Resources and Applications of Biotechnology: The New Wave

To Christine and her parents, Barbara and Eric,
who have made my life in Biotechnology
possible, happy and optimistic

RESOURCES AND APPLICATIONS OF BIOTECHNOLOGY

The New Wave

Edited by
Rod Greenshields
GB Biotechnology
Swansea, Wales

M
STOCKTON PRESS

First published 1989

Published by
THE MACMILLAN PRESS LTD
Houndmills, Basingstoke, Hampshire RG21 2XS
and London
Companies and representatives
throughout the world

Typeset in Great Britain by
TecSet Ltd, Wallington, Surrey
Printed in Hong Kong

British Library Cataloguing in Publication Data
Resources and applications of
biotechnology: the new wave.
1. Biotechnology
I. Greenshields, Rod
660'.6 TP248.2
ISBN 0–333–44358–6

Published in the United States and Canada by
Stockton Press
15 East 26th Street, New York, NY 10010

Library of Congress Cataloging-in-Publication Data
Resources and applications of biotechnology:
the new wave / edited by Rod Greenshields
p. cm.
Bibliography: p.
Includes index.
ISBN 0–935859–22–5: $60.00
1. Biotechnology — Technique. I. Greenshields, Rod, 1933–
TP248.24.R47 1988
660'.6—dc19 88–20108

CONTENTS

FOREWORD

The publication of the Spinks or ACARD report in England in 1980 stimulated the biotechnology boom that has since occurred. In the wake of this intense academic and industrial interest we have seen not only the revamping and updating of traditional research journals but also the appearance of rather expensive newsletters.

However, very few have managed to capture the true message of how research is translated into products and processes. The journal *International Industrial Biotechnology*, guided by the unique experience of Dr Rod Greenshields, was one of the first attempts to marry the more newsy items about the people, institutions and companies involved in biotechnology with the more solid articles on the technologies themselves. The result was an interesting and readable publication crossing the barriers of academia and commerce.

From an academic point of view it presented an up-to-the-minute account of the major thrusts on the commercial side at a sensible price that most could afford; and from the industrial side, a constant review of the state of the art in R&D.

Since many people will not have had the chance to read past issues of *International Industrial Biotechnology*, this book presents a valuable opportunity to study some aspects of the evolution of biotechnology in the UK by publishing revised and edited articles from the journal. I am sure that many academics who are interested in the way that ideas are translated from research to products will find this volume an invaluable reference work to add to their bookshelves.

In these articles is hidden the secrets of the biotechnology of the future—read, mark and inwardly digest!

Dr Mike Gronow
Managing Director
Cambridge Research Laboratories
Cambridge

THE CONTRIBUTORS

Dr John C. Allen
Managing Director,
Newtech (Clwyd) Ltd,
Newtech Innovation Centre,
Deeside Industrial Park,
Clwyd CH5 2NU; Wales

Dr Dennis Allsopp
Head of Culture Collection &
Industrial Services,
CAB International Mycological
Institute,
Ferry Lane,
Kew,
Surrey TW9 3AF, England

Dr D. J. Anderton
Director of New Technology,
Cambridge Life Sciences plc,
Cambridge Science Park,
Milton Road,
Cambridge CB4 4GN, England

Dr David B. Archer
AFRC Institute of Food Research,
Colney Lane,
Norwich NR4 7UA, England

Dr Clifford J. Bailey
Lecturer in Biological Sciences,
Dept of Pharmacy,
Aston University,
Birmingham B4 7ET, England

Dr H. Peter Bennetto
Lecturer in Chemistry,
King's College, London,
Kensington Campus,
Campden Hill Road,
London W8 7AT, England

Mr A. D. Blann
Senior Immunologist,
Dept of Rheumatology,
The Medical School,
Vincent Drive,
Birmingham B15 2TJ, England

Dr C. A. Boulton
Bass plc,
Research Dept,
137 High Street,
Burton-on-Trent,
Staffs. DE14 1JZ, England

Dr M. I. Boxer
Brain Research Group,
Biology Dept,
The Open University,
Walton Hall,
Milton Keynes MK7 6AA, England

Dr A. K. Campbell
Reader in Medical Biochemistry,
University of Wales College of
Medicine,
Heath Park,
Cardiff CF4 4XN, Wales

Professor Roy M. Daniel
Thermophile Research Unit,
University of Waikato,
Hamilton, New Zealand

Dr Alan Doyle
Curator, European Collection of Animal Cell Cultures,
PHLS Centre for Applied Microbiology and Research,
Porton Down,
Salisbury SP4 0JG, England

Dr Peter R. Flatt
Senior Lecturer in Nutrition,
Dept of Biochemistry,
University of Surrey,
Guildford GU2 5XH, England

Dr G. M. Frost
Research Dept,
John & E. Sturge Ltd,
Denison Road,
Selby,
N. Yorkshire, England

Dr J. R. Gallon
Dept of Biochemistry,
University College Swansea,
Singleton Park,
Swansea SA2 8PP, Wales

Professor Tony Godfrey
Managing Director,
Biocatalysts Ltd,
Main Avenue,
Treforest Industrial Estate,
Pontypridd,
Mid Glamorgan CF37 5UT, Wales

Dr Paul Hand
Institute of Horticultural Research,
Worthing Road,
Littlehampton,
W. Sussex BN17 6LP, England

Dr Jayne Heaford
Director,
Futurescan,
17 Beechworth,
Willesden Lane,
London NW6 7YZ, England

Mr Gilbert Jenkins
Cogent,
46 Bishopsgate,
London EC2N 4AJ, England

Dr David B. Johnson
Dept of Biochemistry and Soil Science,
University College of North Wales,
Deiniol Road,
Bangor,
Gwynedd LL57 2UW, Wales

Mr David L. Kemp
Contact Mr. J. Quickenden

Dr Jane Llewelyn
16 Bowden Lane,
High Wycombe,
Bucks. HP11 1DL, England

Dr C. M. Moseley
177 Ridge Valley Road,
Halifax, Nova Scotia,
Canada B3P 2K2

Professor D. F. S. Natusch
Technical Director,
Liquid Fuels Trust Board,
PO Box 17,
Wellington, New Zealand

Dr D. J. Parsons
Manpower Advisor,
National Economic Development Office,
Millbank Tower,
London SW1P 4QX, England

Mr J. Quickenden
Process Design Co-ordinator,
Hamworthy Waste Treatment Systems,
Hamworthy Engineering Ltd,
Poole,
Dorset, England

Dr D. W. Richardson
Programme Manager,
Liquid Fuels Trust Board,
PO Box 17
Wellington, New Zealand

Dr J. J. Rimmer
Lecturer in Biological Sciences,
Dept of Pharmacy,
Aston University,
Birmingham B4 7ET, England

Dr Peter K. Robinson
Lecturer, School of Applied Biology,
Lancashire Polytechnic,
Preston PR1 2TQ, England

Dr P. N. Robinson
Technical Co-ordinator,
Speywood Laboratories Ltd,
Ash Road,
Wrexham,
Clwyd LL13 9UF, Wales

Dr Francisco Rosillo-Callé
Dept of Biology,
King's College,
University of London,
Campden Hill Road,
London W8 7AH, England

Dr I. S. Ross
Dept of Biological Sciences,
University of Keele,
Keele,
Staffs. ST5 5BG, England

Mr Harry Rothman
Reader in Management,
Bristol Polytechnic,
Cold Harbour Lane,
Frenchay,
Bristol BS16 1QY, England

Dr S. W. Shales
Lecturer in Biotechnology,
Queen Mary College,
University of London,
Mile End Road,
London E1 4NS, England

Dr D. A. Stafford
Managing Director,
CLEAR Ltd,
Unit M40, Lewis Road,
East Moors,
Cardiff CF1 5EG, Wales

Dr Tudor H. Thomas
Head, Weed Research Division,
Long Ashton Research Station,
Long Ashton,
Bristol BS18 9AF, England

Dr Brian J. Tighe
Reader in Polymer Science,
Dept of Molecular Sciences,
Aston University,
Birmingham B4 7ET, England

Dr Z. Towalski
Blackpool College of Technology & Art,
Faculty of Academic Business Studies,
Ashfield Road,
Bispham,
Blackpool FY2 0HB, England

Dr Edward J. Yoxen
Dept of Science & Technology Policy,
The University,
Manchester M13 9PL, England

INTRODUCTION

Rod Greenshields

'Give me ae spark o'natures fire,
That's a' the learning I desire.'
From the Epistle to John Lapraik
by R. Burns (1785)

In recent decades much attention has been focused on 'high technology' with its new discoveries, inventions, enterprises and industries mainly in the areas of micro-electronics, nuclear power and aerospace. Since 1970 'biotechnology' has steadily come to everyone's attention. This has been clearly shown, not only by the excitement of the media with the latest genetic engineering discoveries, but also by the business world with global biotechnology ventures and boom biotech companies — a modern 'south sea bubble'.

In academia, the excitement has not been any the less, with new magazines, journals, books, biocentres, institutes, degree courses, diplomas, directives, syllabuses, initiatives, conferences and research grants. These activities have not only been confined to universities, but have also been generated by all levels of education down to schools at junior level.

On the national scene, governments have also taken to biotechnology with initiatives and directives. In the UK, the Department of Trade and Industry, now the Department of Enterprise, has a biotechnology unit and the Science and Engineering Research Council has a Directorate, and there are also initiatives in the other research councils. The European Community also recognises the significance that the technology will have on it and recently the Council passed a major decision adopting a research action programme based on its previous programmes (the Biomolecular Engineering Programme (BEP) and Biotechnology Action (BAP)). Internationally, biotechnology has stirred most nations, particularly in the developing world where its effect will be more significant, although for some nations like Brazil, biotechnology has been all-consuming for over a decade.

Given a reasonable time-span, many of the 'new' high-tech scenes tend to wane and become moments of fashionable interest — not so biotechnology. This is because it is not just another new brilliant technique or one or two new discoveries. It is in fact a fundamental principle of life, a new way of thinking and conceiving ideas, and a fresh application and development of knowledge. To some it even represents the possibility of the dawning of a new age in the next millenium. In fact biotechnology has sustained interest, is growing in its influence and is beginning to affect the very way in which we live.

It would be an immense task to review the whole biotechnology scene because of the many areas it covers, so the purpose of this book is to highlight the concepts, to give some indication of the many resources of biotechnology and to demonstrate some of its major applications. Over the past seven years or so the journal *International Industrial Biotechnology* has been spreading the word of biotechnology and the co-operation between academic researchers and industrial biotechnologists to promote the exchange of ideas and enterprise. The result has been more than a hundred articles written by experienced authors from diverse disciplines who have all been influenced by this new approach. A selection of these articles has been made to illustrate the biotechnology theme and to reflect its concepts and broad understanding, in the hope that it will further stimulate the development of this exciting concept.

WHAT IS BIOTECHNOLOGY? — A PARADOX IN DEFINITIONS!

A question with which all Biotechnology books start is 'What is biotechnology?'. This is because there is considerable misunderstanding, not only in industry but also in academia, as to what the concept really means.

Assessing biotechnology as a new approach poses us with a number of dilemmas or paradoxes as Anita Entebi and David Fishlock have clearly pointed out in their excellent book *Biotechnology — Strategies for Life*. The following illustrate some of these:

1. The major discoveries and developments hailed in biotechnology are not the result of directed research but often answers to questions raised in other fields.

2. Biotechnology is what we are and what we do. It is therefore both very ancient and at the same time extremely modern. It is inescapable in its effects.

3. Biotechnology industry is not so much an industry as a series of enabling technologies.

4. Financiers are putting money into products that not only do not exist at this time, but also that no one can yet specify.

5. Living matter, the basis on which biotechnology exists is subject to necessity, but chance plays an important part in its determination.

The mistake has been to regard biotechnology as a discipline; this has presented a problem to academia because adequate courses with appropriate syllabuses are almost impossible to construct. Some courses with this title do exist but are limited in scope. The consideration of biotechnology as options, or as themes or projects emphasising direction and applications has been more successful. Perhaps the main reason for these difficulties is that many disciplines are involved and while certain of these are fundamental to 'life sciences', such as biology, microbiology, biochemistry, medicine, agriculture and genetics, many others are less obvious because direct human activity is involved, e.g. economics, social science, political science and business studies.

In industry a parallel problem was encountered, namely was biotechnology only genetic engineering involving the medical industries? Where did the traditional industries stand? More importantly, some of the national and international companies, particularly the pharmaceutical companies, did not want to be associated with the nervous developments of the newly formed biotech companies or some of their entrepreneurs, if only to protect their 'blue-chip' shares.

Pressure mounted within scientific circles in the United Kingdom during the 1970s to recognise some of the exciting biotech developments and achievements. It had been exacerbated by the failure to recognise the major discovery of monoclonal antibodies by Cesar Milstein. Shades of the penicillin saga came to haunt the UK again! Nevertheless, the UK was not the only country to fail to see the significance of biotechnology. In 1979, the United Nations held a conference on science and technology without discussing biotechnology.

In 1980, a Joint Working Party under the chairmanship of Prof Spinks, reported to the Royal Society and brought some order to the chaos of different understandings. The organisational framework suggested for co-ordinating national policy on biotechnology was adopted by the British Government.

The working party, now called the Spinks Committee, defined biotechnology as '*the application of biological organism systems and processes to manufacturing and service industries*'. At the time there was a plethora of definitions. This was to be expected because biotechnology is not a discipline and many different facets are seen in the approach by each discipline involved. In some way, and unfairly, the Spinks definition has been interpreted to exclude some people who thought that they might have been involved in this technology. It has also been suggested that the older and more traditional parts of biotechnology were also not included and that biotechnology was genetic manipulation or genetic engineering. Even the simplest definition of '*biology in industry*' seems to exclude what are now essential disciplines for biotechnology to succeed, namely commerce, engineering, economics and the sociopolitical sciences.

A more fundamental definition was sought to enable more people to see the approach and to identify with its direction, particularly at the academic industry interface. When the journal *Industrial Biotechnology* first started as a communication medium, the problem of definition was critical since it wanted not only to bridge academia and industry, and to give top-quality information on the latest discoveries, but also to encourage the traditional areas of this technology. In reality the two were not far apart.

The following definition was adopted: '*biotechnology is the study of the application/exploitation of biological organisms, their parts and activities*'.

HISTORICAL APPRECIATION

Biotechnology appears to be recent but one can immediately see from the above definition that this is not really so — nominally man has always been, and has always had to be a biotechnologist. Ever since he has been on this earth he has used biological materials for his livelihood. First as a hunter trading skins and meat, then some 10 000 years ago when he settled to agriculture and produced cereal foods and bread, and manufactured beer, wine, cider and vinegar, and formulated medicines and dyes, and manufactured yoghurts, cheeses, fats, oils and soaps. All of these were used for barter and trade, giving another definition for biotechnology: '*the utilisation of renewable resources*'. Evidence for this has been found with the Babylonian and Phoenician civilisations in the Tigris and Euphrates Basin, but clearer documentation can be found in the Temple of Thebes in Egypt (4000 years BC). Here on the walls of the temple are cartoons showing how a biotechnology industry was developed to provide the Pharaohs and their dynasty with wine, beer and bread. The Egyptians were the first to tax this activity and from this principle derives the massive global alcohol revenue in which every nation in the world participates.

The religious orders who travelled Europe used biotechnological industries to maintain their existence, and it is from this that the modern fermentation industries of today arc derived. For example, the brewing industry based at Burton-on-Trent, the liqueur distilling of Benedictine and Chartreuse, and many of the famous vineyards with associated vinegar fermentations.

Likewise, the alchemists, with their essential use of herbal plants, also represent biotechnology and have provided the base on which is built the modern pharmaceutical industry. Even today, 25 per cent of our medicines are still plant derived, the antibiotic fermentation industry further amplifying this aspect.

Perhaps the greatest influence was the Industrial Revolution with its dramatic changes on the structure of society. This gave an early impetus to present-day biotechnology, because it was in this period of the nineteenth century that science directed its attention to the bio-world.

In this period, Pasteur must be cited as the biotechnologist *par excellence*, although others before him had initiated the field. Pasteur was a physicist called in by the French wine and beer industry to solve the problem of fermentation 'sickness'. Like many consultants, he ignored the immediate superficiality of the problem and delved in to the fundamental facts. He 'rediscovered' micro-organisms, showing that yeast was responsible for alcoholic fermentation and that 'germs', i.e. bacteria, were responsible for the 'sickness'. The basic facts had been uncovered previously, but were now unequivocally proved and developed as factual science.

Pasteur, who has been called 'the father of microbiology', derived many important concepts from his studies, which were applied not only in the fermentation industry but also in other disciplines such as medicine and farming, which had very direct effects on society. Often his experiments were tried and tested in public. He had a fervent passion for research tempered by a clear enthusiasm for developing the ideas into reality for the benefit of society. Pasteur's life and work clearly exemplified the meaning of the definition of biotechnology by applying the study of organisms, their parts and activities. In this way science became a part of the traditional biotechnology.

Pasteur also presented a problem for biochemistry which delayed its development on the basis of understanding the biological catalyst (enzyme). He claimed that enzymes could not be separated from the 'life force' of the cell, citing the specific case of the enzyme zymase (which fermented sugar to alcohol) from the yeast cell.

Not until 1902 was the Pasteur 'impossibility' resolved, although there was much circumstantial evidence to support the independence of enzymes. This was accomplished by the Buchner brothers who developed a large press which could smash the yeast and release zymase in a cell-free extract which could convert sugar to alcohol. In many ways it was fortuitous that the complex of enzymes which compose zymase were water-soluble and intact, otherwise many more years would have elapsed before this problem was solved.

Biochemistry, as a discipline, came to life at this point, developing rapidly with the understanding of the uniqueness of the bio-catalyst. This too represented a biotechnological 'moment', since it included, for the first time, the interest and influence of engineering in biology and the concept of the industrial bio-process.

The next landmark came in 1911 when a Russian Jewish immigrant by the name of Chaim Weizman came to Britain with a recommendation to Prof Perkin at Manchester University seeking university freedom and a job after his education in Switzerland. He became a demonstrator in the Chemistry Department and researched on Perkin's organic dyes. He paid £6 per annum for the use of his basement laboratory but found the cost of solvents expensive. Undaunted, he developed a biological process for the production of acetone and butanol by bacterial fermentation of starch. The low yield required the production of a pilot plant for which he obtained patents. In 1915 World War I had placed Britain in the position of having a serious shortage of acetone for dissolving gunpowder in the manufacture of cordite for use in her battleships and for the preparation of dope for aeroplane fabric. Lloyd George, then Minister of Munitions, discovered the Weizman process and persuaded Weizman to develop the process to full scale; this became a major industry which helped to win the war.

Weizman's reward was the Balfour declaration which later helped the Zionist movement to form the State of Israel. He became the first President and subsequently the Weizman Institute was formed in his honour. Only recently has the Institute taken up the approach which led to its foundation.

There are two concepts which Weizman's work revealed, both of which are the epitome of biotechnology:

- Firstly, that scientific experiments are not in isolation and have a direct effect on society.
- Secondly (a concept which is part of the definition of biotechnology), the *deliberate use* of organisms for industrial purpose and benefit.

The acetone–butanol industry was untenable post-war when petrochemicals became freely available, but the fermentation industry was assured. A few years later, after the development of yeast production for the baking industry, came antibiotics, initiated by the discovery of penicillin. Even penicillin is not new since there appear to be records of Roman physicians using green mouldy materials to treat septic wounds!

The next milestone in biotechnology came with the technological improvements assisted by the cross-disciplinary activities in World War II. A sound understanding of the biochemical compounds of the cell and their interrelationships (metabolism) was developed, particularly by the new technique of chromatography and its automation. Thus, by 1960, the industrial chemical scene had begun to accept that the bio-industry was reaching maturity and looking at the cell as a means of development for new processes and products. The enzyme, long-recognised as a powerful tool, but neglected in industrial terms, was derived cheaply by microbial fermentation and used initially to enhance the detergent industry. Almost overnight an industry of kilogram size became 'megatonne'. This has not abated, despite the fact that initially only six hydrolytic enzymes were the mainstay of the industry. Its future is assured, with some 3000 enzymes available, and with their synthetic and transforming abilities almost unused.

A decade later, the 1970s brought another concept to light that has since influenced the development of the biotechnological approach. This was the understanding that living organisms were on a 'spaceship' which had a finite volume in which it survived — the 'biosphere'. For the first time, due to pressures of population and well-developed communications, world crises were seen and felt — food, energy, health and pollution, to name just some of the concerns. The global population of 4 000 000 000 was set to become 6 000 000 000 by 2000. For the first time food became a currency of the world, particularly as petrochemicals became critical in supply. The shortages were albeit often local and severely influenced by politics, strategic economies, and religious and social attitudes, but were the first pointers to a situation which is now clearly evident. The crises have now extended to the effect of man's activity directly on his environment in the biosphere by pollution and removal of the natural balancing factors. A 'Green' protectionist attitude has developed which is basically biotechnological. A search for solutions has meant using an approach which is not based on a single discipline or on an attitude which is just academic. It would certainly have to involve biotechnology and have industrial facets, and be one which would involve all parts of human activity and society, and their relationship to the biosphere.

Another dramatic situation occurred about this time from knowledge obtained in the 1960s of the molecular structure of the cell and its biochemical composition. This bore fruit in the ability to manipulate the cell directly by using its nucleic material. Genetic manipulation or 'engineering' revealed an immense potential, but its first products incurred quite a violent reaction in all spheres of life. It was this that highlighted biotechnology and its current significance, showing that the understanding of its effects could not be confined to academics and scientists. Biotechnology had a direct effect on society and was influenced by what society thought about it. Some people consider biotechnology only as genetic engineering, but this is certainly not true; it is a powerful technique which has the potential for involvement in all parts of the wide field of biotechnology.

The history of biotechnology has therefore developed in a number of phases: traditional, scientific, engineering, technological, social, political and genetic. Each has its own historical starting point in a fundamental area of the main activity, yet all phases continue today in parallel and interrelated. It is not always easy to see the connection between the phases, for example, the traditional and the genetic, and yet they can have a significant effect on our everyday lives. The yeast which ferments the beer or wine we drink today, could well provide the nucleotides for enzymes in the latest diagnostic antibody kits for AIDS or the albumen for a pure virus-free blood plasma substitute. Streptomycetes fungi derived from rotting leather hides are assisting in providing the enzymes for sloughing skin in delicate skin operations. The apparently outdated use of leeches has been renewed by mass culture and isolation of their key active enzymes for blood clot removal. Genetic fingerprinting is about to decide on human status and individuality. There are in fact many more examples of this 'traditional–modern' link illustrated in the applications considered by this book.

BIOTECHNOLOGY AS A TECHNOLOGY

David Fishlock has posed the question of Biotechnology as 'Is it a revolution or a renaissance?'. He cites the traditional against the recent history of the biological industry. What was an intellectual puzzle for a few geneticists has not only become a new resurgence for the bio-industry, but has also provided a totally new approach to thinking in fields quite remote from each other. In academia what was the concern of a single discipline now influences a whole structure of disciplines and the concepts of their reasoning. Matter is being explained down to the smallest structure, and the universe to the infinitely large — now is the time to explore ourselves — 'life' in all its aspects.

Perhaps the most dramatic effect of this biotechnological approach is the change in attitudes to the application of knowledge. It is not that academic and industrial relationships did not occur, they did; but neither side was convinced of the other's motives. Biotechnology suddenly allowed industrialists and financiers to see a 'window' into the future, and likewise academics starved of funds could see a route to stable finance and development. Each courted the other until the differences between them were indistinct. Professoriat became directors, research scientists became entrepreneurs, industrialists became teachers, while financiers and bankers were investing in products they did not understand, could not be seen, and had not yet even been developed for what was only a perceived market — an academic approach if ever there was one!

Still, the party was not complete because a further partner for this relationship was necessary. To drive the situation forward required a political will. This was even more so emphasised by the fact that biotechnology was affecting society directly. No longer were research-laboratory discoveries scientific curiosities, but now of immediate and considerable concern to the knowledgeable layman and his pressure groups. Application of the knowledge was now a matter for society's desire or repression, and this dictated the political will.

Biotechnology thus depends on a triangle relationship, with politics providing the incentive and investment direction, academia researching and developing the ideas, and industry investing in the application for society. Each is dependent and driven by the other. Biotechnology can only exist when this triangle is complete.

As biotechnology is an approach to life by life, its effects are multi-faceted and all-embracing. As a technology it is irresistible since it is what the human organism does.

CONCLUSION

A short while ago the 4 000 000 000th child was born and, barring a global catastrophe, it is anticipated that by the year 2070 the population will rise to 10 000 000 000. This is providing a considerable pressure on the biosphere and causing many problems, some now on a massive natural scale. Food and drink, resources and energy, health and population, environment and pollution, are the main areas for concern, all of which are key areas for biotechnology and its approach. Problems like these are *understandable* by many people at all levels and are *appropriate* to the exercise of the many disciplines involved in biotechnology; they are *approachable* as problems from the many aspects of biotechnology and *solvable* by biotechnology in an *applicable* way. Fortunately the biosphere operates cyclically and therefore the situation is *renewable*. These are the tenets of biotechnology which are amply illustrated by the examples of the biotechnological resources and applications described in this book.

Part I
Resources of Biotechnology

1
THE INDUSTRIAL POTENTIAL OF CYANOBACTERIA — MULTIPURPOSE ORGANISMS?

John R. Gallon

INTRODUCTION

Cyanobacteria are unusual organisms. Although they possess a photosynthetic system that functions like that of higher plants and green algae (for example, it produces oxygen), they nevertheless resemble bacteria in their prokaryotic cellular structure. Because of their photosynthetic properties, cyanobacteria were, until very recently, almost universally referred to as blue-green algae. Even now, this name appears in the literature at least as frequently as does the name 'cyanobacteria'. Indeed, over the years, cyanobacteria have been given a variety of names. In addition to the terms 'cyanobacteria' and 'blue-green algae', names include 'cyanochloronta' (referring to their blue-green appearance), 'myxophytes' (referring to their plant-type photosynthesis and to the fact that many species are surrounded by mucus), 'cyanophytes' (blue plants) and 'blue-green bacteria'.

The term 'cyanobacteria' seems to be gaining popularity. For example, Carr and Whitton[8] entitle their book *The Biology of Cyanobacteria*, thereby replacing the term 'blue-green algae' which they used in the title of an earlier edition. On the other hand, Fay[11] in his monograph *The Blue-Greens* neatly avoids the controversy as to whether the cyanobacteria should be classified as bacteria or as plants. In truth, as he says, cyanobacteria are neither typical bacteria nor typical algae.

Cyanobacteria are ancient organisms. They appear to have been widespread at the beginning of the Proterozoic age, 2.5×10^9 years ago, so are probably older than that. Unfortunately, in older rocks it is very difficult to distinguish between the fossils of cyanobacteria and those of other prokaryotes, but current thinking has it that cyanobacteria evolved between 3.5 and 2.8×10^9 years ago. In geological terms this means they appeared not long after the supposed origin of life itself, about 3.8×10^9 years ago. Mineralogic studies indicate that the atmosphere of the earth, which was originally reducing, became oxygenic about 2.3×10^9 years ago, and it seems very likely that the source of this oxygen was cyanobacterial photosynthesis. Cyanobacteria are, therefore, probably largely responsible for the oxygen found in the present-day atmosphere of the earth.

Cyanobacteria are microscopic organisms. Their cells measure up to about 5 μm in diameter. They may be unicellular, occurring singly or in colonies surrounded by slime, or they may be filamentous, in which case the cells are organised into long chains that may be straight, spiral, branched or unbranched. In some species of filamentous cyanobacteria, differentiated cells appear within the chains. These differentiated cells may be akinites, which function like bacterial spores, or heterocysts, which are produced under nitrogen-deficient conditions and are the site of N_2 fixation in these cyanobacteria.

Cyanobacteria are very widespread, and most people are probably familiar with them, even though they may not realise it. For example, cyanobacteria often accumulate or 'bloom' in lakes at certain times of the year, particularly during summer and early autumn. They can then form a slimy green scum on the surface of the lake — something that almost everyone will have seen at one time or another. Similarly, cyanobacteria occur in terrestrial habitats, forming, for example, green patches on damp walls and moist rocks, and, once again, these stains will probably have come to the attention of most people.

Cyanobacteria exist in both aquatic and terrestrial forms. Aquatic species may be either fresh-water or marine, and, although they are found in cold lakes and oceans, are most abundant in the tropics. Some species are found in hot springs. Similarly, terrestrial species are found almost everywhere, from Antarctica to some of the hottest deserts. The slime that surrounds many of these cyanobacteria protects them from desiccation. Like aquatic species, terrestrial cyanobacteria are most abundant in the tropics, although they are by no means exclusive to these regions. They occur on rocks, in soil and in more unusual environments such as the bark of trees, on buildings and carved stone (where they can cause considerable damage as well as unpleasant, disfiguring stains); and cells of cyanobacteria have even been found in clouds. Cyanobacteria, therefore, inhabit almost every environment, a tribute to their diverse range of physiological and metabolic capacities.

As well as leading an independent existence, cyanobacteria are found in association with a wide range of other organisms. Among these are fungi, bryophytes, pteridophytes, higher plants and animals. The symbiotic association of cyanobacteria with plants is of special significance because, in many respects, cyanobacteria resemble chloroplasts, the photosynthesising organelles of higher plants. Indeed, it is currently considered that chloroplasts may have arisen from the invasion of an ancestral plant cell by a photosynthetic cyanobacterium-like organism, which eventually became incapable of survival in a free-living form.

It is largely the ability of many cyanobacteria simultaneously to photosynthesise and to fix atmospheric N_2 that has led to their intensive study in recent years. In our own laboratory, we have a particular interest in N_2 fixation by these organisms, but, like many other workers, we have gradually extended our interests into most aspects of cyanobacterial metabolism and we now routinely grow several genera on both a small and a large scale. Our interest in the commercial potential of these organisms stems from the recent general shift of emphasis towards the application of biological systems to industry, coupled with the feeling that there must be a role in industry for organisms that, in effect, live on fresh air, requiring only illumination and a dilute inorganic salts medium for their growth. However, our expertise is in pure research, and our experience of industrial processes is limited. In consequence, some of the possible applications of cyanobacteria which are listed below, and which are based on our own observations rather than on proven usefulness, may not be feasible at present. On the other hand, in some areas, cyanobacteria are already being used and have demonstrated their worth.

The aims of this chapter are, therefore, to paint as wide a picture as possible of the potential exploitation of cyanobacteria in industry, to record our own interest in this area and also, perhaps, to sow a few ideas in the minds of others.

CYANOBACTERIAL PROCESSES AND PRODUCTS

Nitrogen Fixation

Many species of cyanobacteria can fix nitrogen. However, this ability is not a universal property of these organisms. Furthermore, not all nitrogen-fixing cyanobacteria do so under aerobic conditions of growth. Nitrogen fixation is an oxygen-sensitive process. Thus, the nitrogen-fixing enzyme, nitrogenase, from all nitrogen-fixing organisms is rapidly and irreversibly inactivated by exposure to air. Nevertheless, many cyanobacteria can fix nitrogen aerobically and are therefore capable of protecting their nitrogenase from both atmospheric and photosynthetic oxygen. The mechanism of this protection is of great interest to biochemists and microbiologists, but from the industrial standpoint, its significance is that these species can be grown without the provision of a special anaerobic gas mixture.

Nitrogen-fixing cyanobacteria are grown on a large scale in the Far East and are applied to rice paddy fields, which provide an ideal environment for the growth of these organisms[2]. In the rice paddies, the cyanobacteria excrete a portion of the nitrogen that they fix and, when they die, their cellular nitrogen serves as an excellent source of fertiliser for the growing rice plants. As much as 40 kg nitrogen per hectare can be added annually to the soil in this way. The use of nitrogen-fixing cyanobacteria is not confined to rice paddies, however. For example, it has recently been shown that inoculation of sugar cane with soils containing various species of nitrogen-fixing cyanobacteria increased the yield of sugar by as much as 20 per cent more than was obtained by application of conventional fertiliser[15].

In the case of rice paddies, the water fern *Azolla*, which contains symbiotic nitrogen-fixing cyanobacterial cells within its leaves, is more efficient as 'green' manure than are free-living cyanobacteria[40]. *Azolla* has been applied to rice paddies for centuries in Vietnam and can provide as much as 300 kg nitrogen per hectare per year, improving rice yield by almost 40 per cent.

Thus, nitrogen-fixing cyanobacteria make a significant contribution, as fertiliser, to tropical agriculture. However, their use could be extended. For example, cyanobacteria are primarily colonisers of new land and, in many desert soils, they may comprise the bulk of the vegetation. Their mucus sheaths protect them from desiccation, although, under really unfavourable conditions, they produce akinites which will germinate later upon wetting of the soil. It may be good practice to introduce nitrogen-fixing cyanobacteria into irrigation water and into waste-water outflows from industry[5] and thereby improve soil fertility as well as providing water.

There is good evidence that nitrogen-fixing cyanobacteria improve the nitrogen status of soil. They release amino acids and polypeptides during their growth and also after their death. Furthermore, cyanobacteria excrete substantial quantities of carbohydrate which can improve soil structure (and thereby limit erosion) as well as soil fertility.

In a world in which millions of people are starving, better land management is a dire necessity. Part of such management must include better utilisation of biological nitrogen fixation, including that brought about by cyanobacteria. However, considerable research, both pure and applied, remains to be done. For example, there has been no systematic survey of cyanobacterial strains for their ability to survive under the conditions prevailing in the soil where they may be used, nor do we have many details on cyanobacterial predators. In addition, it may be worth investigating whether the proven usefulness of cyanobacteria in tropical agriculture could be extended to temperate soils.

Biomass

Biomass is, basically, everything that grows. It can serve as a feedstock to several industrial processes, particularly those associated with fuel production (e.g. methane, ethanol). At present, utilisation of biomass is concentrated on the waste products of agriculture (e.g. straw), but it is worth bearing in mind that, should the demand for biomass exceed its availability from these sources, it may become necessary to grow organisms specifically to provide biomass. The simple growth requirements of cyanobacteria may make them promising candidates for any such process. Cyanobacteria grown in aquatic systems show a relatively efficient conversion of solar energy into biomass (3–5 per cent, compared with 6 per cent for higher plants in conventional agriculture) and some species show a fairly high growth rate, with doubling times of about 2 h, although 5–24 h is more usual in nitrogen-fixing strains (see Ref. 39).

Food

Certain species of cyanobacteria are edible and, indeed, have been eaten since ancient times. Perhaps the best-known examples of such organisms are *Spirulina platensis*, which is used as food by tribes in the region of Lake Chad (Africa), and *Spirulina geitleri*, which was eaten by the Aztecs in Mexico.

There has recently been considerable interest in these organisms as a source of single-cell protein for both man and animals, and *Spirulina* is now grown on a large scale in several countries[36]. However, most *Spirulina* is grown under semi-natural conditions, in open-air ponds rather than in industrial plants. Such exploitation of this cyanobacterium demonstrates the potential of cyanobacteria as alternative sources of food. *Spirulina* is an excellent source of protein, although some of the claims as to its health-giving qualities may be exaggerated. Nevertheless, as a foodstuff, *Spirulina* has passed the sanitary regulations of many countries, and its consumption is expanding.

Spirulina cannot fix nitrogen, but edible nitrogen-fixing cyanobacteria are known. For example, *Nostoc flagelliforme* is a traditional food in China and is by no means unpleasant to eat[42]. It appears possible to grow this cyanobacterium under nitrogen-fixing conditions in the laboratory, although to date it has not been commercially exploited.

As is the case with single-cell protein derived from bacteria, there may be consumer resistance to the consumption of cyanobacteria or cyanobacterial protein by humans. Despite this, however, there has been a report of attempts to preserve cyanobacteria by canning[10]. Perhaps a more realistic approach would be to use cyanobacteria as animal fodder, or in aquaculture[25]. A word of caution, though: several species of cyanobacteria are toxic (see below). Indiscriminate sampling of these organisms as potential foodstuffs, therefore, is not recommended and it may be necessary to subject even those species traditionally consumed by man to rigorous testing before they can be used as food on a commercial, large-scale basis.

Production of Ammonia

Nitrogen-fixing cyanobacteria reduce nitrogen to ammonia. This ammonia is usually rapidly assimilated into organic material, but treatment with certain chemicals, or mutation, can produce cyanobacteria that excrete ammonia into the growth medium. Musgrave *et al.*[29] describe ammonia production for up to 130 h by cultures of *Anabaena* spp., immobilised in beads of calcium alginate. However, it is unrealistic to expect such preparations to replace the Haber process of industrial nitrogen-fixation, not least because recovery of ammonia from its dilute solution would be difficult and expensive. Nevertheless, it is possible that, in rural agriculture, small plants, containing columns of immobilised cyanobacteria, may be useful in providing ammonia in irrigation waters, at the expense of solar energy.

Production of Hydrogen Solar Energy Conversion

The dual ability of many cyanobacteria to fix nitrogen and to photosynthesise has led to their investigation as organisms for solar energy conversion on an industrial scale (see, for example, Refs. 7 and 23). The literature on this topic is extensive. The basis of the interest in cyanobacteria is that their photosynthesis generates oxygen from water, while nitrogen fixation produces hydrogen simultaneously with ammonia as an obligatory part of the nitrogenase reaction. In the absence of atmospheric nitrogen, hydrogen evolution is considerably increased. Cyanobacteria, therefore, can effect a splitting of water into hydrogen and oxygen, at the expense of light energy.

Hydrogen has considerable advantages as a fuel. It is clean-burning and, as a liquefied gas, has an energy content 250 per cent of that of petrol per unit weight. Furthermore, in most industrial countries, there already exist pipelines for gas distribution which could easily be adapted to carry hydrogen.

Hydrogen evolution is not confined to cyanobacteria. Green algae can also photoevolve hydrogen. However, in these organisms, hydrogen evolution is not catalysed by nitrogenase, but by a hydrogenase enzyme. Both hydrogen evolution and hydrogenase itself are inhibited by exposure to oxygen, and green algae will only generate hydrogen under atmospheres containing less than 1 per cent oxygen and under relatively low light intensities, to limit photosynthetic evolution of oxygen. In contrast, certain cyanobacteria can photoevolve hydrogen aerobically and at relatively high light intensities.

Photosynthetic bacteria can also photoevolve hydrogen, but they are strict anaerobes and have rather more exacting growth requirements than do cyanobacteria. In addition, they do not simultaneously evolve oxygen, although this could be an advantage by obviating the subsequent need to separate hydrogen from oxygen. However, this could also be avoided by using cyanobacteria such as *Gloeothece* which, when grown under alternating 12 h light and 12 h darkness, effect a temporal separation of photosynthesis (oxygen evolution) in the light and nitrogen fixation (hydrogen evolution) in the dark[27].

On an efficiency basis, cyanobacteria have a distinct lead as photoevolvers of hydrogen (although 3 per cent may be the practical upper limit). Unfortunately, many cyanobacteria also possess a hydrogen-uptake system, and optimum rates of hydrogen evolution require not only absence of nitrogen, but also inhibition of hydrogen uptake. Fortunately, the latter is relatively easily achieved by inhibition (e.g. nickel deficiency: Ref. 9) or by mutation. Recently, absence of nitrogen has been achieved by using, as a gas phase for these cyanobacteria, a methane/CO_2 mixture produced from an industrial methane-fermenting system[24].

Calculations of the potential of cyanobacteria to provide hydrogen as a fuel have been many. One such[30] estimates that an area of about 17.5 km^2 filled with hydrogen-evolving *Anabaena cylindrica* would provide enough fuel to provide the energy needs of a city of one million people, at a power rating of 10 kW per person.

Immobilisation of cyanobacterial cultures has also been employed in the establishment of hydrogen-evolving systems[19, 22, 26].

Cyanobacteria have also been employed as components of photosynthetic solar cells, by which light energy can be converted directly to electrical energy. Higher plant chloroplasts have also been used in this way, but cyanobacteria are much more stable, lasting for as long as 20 days. Cells of *Phormidium* spp., immobilised on a SnO_2 semiconductor electrode and dried for 60 min at 50°C, can produce an electrical current of as much as 10 μA per cm^2 of electrode (containing 10 μg of chlorophyll) under illumination at an intensity of 250 J/m^2 — a power efficiency of about 0.1 per cent[32]. Clearly, this area of research promises well for the future.

Possible Bulk Products from Cyanobacteria

Most cyanobacteria are surrounded by mucus or slime which consists largely of free polysaccharides and lipopolysaccharides. The sugar composition of these extracellular polysaccharides varies among different species and, as well as the more usual glucose and glucuronic acid, sugars such as xylose, ribose and fucose are often found (see Drews and Weckesser in Ref. 8). In the laboratory, it is relatively easy to recover polysaccharides from medium in which cyanobacteria were grown. Bulk harvesting of this material for industrial purposes is possible.

Cyanobacteria also excrete considerable amounts of peptides and amino acids into the growth medium[12], along with sterols, free fatty acids, hydrocarbons, phenolics[41] and various volatile materials[38]. Any, or all, of these materials may be suitable for commercial exploitation.

As well as extracellular products, certain intracellular products of cyanobacteria may have commercial potential. Among these are cyanophycin (a unique polymer of arginine and aspartic acid which can be easily isolated and is very resistant to hydrolysis) and low molecular weight carbohydrates which accumulate in certain cyanobacteria in response to osmotic stress[35]. Furthermore, some cyanobacteria are relatively tolerant of X- and Y-radiation[3, 4, 37]. They may therefore be ideal organisms from which to isolate radiochemically labelled intracellular or extracellular products.

Specialised Products

Cyanobacteria contain a variety of unusual products, some of which occur in all species, others in only one or two organisms. Several of these products may have industrial applications, although, in most cases, these remain to be assessed.

Pigments and Food Additives

Cyanobacteria contain a variety of brightly pigmented compounds. In addition to chlorophyll and carotenoids, which also occur in other photosynthetic organisms, cyanobacteria contain the protein pigments phycocyanin (blue) and phycoerythrin (red). It has been suggested that these pigments may have applications in the food industry.

Several species of, in particular, unicellular cyanobacteria contain large quantities of polyunsaturated fatty acids[20]. This may eventually be commercially significant, given the current interest in these compounds in the food industry. In addition, certain cyanobacteria contain tocopherols[31], which may have a role as antioxidants in food preservation.

Toxins

At the opposite end of the spectrum, some cyanobacteria are extremely toxic to animals[13]. The nature of the toxic constituent varies from species to species. For example, in *Microcystis aeruginosa* the toxin is a peptide, in *Anabaena flos-aquae* it is an alkaloid, while *Lyngbya majuscula* contains both an alkaloid and a toxic phenolic compound. Some toxins contain halogens, e.g. bromoaplysiatoxin from *Oscillatoria nigroviridis* and *Schizothrix calcicola* in mixed culture, and cyanobacterin, a chlorine containing anti-algal agent, from *Scytonema hofmanni*[33].

Blooms of toxic cyanobacteria have been responsible for the death of livestock and are occasionally a significant factor in agriculture. On the other hand, toxins can often have pharmacological applications. However, in the case of the cyanobacterial toxins, these remain to be investigated.

Enzymes

Like other micro-organisms, cyanobacteria contain restriction endonucleases, enzymes that have proved indispensable in modern recombinant DNA technology. Restriction endonucleases recognise specific sequences within a DNA molecule and cleave both strands of the DNA within this sequence in such a way that it is relatively easy to splice the fragments together in an artificial fashion. In this way a fragment of DNA from one organism may be introduced into another.

Cyanobacterial restriction endonucleases are already available commercially. Perhaps the best examples are *Ava* I and *Ava* II from *Anabaena variabilis*[16, 28], but there are, of course, others.

In addition to restriction endonucleases, other cyanobacterial enzymes, particularly those associated with the biosynthesis of products unique to cyanobacteria, may have a commercial value by virtue of their individuality.

Cyanobacteria in Microbial Mining

There are reports that cyanobacteria may have a role in the concentration and extraction of minerals from sea-water and terrestrial ores. Among these minerals are iron[14] and uranium[21].

Water Purification

The use of cyanobacteria for the removal of phosphorus, nitrogen, toxic metals and suspended matter from water has been investigated (see, for example, Ref. 34) and these organisms may therefore have a role in water purification. In addition, the cyanobacteria themselves may be subsequently used as nitrogen fertiliser[6].

LARGE-SCALE GROWTH OF CYANOBACTERIA

In the laboratory, growth of cyanobacteria in batch cultures up to 20 litres in volume (which can yield about 50 g, wet weight, of cells) is simple. The only requirements are a dilute salts solution, aeration, temperature control and a source of illumination. Our own findings are that many species of cyanobacterium grow just as well under alternating 12 h light and 12 h darkness as under constant illumination. This raises the possibility of using natural illumination for the large-scale culture of these organisms, thereby saving in energy costs. Growth of cyanobacteria on a larger scale (up to 1000 litres) has also been successful but yields may not be optimal, since the equipment used is usually designed primarily for growth of heterotrophic micro-organisms. Jüttner[18] describes some of the problems associated with the use of such equipment and also describes a purpose-built fermenter for culture of photosynthetic micro-organisms on the 110 litre scale. Cyanobacteria have also been grown on a laboratory scale in continuous culture (see Van Liere and Walsby in Ref. 8).

However, although cyanobacteria have been grown on a large scale under entirely artificial conditions, the technical and economic problems encountered have limited industrial production of these organisms (see Ref. 36). Although there is no reason to believe that these problems cannot be overcome, it seems that seminatural cultivation of cyanobacteria in ponds, in saline and alkaline tropical waters and on arid land offers the best prospects for bulk production of these organisms at present[1]. Nevertheless, in the future, if the value of the potential products warrant it, construction of plants for the large-scale industrial culture of cyanobacteria will become economically feasible. Meanwhile there is clearly a need for more research into reactor design and operation in a cost-effective manner, probably utilising natural illumination.

Harvesting of cyanobacterial cells and isolation of cyanobacterial products should present no serious problems over and above those encountered with other organisms. However, the cells of some cyanobacteria are particularly resistant to disruption, and this should be taken into account in assessing the feasibility of these particular organisms as sources of commercial products.

The genetic potential of cyanobacteria is only beginning to emerge. Recently, cyanobacterial genetics have made great strides and the possibility of tailoring a particular organism for a particular process (e.g. overproduction of a product, ease of cell disruption) is already with us. Furthermore, the isolation of cyanobacterial plasmids and the demonstration of transformation in cyanobacteria offer the prospect of genetically manipulating these organisms. Since cyanobacteria grow at the expense of light energy and have minimal nutritional requirements, it may be economically worth while to introduce into a cyanobacterium foreign genes coding for a particularly high-value product. For a recent survey of the genetic properties of cyanobacteria, see Ref. 17.

In conclusion, therefore, it is apparent that in some directions, e.g. food production and solar energy conversion, cyanobacteria already show immense promise for industrial exploitation. In other areas the commercial prospects may be less immediately apparent but should not be ignored. Who knows how useful to man's industrial economy in the twenty-first and twenty-second centuries will be organisms that grow on fresh air and directly convert solar energy into a whole array of potentially useful materials? Certainly, the diversity of cyanobacterial metabolism means that they may have applications in many different directions: truly multipurpose organisms.

EDITOR'S NOTE

The developments in space travel open yet another area where the cyanobacteria could have a big potential. The availability of sunlight but restriction of other materials could make their multipurpose nature a crucial factor in operating space stations or in lengthy space flights, where renewal of all organics is essential.

REFERENCES

1. Aaronson, S. and Dubinsky, Z. (1982). Mass production of microalgae. *Experientia,* **38**, 36–40
2. Agarwal, A. (1979). Blue-green algae to fertilize Indian rice paddies. *Nature, Lond.,* **279**, 181
3. Asato, Y. (1971). Photorecovery of γ-irradiated cultures of blue-green alga *Anacystis nidulans. Radiation Botany,* **11**, 131–136
4. Astier, C., Joset-Espardellier, F. and Meyer, I. (1979). Conditions for mutagenesis in the cyanobacterium, *Aphanocapsa* 6714. Influence of repair phenomena. *Archives of Microbiology,* **120**, 93–96
5. Balloni, W., Materassi, R. and Florenzano, G. (1983). Wastewater treatment

with photosynthetic bacteria and direct utilization of their biomass in agriculture. In *Fitodepuratori Impieghi Biomasse Produzione Convengno Internazionale* (1981). Ed. P. F. Gretti, pp. 289–298. Centre Ricerche Produzione Animali, Reggio Emilia, Italy

6. Benemann, J. R. (1979). Production of nitrogen fertilizer with nitrogen-fixing blue-green algae. *Enzyme and Microbial Technology,* **1**, 83–90
7. Bothe, H. (1982). Hydrogen production by algae. *Experientia,* **38**, 59–64
8. Carr, N. G. and Whitton, B. A. (Eds.) (1982). *The Biology of Cyanobacteria.* Blackwell, Oxford
9. Daday, A. and Smith, G. D. (1983). The effect of nickel on the hydrogen metabolism of the cyanobacterium *Anabaena cylindrica. FEMS Microbiology Letters,* **20**, 327–330
10. Fang-Yung, A. F., Istkovskaya, L. S., Biryukova, S. N. and Davydova, E. V. (1976). Canning of blue-green algae. *Rybn. Khoz (Moscow),* **1976**, 72–73
11. Fay, P. (1983). *The Blue Greens.* Arnold, London
12. Fogg, G. E. (1952). The production of extracellular nitrogenous substances by blue-green algae. *Proceedings of the Royal Society, Series B,* **139**, 372–397
13. Gorham, P. R. and Carmichael, W. W. (1979). Phycotoxins from blue-green algae. *Pure and Applied Chemistry,* **52**, 165–174
14. Hou, Q., Chen, Z. and Yu, J. (1983). Ore fabric and effect of blue-green algae on iron richment in Xuanlong iron mine, Hebei (China). *Dizhi Kexue,* **1983**, 246–250
15. Huang, C-M., Chang, C-H. and Wang, C-C. (1982). Effect of soil inoculant application on sugar cane growth. *Report of the Taiwan Sugar Research Institute,* **98**, 15–24
16. Hughes, S. G. and Murray, K. H. (1980). The nucleotide sequences recognized by endonucleases *Ava* I and *Ava* II from *Anabaena variabilis. Biochemical Journal,* **185**, 65–67
17. Joset, F. (1983). Les cyanobactéries ont-elles un avenir en biotechnologie? Analyse de leurs propriétés génétiques. *Biosciences,* **2**, 214–218
18. Jüttner, F. (1982). Mass cultivation of microalgae and phototrophic bacteria under sterile conditions. *Process Biochemistry,* **17**, 2–7
19. Kayano, H., Karube, I., Matsunaga, T., Suzuki, S. and Nakayama, O. (1981). A photochemical fuel cell system using *Anabaena. European Journal of Applied Microbiology and Biotechnology,* **12**, 1–5
20. Kenyon, C. N. (1972). Fatty acid composition of unicellular strains of blue-green algae. *Journal of Bacteriology,* **109**, 827–834
21. Kernforschungsanlage Jeulich GmbH (1978). Microorganism cultures for extracting uranium from seawater. *British patent* 1,507,003 (C1.A01H13/00), 12 April 1978
22. Lambert, G. R., Daday, A. and Smith, G. D. (1979). Hydrogen evolution from immobilized cultures of the cyanobacterium *Anabaena cylindrica* B629. *FEBS Letters,* **101**, 125–128
23. Lambert, G. R. and Smith, G. D. (1981). The hydrogen metabolism of cyanobacteria (blue-green algae). *Biological Reviews,* **56**, 589–660
24. Mitsubishi Heavy Industries Ltd (1983). Hydrogen production of a blue-green algal photochemical system. Japanese Patent 58 76, 094 [83 76m 094] (C1.C12P3/00), 9 May 1983

25. Mitsui, A. (1980). Saltwater based biological solar energy conversion for fuel, chemicals, fertilizer, food and feed. In *Proceedings of Bio-Energy,* **80**. *World Congress and Exposition*, pp. 486–491. Bioenergy Council, Washington
26. Muallem, A., Bruce, D. L., and Hall, D. O. (1984). Photoproduction of hydrogen and $NADPH_2$ by polyurethane-immobilized cyanobacteria. *Advances in Photosynthetic Research,* **2**, 797–800
27. Mullineaux, P. M., Gallon, J. R. and Chaplin, A. E. (1981). Acetylene reduction (nitrogen fixation) by cyanobacteria grown under alternating light–dark cycles. *FEMS Microbiology Letters,* **10**, 245–247
28. Murray, K., Hughes, S. G., Brown, J. S. and Bruce, S. A. (1976). Isolation and characterization of two sequence-specific endonucleases from *Anabaena variabilis. Biochemical Journal,* **159**, 317–322
29. Musgrave, S. C., Kerby, N. W., Codd, G. A. and Stewart, W. D. P. (1982). Sustained ammonia production by immobilized filaments of the nitrogen-fixing cyanobacterium *Anabaena* 27,893. *Biotechnology Letters,* **4**, 647–652
30. Neil, G., Nicholas, D. J. D., Bockris, J. O'M. and McCann, J. F. (1976). The photosynthetic production of hydrogen. In *Proceedings of the First World Hydrogen Energy Conference*. Ed. T. N. Veziroglu. Vol. 2, pp. 69–76. University of Miami, Coral Gables, Florida
31. Newton, R. P., Walton, T. J. and Moyse, C. D. (1977). Non-α-tocopherols in the unicellular blue-green alga *Gloeocapsa. Biochemical Society Transactions,* **5**, 1486–1489
32. Ochiai, H., Shibata, H., Sawa, Y., Shoga, M. and Ohta, S. (1983). Properties of semiconductor electrodes coated with living films of cyanobacteria. *Applied Biochemistry and Biotechnology,* **8**, 289–303
33. Pignatello, J. J., Porwoll, J., Carlson, R. E., Xavier, A., Gleason, F. K. and Wood, J. M. (1983). Structure of the antibiotic cyanobacterin, a chlorine containing gammalactone from the freshwater cyanobacterium *Scytonema hofmanni. Journal of Organic Chemistry,* **48**, 4035–4038
34. Pushparaj, B., Caroppo, S., Sili, C. and Miclaus, N. (1976). Preliminary tests for waste water purification by algal filters. *Nuovi Annali Ig. Microbiologia,* **27**, 407–417
35. Reed, R. H., Richardson, D. L., Warr, S. R. C. and Stewart, W. D. P. (1984). Carbohydrate accumulation and osmotic stress in cyanobacteria. *Journal of General Microbiology,* **130**, 1–4
36. Santillan, C. (1982). Mass production of *Spirulina. Experientia,* **38**, 40–43
37. Shestakov, S. V. and Jevner, V. D. (1968). Study of mutagenesis in blue-green alga *Anacystis nidulans. Proceedings of the XIIth International Congress on Genetics,* **1**, 84
38. Slater, G. C. and Block, V. C. (1983). Odorous compounds from cyanobacterial blooms. *Water Science Technology,* **15**, 229–240
39. Stacey, G., Van Baalen, C. and Tabita, F. R. (1977). Isolation and characterization of a marine *Anabaena* sp. capable of rapid growth on molecular nitrogen. *Archives of Microbiology,* **114**, 197–201
40. Talley, S. M., Lim, E. and Rains, D. W. (1981). Application of *Azolla* in crop production. In *Genetic Engineering of Symbiotic Nitrogen Fixation and Conservation of Fixed Nitrogen*. Ed. J. M. Lyons *et al*. pp. 363–384. Plenum, New York

41. Tambiev, A. K., Chitao, S. I., Kirikova, N. N., Shelyastina, N. N. and Chang, D. Q. (1984). Extracellular products of cyanobacteria and their biological activity. *Samoochishchenie Vody MIgr. Zagryaz. Troficheskoi Tsepi (Sb. Tr. Vses. Soveshch. Sanit. Gidrobiol.)*, 4th. pp. 70–74
42. Wang, F.-C., Zhung, Z. and Hu, Z. (1981). Nitrogen fixation by edible terrestrial blue-green algae. In *Current Perspectives in Nitrogen Fixation*. Ed. A. H. Gibson and W. E. Newton, p.455. Australian Academy of Science

2
PHYCOTECHNOLOGY

Peter K. Robinson

INTRODUCTION

Algae are non-vascular, photosynthetic plants which contain chlorophyll *a* and have simple reproductive structures. The group contains some 1800 genera and 21 000 species, and includes unicellular forms (such as *Chlorella*), small colonial forms (such as *Volvox*) and large multicellular forms (such as the seaweeds). The group has also historically included the prokaryotic blue-green forms, although these are now thought to be more allied to the bacteria than to the rest of the (eukaryotic) algae.

The biotechnological use of algae (phycotechnology) has until recent times almost exclusively involved use of the marine macroalgae (i.e. the seaweeds). The earliest record of seaweed utilisation dates back to 2700 BC in a Chinese herbal compiled according to the directions of the Emperor Shen Nung, while in Europe Roman women are known to have used a red algal extract as a dye for cosmetic purposes around 50 BC.

In recent years advances in biotechnological processes employing algae have been much slower than in those employing bacteria or fungi. Many phycotechnological processes are still essentially 'low-tech' concerns, perhaps being of local importance but with little global impact. It is, perhaps, not surprising that a recent review of targets in biotechnology[29] almost overlooked phycotechnology. This chapter outlines current uses of algae and algal products, and introduces some of the potential uses of the algae.

The current usage of algae and algal products almost exclusively involves the marine macroalgae. Natural ('wild') microalgal populations are not particularly amenable to human exploitation, since they are usually too dispersed to be economically harvested, are often ephemeral and are rarely homogeneous. Thus, exploitation of microalgal biomass generally necessitates deliberate cultivation, resulting in a costly product. Using current technology, algal protein is 10 times more expensive to produce than is its soya protein equivalent. Growth of microalgae on a commercial basis is therefore not economically viable in Europe or North America, although in the Far East 1000 tons of microalgae are produced annually[33], the high production cost being offset by the high price commanded by the product as a health food. The marine macroalgae, however, are much more amenable to human exploitation and have historically been used as fertilisers, as food for both animals and man, and as raw materials for various industrial processes. The current economic value of seaweed production is shown in Table 1.

Table 1 Estimate of world seaweed production (after Naylor[23])

Location	10^3 t *wet weight*	*Value* *($ million)*
Japan	654	563.0
China	700	130.0
Korea (Republic)	224	45.0
USA	126	1.9
USSR	100	5.8
UK	24	0.4
Others	572	18.9
Total world	2400×10^3	765.0

In Europe seaweeds such as *Laminaria, Macrocystis* and *Ascophyllum* have been used as fertilisers since the twelfth century. For such use, seaweeds have been used directly (dug into the soil), composted, or reduced to dry or liquid preparations. Liquid products are nowadays increasingly preferred, two well-known brands being 'Maxicrop' and 'Baby-Bio'.

Seaweeds have been used for centuries as fodder (coarse food for animals) in numerous countries, including the UK, France, Canada and Iceland. Although in earlier times the use of seaweeds was restricted to coastal areas near the source of supply, modern practice is to use dried seaweed meal, which may be easily transported for use in coastal or inland areas. Such meal is usually fed to livestock in the proportion of 10 per cent of the total ration by weight. Figures for seaweed meal production are difficult to find; Jensen[15], however, estimated that seaweed meal production in Norway alone was 12 000–15 000 tonne/annum.

In Europe the general use of seaweed as human food was discontinued about 50–100 years ago, although small quantities of *Porphyra* are still eaten in Iceland and Wales. Annual production of laver (a *Porphyra* product) in Great Britain has been estimated as 200 tons[27]. Probably only distress and/or poverty would bring about a wider usage comparable to that of a century or so ago.

Only in the Far East have algae been regularly used for human food, the most prolific users being the coastal populations of China, Japan and the tropical Pacific Islands, where numerous genera are used, the two most common being *Porphyra* and *Laminaria*. In the Pacific Islands chopped raw algae are added to many dishes. In Japan *Porphyra* is washed, chopped and spread on frames to dry. These dried sheets ('hoshi-nori') may be rolled around rice and seafood as 'sushi', toasted and sprinkled on rice, or used in various other ways.

In Europe the industrial use of seaweed historically comprised three phases. The first began in the seventeenth century in France, where brown rockweeds (species of *Fucus* and *Ascophyllum*) and then later oarweeds (species of *Laminaria*) were collected for the production of soda, which was subsequently used in the glazing of pottery and in glass manufacture. This industry declined after 1810 as the importation of Barilla soda began, and by 1845 the industry was almost dead. The discovery of the element iodine in algae by Courtois in 1811 and the subsequent use of it in medicine revived the industry, but with the discovery of the Chilean mineral deposits of iodine the industry entered a second decline. The third revival took place mainly during World War II, with the rise of the alginate industry.

Nowadays the algal soda industry is apparently dead. So far as the algal iodine industry is concerned, Chapman[5] suggested that Russia was possibly the sole producer, while Bonotto[4] suggested that only China and Japan produce seaweed iodine for commercial purposes, and Volesky *et al.*[34] report that the industry is still active in Japan, where it contributes 5–7 per cent of world production.

Today much of the macroalgae industry is devoted to the production of various algal polysaccharides. Such polysaccharides may make up 10–65 per cent of the algal dry weight. The polysaccharides of greatest importance are from the brown algae (and include alginic acid derivatives, fucoidans and laminarans) and the red algae (including agar and carrageenin), while other polymers, such as xylans and mannans, are currently of more academic interest. The source, composition and uses of some of the economically important algal polysaccharides are shown in Table 2.

ALGAE FOR FOOD

The nutritional value of algae at present lies primarily with their vitamin and mineral contents. Even in countries where algae are regularly consumed, they are considered a delicacy or luxury rather than part of the basic necessary diet. In Japan, for example, average consumption of algae is only 10 g per person per day[16].

Table 2 Algal polysaccharides

Polysaccharide	*Source*	*Composition*	*Uses*	*Cost*[a]
Agar	Red algae—*Gelidium, Gracilaria, Gigartina*	Agarose combined with agaropectin	Microbiology, food preparation and canning, manufacture of processed cheese, mayonnaise, jelly and ice-cream, stabiliser and emulsifier, 'carrier' for drugs	£30/kg
Alginate	Brown algae, mainly *Macrocystis*	Mannuronic and guluronic acid residues	Ice-cream, paper products and adhesives, thickener in paints, 'filler' in drugs	£20/kg
Carrageenin	Red algae—*Chondrus, Gigartina, Iridae*	Galactose residues	Stabiliser for emulsions in food, drug and brewing industries	£30/kg
Fucoidan	Brown algae	L-Fucose residues		£12/kg
Laminaran	Browm algae—*Laminaria, Ascophyllum, Fucus*	Glucose residues		£15/kg

[a]All costs are approximate and obviously vary according to purity of compound required.

Table 3 Nutritional composition of algae

Dried material	% *Water*	% *Raw protein*	% *Fat*	% *Carbohydrate*	% *Fibre*	% *Ash*
Enteromorpha	13.6	12.4		53.0	10.6	10.4
Ulva	18.7	14.9	0.04		0.2	15.6
Laminaria	23.5	5.85	1.15		6.7	21.1
Porphyra	17.1	25–30	0.8	40.1		10.3
Gracilaria	12.85	7.9	0.05		3.0	17.8

Table 3 contd.

Source	*Essential amino acids*[a]	*Non-essential amino acids*[a]	*Total N*[a]
Chlorella	53.6	47.7	101.3
Ulva	21.7	27.8	49.5
Laminaria	27.4	31.2	58.6
Chondrus	27.2	27.5	54.7

[a]Amounts given as g amino acid N/100g protein N

Table 3 cont'd.

Source	*Ascorbic acid (C)* (ppm)	*Thiamine (B1)* (ppm)	*Riboflavin (B2)* (ppm)	*Niacin* (ppm)	*Pantothenic acid* (ppm)	*Cobalamin (B12)* (ppm)
Chlorella						
Ascophyllum	1800–4000	3.7–22	19.5–55	111–220	18.5	0.021
meal	500–2000	1.4–5.4	5–10	10–30	0.2	0.004–0.08
Laminaria meal	2094	1.3–7.2	2.4	19.4	0.28	0.06–0.12

Table 3 cont'd.

Source	*Fe* (ppm)	*Mn* (ppm)	*Cu* (ppm)	*Zn* (ppm)	*Mo* (ppm)	*Co* (ppm)	*Sr* (ppm)
Laminaria	138–293	9–10	3–14	62–76	0	0.25–0.92	4000
Ascophyllum	283	27	1.1–4	60	0.29	0.73	2600
Fucus	221–638	104–155	3.4–17	60–70	0.29–0.65	0.65–1.39	2800

Consideration of the 'nutrient profile' of algae (Table 3) reveals that they do contain substantial amounts of carbohydrate, together with small quantities of fat and protein. However, since many of the algal carbohydrates cannot be assimilated, their nutritive value is less than may be expected, perhaps making them of some use in low-calorie diets (as suggested by Takagi[31]). Algal polysaccharides may also be subjected to enzymatic hydrolysis to yield simple carbohydrates assimilable by man, but as yet this process is not economically feasible.

World stocks of macroalgae are large. Michanek[20] estimated the current annual world harvest of seaweed to be 2 million tonnes (fresh weight), with a potential harvest of about an order of magnitude greater. The deliberate cultivation of macroalgae may increase this figure. However, areas useful for seaweed cultivation are limited to those coastal waters free from pollution, of open aspect (allowing easy, preferably mechanical, harvesting) and free from major climatic fluctuations (especially storms). The availability of such areas limits the further development of the seaweed cultivation industry, which in Japan already covers 70 000 ha for the cultivation of *Porphyra* alone. Land-based ponds may also be used for seaweed cultivation. In Taiwan, for example, 12 000 t (dry weight) *Gracilaria* are produced annually from 300 ha of ponds; but while this 'low-tech' culture may be further developed in the Third World (where land and labour are cheap), it is unlikely to gain widespread usage in the more developed countries (where both land and labour are expensive).

While growth rates of macroalgae can be spectacular (giant kelp fronds growing at a rate of 65 cm/day, for example), *specific* growth rates of microalgae are much higher. Maximum specific growth rates of macroalgae range from 0.02 to 3.30/day, and average about 0.14/day[14], while specific growth rates of microalgae may be as high as 5.5/day in laboratory-scale cultures[10]. Developments in microalgal culture towards commercial exploitation go back over 30 years. The first attempts at large-scale outdoor culture of microalgae involved the use of systems in which the culture was partitioned from the environment by a transparent barrier so as to enable control of as many environmental variables as possible and to reduce the risk of non-algal contamination. But, once the feasibility of open systems was demonstrated, subsequent developments in this technology have almost exclusively featured open cultures. Closed systems do, however, still possess two major advantages over their open counterparts, the first being their reduced rate of water loss through evaporation (useful in semi-arid areas) and the second being the possibility of harnessing released oxygen.

The productivities and costs of microalgal production have been variously quoted. Tamiya[32] considered that microalgae could be produced in outdoor ponds at a cost of $0.26/lb, Soeder[28] suggests $1/kg and

Oswald[25] \$0.15–0.25/kg. White and Plaskett[35] consider 60–90 dry tonnes per hectare per annum as a realistic productivity in such open pond systems. In order to reduce the costs of mass culture of microalgae, it has been suggested that the process may be used in the latter stages of sewage treatment such that the algae are able to metabolise the materials in the waste-water (thereby obviating the need for supplied organic or inorganic nutrients) and at the same time purify the waste-water. Oswald[24] estimated that the cost of algae produced in this way would be only \$0.01/kg (dry weight). Consideration of this aspect of algal culture has been made in Ref. 26; suffice it to say that problems associated with the presence of pathogens and parasites and with bioaccumulation of various substances currently discourage the use of algae grown in such a fashion for human consumption. Algal growth in waste-water treatment systems is, however, perhaps the only way (using currently available technology) that a microalgal product can be grown at a cost such that it could be used as an animal (rather than human) feedstuff, and preliminary trials have shown that waste-grown algal biomass may be successfully used to supplement the diets of poultry, fish and livestock without adverse effect (see, for example, Ref. 22). Algal biomass produced in this way may also be useful for the generation of available energy; this will be discussed in the next section.

Before leaving the use of algae for food, one point of interest is the indirect use of algae to improve the growth of other plants. Certain blue-green algae (both free-living and endophytic) are known to be capable of nitrogen fixation, i.e. they are able to live without any supply of combined nitrogen and use gaseous (atmospheric) nitrogen in their metabolism. Growth of these organisms alongside crop plants is thus beneficial to crop productivity in that it allows atmospheric nitrogen to enter the ecosystem. It has been reported, for example, that rice from these inoculated fields contained 7.5 lb more nitrogen per acre than the controls.

ALGAE FOR ENERGY

There are various ways by which available energy can be derived from algal cells. Cells may be dried and incinerated to generate electricity, but this seems neither energetically nor economically feasible, and has been little investigated. Algal cells may also be anaerobically digested to form biogas, rich in methane. Work indicates that yields of 0.4–0.56 m^3/kg are theoretically possible from the digestion of *Macrocystis*, with over 80 per cent conversion of the organic matter content. In practice, however, large-scale digestion is likely to result in yields of about 0.28 m^3/kg, which is still high when compared with other green vegetable feedstock. The advantage of many algae is that their organic matter is algin and cellulose, which is almost completely biodegradable by anaerobic digestion. There may be problems, however, with the digestion of marine algae because of the high chloride content of the associated water, and for this reason it may be that thermal conversion may offer an alternative route particularly towards liquid fuel generation with species with high oil contents.

Algae may also be of use in energy production, in that they are capable of producing hydrogen by the photolysis of water. Photolysis of water is a natural process taking place in the chloroplasts of all plants to produce free oxygen and organic matter from carbon dioxide and water. In 1974, however, it was first shown that the blue-green alga *Anabaena*, cultured under strict anaerobic conditions and in the absence of dinitrogen, could produce free hydrogen through a side reaction of the nitrogenase-catalysed reduction of atmospheric nitrogen to ammonia. It was subsequently shown that this process is not necessarily an anaerobic one, since when algae were incubated in air with low concentrations of carbon monoxide (an inhibitor of all but the hydrogen-producing reaction of nitrogenase) and acetylene (an inhibitor of hydrogenase), the rate of hydrogen evolution was comparable to that produced when cultured under argon. Subsequent studies have shown that the process is widely distributed among the blue-green algae, being carried out by fresh-water and marine, mesophilic and thermophilic, species. Experimental work (e.g. Ref. 3) has shown that algal cultures are able to produce hydrogen at the rate of 0.4 per cent of the solar conversion efficiency, and possibly to as high as 2 or 3 per cent, and it has been calculated that under favourable conditions 0.65 litres hydrogen gas/m culture surface could be produced. Advantages of the process are that it operates at low temperature, inputs (solar energy and water) are cheap and it results in a clean burning fuel which produces no pollutants. Mitsui and Kamazawa[21] calculated that a 214 km^2 algal culture area could supply the total US needs for hydrogen in the year 2000.

ALGAE FOR BIOCHEMICALS

The commercial usage of algae as sources of biochemicals is currently largely restricted to algal polysaccharides. However, the algae are known to contain (and in some cases actively excrete) a wide range of substances from simple organic acids to complex peptides, from gamete attractants to toxins. Rather than include a lengthy discussion of this wide field of interest, much of the relevant work on algal products is summarised in Table 4 (although this is in no way meant to be comprehensive) and discussion is restricted to a few of the more interesting findings.

One of the greatest biotechnological advances utilising algae is that of the production of glycerol from the salt-tolerant green alga *Dunaliella*. In high-salinity environments this alga naturally produces large amounts of intracellular glycerol to counterbalance the high external osmotic stress,

Table 4 Examples of algal products

Source	*Product*	*Effect*
Laminaria	Laminine	Hypotensive activity
Fucus/Sargassum	Unsaturated fatty acids	Hypocholesterolaemic activity
Digenea simplex	L-alpha-kainic acid	Antihelminthic
Chondria armata	Domoic acid	Antihelminthic
Anbaena cycadea	Whole cells	Tuberisation of cycadea roots
Nostoc rivulare	Culture filtrates/extracts	Tumours in mice after intraperitoneal or subcutaneous injection
Fucus	Volatile	Male gamete attractant
Chlamydomonas eugametos	Gamone—high molecular weight Glycoprotein	Agglutination of gametes of opposite mating types
Various	Endotoxins/exotoxins	
Various	Growth inhibitors	
Various	Enzymes, including those which hydrolyse sucrose, starch and proteins, acid and alkaline phosphatases, glutaminase and enzymes which depolymerise agar	
Various	Vitamins, including thiamine, biotin, pyridoxine, ascorbate, nicotinic acid, pantothenate, riboflavin and tocopherol	

such that 50 per cent of the algal dry weight may be glycerol. Chen and Chi[6] evaluated the cost of producing such glycerol and revealed that while the traditional petrochemical process manufactured glycerol at $0.89/kg, the algal process could manufacture it at $0.75/kg; and since the algal process is capital- and labour-intensive, whereas the petrochemical process is energy-intensive, production costs for the petrochemical process would be expected to rise more rapidly in the future than costs for the algal process. Ben-Amotz[2] suggested that technical-grade glycerol costs $1.20/kg, while 99 per cent glycerol now costs about $7/kg. The market for the product is large; the USA alone used 510 million lb in 1978, major uses being drugs and cosmetics (19 per cent), alkyds (17 per cent), tobacco (12 per cent), food and beverages (10 per cent), cellophane (10 per cent), urethane polyols (7 per cent) and explosives (4 per cent). The high labour demand of the algal process makes it particularly suitable for use in less developed countries where labour costs are low, such countries often also having favourable climates (warm and sunny). Most current interest in the algal glycerol process seems to be in the Middle East, where already Koor Food Ltd (Israel) have announced intentions to construct a 2 km^2 algae farm which is expected to produce 11 000 t of glycerol, 550 t of beta-carotene and 19 800 t of animal feed per year.

Algae also contain various other intracellular components which may be extracted economically. For example, Ahern *et al.*[1] report that up to 36 per cent of the lipids of the red alga *Porphyridium cruentum* may be arachidonic acid, which is traditionally isolated from the adrenal gland and liver tissues. An essential human dietary constituent, arachidonic acid is also the natural precursor of a large group of C20 compounds (i.e. prostaglandins, thromboxanes, leukotrienes and prostacyclin) which are at present the subject of intensive medical research. The current cost of arachidonic acid is about $85/g (free acid, 99 per cent purity) and it is claimed that the cost of producing the *Porphyridium* for the purpose of arachidonic acid extraction would be only $0.16–1.00/g arachidonic acid.

One of the most fundamental chemicals produced by algae is oxygen, and it has been suggested by several authors that in enclosed spaces (particularly in space flight) illuminated algae may be able to supply the oxygen demands of the crew (probably first suggested by J. B. S. Haldane and others connected with the early years of the British Interplanetary Society). While the problem of man's respiratory needs during short-duration flights is solvable by carrying enough oxygen for the voyage, with prolonged voyages this approach becomes logistically suboptimal and photosynthetic systems become more attractive.

Algae are also known to produce and excrete a wide range of antibiotic/biostatic agents. However, as yet, no antibiotic has been developed to the commercial scale, since either the activity is insufficiently potent or the production excessively costly (or both) when compared with fungal antibiotics.

Algal products have been used in herbal medicine in the Far East for centuries. One example of more conventional usage was the use of phycocolloid (Algasol T 331), which proved efficacious in the treatment of oncologic patients, producing a good recovery in 68 per cent of the 162 patients treated[8]. While various drugs have made use of algal products, there is at least one record of the surgical use of algae: short pieces of the stems of *Laminaria* have been used in surgery for widening fistulae and wound entrances, their use relying on their great ability to swell when moistened[5].

Thus, the list of useful algal products is long, and includes a great diversity of biochemicals having a great diversity of uses/effects, only a few of which have been mentioned. A more extensive coverage of the subject of algal products (though including little of their biotechnological uses) is given in the texts of Jackson[13], Lewin[17] and Stewart[30].

CONCLUSION

It is clear, therefore, that while we are just beginning to exploit the algae for biotechnological uses, certain processes (e.g. glycerol from *Dunaliella*) are already economically competitive with existing processes. However, it is questionable whether we should really expect major financial rewards at such an early stage of exploitation. It has taken many years of selective breeding of terrestrial plants to produce the 'super-productive' crops we have today, whereas exploitation of the algae still largely uses only the wild types; and it seems certain that the processes considered in this chapter would be further improved by some type of genetic manipulation.

The Chinese are currently the world leaders in the practical breeding of algae — in particular, *Laminaria*. Through a selection and inbreeding programme carried out over 15 years the Chinese have developed strains superior to wild types in growth rate, iodine content and temperature tolerance. As well as this conventional selection programme, the Chinese are also experimenting with a monoploid breeding programme[11]. Here monoploid female *Laminaria* plants are induced to develop parthenogenetically into large sporophytes which then undergo spontaneous or induced chromosome doubling to form completely homozygous diploid sporophytes, thereby allowing easy comparison of the desirability of individual genomes.

Some groups of algae have also been found to readily hybridise across specific and even generic boundaries (see, for example, Ref. 19), and though it is impossible to predict the characteristics of such hybrids, it does not seem unreasonable to suppose that some of the progeny will be of use for biotechnological exploitation.

The prokaryotic blue-green algae lack sexual reproduction, but it is possible to incorporate desirable traits into the genotype by transformation or transduction, as has been demonstrated by Delaney[9] and Hondel *et al.*[12]. Lewin[18] suggests that eventually such techniques may be applicable to the eukaryotic algae also.

Thus, the application of such genetic manipulation techniques to potentially useful algae is likely to be a major objective for biotechnologists over the coming years. With our increasing understanding of basic algal physiology and biochemistry, this should encourage a more vigorous exploitation of algae at the commercial level than has been so far apparent.

The algae, then, are a biotechnologically useful group of organisms with considerable potential in the future. Like many biotechnologies, much of the current interest in the use of algae is related to the production of high-value, low-volume, products. For the future, therefore, it would seem likely that while the scale of algal protein (food) production would remain globally small, the scale of algal-derived fine-chemical production may well become globally significant.

REFERENCES

1. Ahern, T. J., Katoh, S. and Sada, E. (1983). Arachidonic acid production by the red alga *Porphyridium cruentum. Biotechnol. Bioeng.,* **25**, 1057–1070
2. Ben-Amotz, A. (1980). Glycerol production in the alga *Dunaliella.* In *Biochemical and Photosynthetic Aspects of Energy Production* (Ed. Pietro, A. San). pp. 191–208. Academic Press, New York
3. Benemann, J. R. (1978). Recent developments in hydrogen production in micro-algae. In *Energy from Biomass and Wastes* (Eds. Klass, D. L. and Waterman, W. W.). Institute of Gas Technology, Chicago
4. Bonotto, S. (1976). Multicellular plants. In *Marine Ecology,* Vol. 3: *Cultivation, Part 1* (Ed. Kinne, O.). pp. 467–530. Wiley, London
5. Chapman, V. J. (1970). *Seaweeds and Their Uses*, 2nd edn. Methuen, London
6. Chen, B. J. and Chi, C. H. (1981). Process development and evaluation for algal glycerol production. *Biotechnol. Bioeng.*, **23**, 1267–1287
7. Chynoweth, D. P., Klass, D. A. and Ghosh, S. (1978). Biomethanation of giant brown kelp, *Macrocystis pyrifera.* In *Energy from Biomass and Wastes* (Eds. Klass, D. L. and Waterman, W. W.). Institute of Gas Technology, Chicago
8. Claudio, F. and Stendardo, B. (1966). In *Proc. 5th Int. Seaweed Symp.* pp. 369 *et seq*. Pergamon Press, Oxford
9. Delaney, S. P., Herdman, M. and Carr, N. G. (1976). Genetics of blue-green algae. In *The Genetics of Algae* (Ed. Lewis, R. A.). pp. 7–12. Blackwell, Oxford
10. Droop, M. R. (1974). Heterotrophy of carbon. In *Algal Physiology and Biochemistry* (Ed. Steward, W. D. P.). pp. 530–559. Blackwell, Oxford
11. Fang, T. C., Tai, C. H., Ou, Y. L., Tseui, C. C. and Chen, T. C. (1978). Some

genetic observations on the monoploid breeding of *Laminaria japonica*. *Sci. Sin.*, **21**, 401–408
12. Hondel, C. A. M. J. J. Van Der, Verbeck, S., Ende, A. Van Der, Weisbeck, P. J., Borrias, W. E. and Arkel, G. A. Van (1980). Introduction of transposon Tn901 into a plasmid of *Anacystis nidulans*. Preparation for cloning. *Proc. Natl Acad. Sci. U.S.A.*, **77**, 1570–1574
13. Jackson, D. F. (Ed.) (1964). *Algae and Man*. Plenum Press, Oxford
14. Jackson, G. A. (1980). Marine biomass production through aquaculture. In *Biochemical and Photosynthetic Aspects of Energy Production* (Ed. Pietro, A. San). pp. 31–58. Academic Press, Oxford
15. Jensen, A. (1966). Norsk. Inst. Tang-og Tareforsk. Rept. 31
16. Kirby, R. H. (1953). *Seaweeds in Commerce*. HMSO, London
17. Lewin, R. A. (Ed.) (1962). *Physiology and Biochemistry of Algae*. Academic Press, New York
18. Lewin, R. A. (1983). Phycotechnology — how microbial geneticists might help. *Bioscience*, **33**, 177–179
19. Luning, K., Chapman, A. R. O. and Mann, K. H. (1978). Crossing experiments in the nondigitate complex of *Laminaria* from both sides of the Atlantic. *Phycologia*, **17**, 293–398
20. Michanek, G. (1975). Seaweed resources of the ocean. FAO Fish. Tech. Pap. No. 138
21. Mitsui, A. and Kamazawa, S. (1977). Hydrogen production by marine photosynthetic organisms as a potential energy resource. In *Biological Solar Energy Conversion* (Eds. Mitsui, A. *et al.*). Academic Press, New York
22. Moraine, R., Shelef, G., Meydan, A. and Levi, A. (1979). Algal single cell protein from wastewater treatment and renovation process. *Biotechnol. Bioeng.*, **21**, 1191–1207
23. Naylor, J. (1976). Production, trade and utilisation of seaweeds and seaweed products. FAO Fish. Tech. Pap. No. 159
24. Oswald, W. J. (1973). In *Progress in Water Technology*, Vol. 3: *Water Quality: Management and Pollution Control Problems* (Ed. Jenkins, S. H.). pp. 153–163. Pergamon Press, Oxford
25. Oswald, W. J. (1979). Complete waste treatment in ponds. In *Handbook of Microbiology*. CRC Press, Cleveland, Ohio
26. Robinson, P. K. (1984). Biotechnological involvement and advances in wastewater treatment. Part I: *Industrial Biotechnology Wales*, **3** (issue 3), 5–15. Part II: *Industrial Biotechnology Wales*, **3** (issue 4), 1–6
27. Scmid, O. J. and Hoppe, H. A. (1962). *Bot. Marina, Suppl.*, **101**
28. Soeder, C. J. (1978). Economic considerations concerning the autotrophic production of microalgae at the technical scale. *Arch. Hydrobiol.*, **11**, 259–273
29. Spinks, A. (1982). Targets in biotechnology. *Proc. R. Soc. Lond. (B)*, **214**, 289–303
30. Stewart, W. D. P. (Ed.) (1974). *Algal Physiology and Biochemistry*. Blackwell, Oxford
31. Takagi, M. (1975). Seaweeds as medicine. In *Advances in Phycology in Japan* (Eds. Tokida, J. and Hirose, H.). pp. 321–325. Junk, The Hague
32. Tamiya, H. (1975). Mass culture of algae. *Ann. Rev. Plant Physiol.*, **8**, 309–334
33. Tsudaka, O., Kawahara, T. and Miyachi, S. (1978). Mass culture of *Chlorella*

in Asian countries. In *Biological Solar Energy Conversion* (Eds. Mitsui, A. *et al.*). Academic Press, London
34. Volesky, B., Zajic, J. E. and Knettig, E. (1970). Algal products. In *Properties and Products of the Algae* (Ed. Zajic, J. E.). pp. 49–82. Plenum Press, New York
35. White, L. P. and Plaskett, L. G. (1981). *Biomass as Fuel*. Academic Press, New York

3
THE INDUSTRIAL POTENTIAL OF EXTREME THERMOPHILES

Roy M. Daniel, H. W. Morgan and A. M. Donnison

INTRODUCTION

New Zealand has extensive thermal regions which, unlike those of other parts of the world, are easily accessible at all seasons of the year. The hot springs of these regions are diverse in mineral composition, acidity and temperature. Only bacteria can live at temperatures higher than 65°C, and those growing optimally at 65°C or higher are called extreme thermophiles. Currently biochemical and physiological studies of extremely thermophilic bacteria are being made by the Thermophile Research Unit at the University of Waikato alongside studies of the ecology of the hot springs from which the bacteria are isolated.

BACTERIAL TYPES

A large number of bacteria have been isolated from the New Zealand hot springs (and from hot springs of other parts of the world), but apart from the factors enabling the organisms to grow at high temperatures (i.e. the stability of their proteins and the high melting point of their membrane lipids) they do not differ in physiology from other bacteria. Table 1 sets out the current nomenclature of 'thermophily'[1].

Table 1 Current Nomenclature of Thermophily

Technical term	*Representative*	*Optimal growth temperature* (°C)
Mesophilic	*Escherichia coli*	37
Thermophilic	*Bacillus stearothermophilus*	60
Extremely thermophilic	*Thermus thermophilus*	
	Thermoproteus tenax	88

THE BIOCHEMISTRY

It was originally thought that growth and survival of the thermophilic bacteria at high temperatures was achieved by rapid enzyme turnover rather than enzyme stability[2, 3], but it is now well accepted that most enzymes isolated from extreme thermophiles are stable at the optimum growth temperature of the organism, even in the purified state[4]. There have been many theories proposed as to the mechanisms of this thermostability (Refs. 4, 5, 6–8 review this work) but no general cause seems to apply. Current evidence suggests that since the net free energy of stabilisation of proteins is small (this free energy is the result of forces which hold it together), one or more quite minor changes in the structure of a protein could lead to large changes in thermostability[9]. It was found that by changing only one of the many amino acids which make up the lysozyme from bacteriophage T4 the functional temperature was lowered by 14°C. The structure and properties of the enzymes from extreme thermophiles is naturally the major research interest of the Thermophile Research Unit.

ENZYMES IN INDUSTRY

There are a number of advantages in using enzymes as industrial catalysts. They are usually highly specific and thus undesirable side-reactions which occur in chemically catalysed reactions are rarely a problem. Enzymes are generally more active and efficient than chemical catalysts, and the effluent from such processes is less toxic.

An example of the advantages of using enzymes as compared with chemical catalysts is the conversion of glucose to fructose[10]. The chemical conversion is carried out under alkaline conditions at high temperatures. This results in the production of contaminating substances (ketohexose and coloured by-products) which are costly to remove[11]. Chemical conversion techniques are not economically viable. A number of glucose isomerase enzymes can convert 50 per cent of the glucose to fructose without undesirable by-products, although production costs are still high at the present time.

The rapid growth of the enzyme industry reflects their advantages. Enzymes are now used in the food and beverage industry (having a market value of about $500 million in 1985), in detergents and in medical diagnostics. Current growth rates of 10–15 per cent per year can be expected to rise as further new processes are developed. This occurred in the development of enzymes to produce high fructose corn syrup from starch, with the enzymes amylase, amyloglucosidase and glucose isomerase. The value of the enzymes in the USA rose from $26 million in 1975 to $76 in 1980.

The most significant new developments are likely to be in the fields of clinical diagnostics, cellulose hydrolysis and immobilisation of enzymes.

In diagnostics the absolute specificity of many enzymes is used to measure the concentration of important metabolites — for example, glucose, urea and cholesterol in biological fluids such as blood and urine. One such kit measures the level of creatine kinase — a substance which increases in the blood after a heart attack.

The second important growth area is enzymatic hydrolysis of cellulose to glucose. In this process a waste product which is readily and cheaply available (e.g. corn stover, wood chip or straw) can be converted to glucose which can, in turn, be fermented to produce single-cell protein or ethanol.

The recent development of enzyme immobilisation technology (i.e. the linking of enzymes to insoluble support matrices) is likely to further widen the application of enzymes to industrial catalytic processes. Immobilisation often increases the stability of an enzyme and may modify its properties in ways which make it more efficient in an industrial process. The use of immobilised enzymes has permitted the development of highly efficient continuous-flow reactor systems.

ADVANTAGES OF ENZYMES FROM EXTREME THERMOPHILES

Although enzymes have many advantages, those derived from mesophilic organisms are generally unstable and expensive. The high cost of commercial enzyme production arises from low yields of enzyme per unit of biomass, the expense of extraction and purification, and losses of active enzyme during purification, storage and handling. Running costs are often high because of the short life-span of most enzymes.

A number of proccsses are considered unsuitable for enzyme catalysis because of the cost and/or instability of currently available enzymes, and, indeed, enzymes have traditionally been regarded as rather thermolabile.

It has been shown, however, that enzymes from thermophilic bacteria are more stable than those from mesophiles and that those from extreme thermophiles are more stable still: that is, the resistance of the enzyme to thermal denaturation is what one might expect from the growth temperature of the organism.

The differences in thermostability are often so large that it is difficult to obtain strictly comparable data. This is due to the fact that for convenience enzyme thermostabilities are usually measured at temperatures which give half-lives of 5–200 min, but the half-life of a typical enzyme from a mesophile will be of the order of seconds at 70°C, while the half-life of an enzyme from an extreme thermophile may be of the order of days.

Some typical stabilities (half-lives) of enzymes from extreme thermophiles are shown in Table 2. These enzymes were not specifically selected for stability, so it seems likely that selection among extreme thermophiles for particularly stable enzymes or the use of extreme thermophiles which have higher optimum growth temperatures will undoubtedly yield enzymes of even greater thermostability.

Table 2 Thermal stability (half-lives) of some enzymes from mesophiles and from extreme thermophiles

Enzyme	*Extreme thermophile half-lives* (min)		*Mesophile half-lives* (min)	
L-Asparaginase	85°C/40	95°C/2.5	60°C/40	60°C/
NADH dehydrogenase	85°C/7300	95°C/40	60°C/5	70°C/3
β-Galactosidase	85°C/7300	95°C/2	50°C/36	50°C/3
α-Glucosidase	75°C/100	85°C/1.4	40°C/5	60°C/5
Carboxymethyl-cellulase	85°C/760	—	—	—
α-Amylase	—	95°C/60	55°C/15	60°C/60
Caldolysin	85°C/7300	95°C/60	—	—
Protease	100°C/60	—	50°C/15	60°C/15

It is now clear that factors which render an enzyme thermostable confer a general robustness on the enzyme.

Thermostable enzymes resist the denaturing action of solvents and detergents, and proteolytic attack. Arising from this, enzymes derived from extremely thermophilic bacteria have a number of economic advantages.

(1) The large-scale growth of thermophilic bacteria is likely to be cheaper than that of mesophilic bacteria. This is partly as a result of reduced capital outlay, since the heat exchange and refrigeration equipment needed for the cooling of large-scale mesophilic cultures will not be needed. (The additional costs of insulation and heating and of heat

recovery equipment are not likely to outweigh this saving.) Furthermore, more of the cost of mesophile growth vessels is associated with preventing or reducing contamination. Since such contamination is less likely above 70°C, much simpler and cheaper vessels can be used to grow thermophilic bacteria.

(2) There is evidence that higher yields of purified enzymes can be obtained from isolation procedures when thermophilic micro-organisms are used as the source. This is a function of the greater stability of their enzymes[12].

(3) Being more stable, thermophilic enzymes have a longer useful life in industrial enzyme reactors than their mesophilic equivalents, which increases their cost-effectiveness.

(4) Reactors using thermophilic enzymes can be operated at temperatures sufficiently high to prevent microbial contamination. This would permit reactors to operate for longer periods between sanitising procedures and without the necessity of presterilisation or any addition of antimicrobial agents. The resistance of thermophilic enzymes to proteolytic attack will tend to negate the effect of any microbial contamination which does occur.

(5) The resistance to detergents and solvents will enable the use of these to improve the solubility of enzyme substrates or products. It will also minimise losses during the cleaning of reactors using immobilised enzymes.

(6) In high-temperature bioreactors, volatilisation of reaction end-products could occur. This could be valuable either as a recovery step or as a means of removing a potentially inhibitory end-product.

(7) There are a number of general advantages in operating industrial processes at higher temperatures. Viscosity is reduced, solubilities are higher, diffusion is accelerated and less effort needs to be taken to dissipate the waste heat inevitably generated during industrial processes.

Many of the advantages referred to here for thermophilic enzymes, and some additional ones, apply to the use of whole extremely thermophilic bacteria to replace mesophilic micro-organisms in existing processes. The advantages of a fermentation carried out near the boiling point of ethanol are fairly obvious, for example. Extreme thermophiles are, of course, non-pathogenic, since by definition they will not grow at body temperatures.

All the techniques, such as immobilisation and genetic engineering, currently being developed to enhance the usefulness of enzymes from mesophiles are also applicable to enzymes from extreme thermophiles.

The carboxymethyl cellulase and cellobiase genes of *Thermus* have been cloned and expressed in *E. coli*. As a result, a mesophilic fermentation can now be conducted to yield specific enzymes which are stable at up to 80°C. While the enzymes have their unique value, such a system also provides an ideal method of purification, in that the simple process of heating the fermentation broth will inactivate all enzymes except the thermostable species required. Because extreme thermophiles cannot be pathogenic, the use of recombinant DNA techniques carries none of the potential hazards normally associated with these.

The stability of an extracellular protease and an intracellular asparaginase from isolate *Thermus* T351 has been further increased by immobilisation (Table 3).

Table 3 Effect of immobilisation on enzymes from *Thermus* T351

Property	D-*Asparaginase*[13]		*Protease*[14]	
	Free enzyme	*Immobilised enzyme*	*Free enzyme*	*Immobilised enzyme*
Half-life (min) at 85°C	30	600	360	1060
pH optimum	No change	No change	7.0	6.0
Substrate inhibition	—	—	Present	Absent

CONCLUSION

As well as the advantages referred to above, enzymes from extreme thermophiles should make good subjects for enzyme engineering, and the 'heat purification' method available when the enzymes are cloned into mesophilic bacteria should enable pure enzymes to be made available at relatively low cost[15]. Enzymes from extreme thermophiles have a promising future in replacing many enzymes used in existing processes. However, some enzyme processes are designed around the properties of particular enzymes, and process redesign will not always be worth-while. It may be that the full potential of these enzymes will only be realised with the development of new processes — for example, using organic solvents or at high temperatures — which are dependent upon their stability[16, 17].

REFERENCES

1. Sonnleitner, A. and Fiechter, A. (1983). Advantages of using thermophiles in biotechnological processes: expectations and reality. *Trends in Biotechnology,* **1** (3), 74
2. Gaughran, E. R. L. (1947). The thermophilic micro-organisms. *Bacteriological Reviews,* **11**, 189
3. Koffler, H. (1957). Protoplasmic differences between mesophiles and thermophiles. *Bacteriological Reviews,* **27**, 227
4. Williams, R. A. D. (1975). Caldoactive and thermophilic bacteria and their thermostable proteins. *Science Progress Oxford,* **62**, 373
5. Singleton, R. (1976). A comparison of the amino acid composition of proteins from thermophilic origins. In *Extreme Environments; Mechanisms of Microbial Adaptation*. Academic Press, New York, p. 189
6. Zuber, H. (1978). Comparative studies of thermophilic and mesophilic enzymes: objectives, problems, results. In *Biochemistry of Thermophily* (Ed. Friedman, S. M.). Academic Press, New York, p. 267
7. Zuber, H. (1979). Structure and function of enzymes from thermophilic micro-organisms. In *Strategies of Life in Extreme Environments* (Ed. Shilo, M.)
8. Amelunxen, R. E. and Murdock, A. L. (1978). Microbial life at high temperatures: mechanisms and molecular aspects. In *Microbial Life in Extreme Environments* (Ed. Kushner, D. J.). Springer-Verlag, Berlin, p. 217
9. Perutz, M. F. and Faidt, H. (1975). Stereochemical basis of heat stability in bacterial ferrodoxins and in haemoglobin. *Nature, Lond.,* **255**, 256
10. Bucke, C. (1977). Industrial glucose isomerase. In *Topics in Enzyme and Fermentation Biotechnology* (Ed. Wiseman, A.). Ellis Horwood, Chichester, p. 147
11. Pintauro, N. D. (1979). *Food Processing Enzymes: Recent Developments*. Noyes Data Corporation, New Jersey, p. 11
12. Doig, A. R. (1974). Stability of enzymes from thermophilic micro-organisms. In *Enzyme Engineering*, Vol. 2 (Eds. Pye, E. K. and Wingard, L. B.). Plenum Press, London, p. 17
13. Guy, G. R. (1980). Asparaginases from thermus T-351. M.Sc. Thesis, University of Waikato, New Zealand
14. Cowan, D. A. (1980). A protease from an extreme thermophile. D.Phil. Thesis, University of Waikato, New Zealand
15. Daniel, R. M., Donnison, A. M., Bragger, J. and Morgan, H. W. (1986). The stability of enzymes from extreme thermophiles. In *Proceedings of 7th Australian Biotechnology Conference*, University of Melbourne, p. 158
16. Daniel, R. M. (1986). The immune response to proteins from extreme thermophiles. *Journal of Theoretical Biology,* **120**, 125
17. Daniel, R. M. (1986). The stability of proteins from extreme thermophiles. *Journal of Cellular Biochemistry,* **32**

4
EARTHWORM BIOTECHNOLOGY (VERMICOMPOSTING)

Paul Hand

INTRODUCTION

The contribution of earthworms to soil fertility through their burrowing and feeding activities was recognised by Gilbert White as long ago as 1789[18]. In 1881 Charles Darwin[2] showed that earthworms play an important part in the disappearance of surface plant litter and its breakdown and incorporation into the soil. More recent work on earthworm ecology has shown the precise importance of earthworms in the cycling of organic matter in terrestrial ecosystems.

Modern society, with its high population densities, sophisticated industries and intensive methods of agriculture, produces ever-increasing quantities of wastes. Existing methods of waste treatment and disposal tend to be costly, and recent interest has been shown in the development of novel processes which are based upon the utilisation of biological systems[10]. One such system involves the harnessing of earthworms for the stabilisation of a variety of organic wastes: a process known as 'vermicomposting'. This has been the subject of intensive research in many countries, and is the most important aspect of earthworm biotechnology at present. Its importance is reflected in the recent *Pollution Abatement Technology* award to one group of researchers in this field in the United Kingdom[1]. This chapter will therefore deal chiefly with the vermicomposting process, although the actual and potential application of earthworms in other areas will also be discussed.

VERMICOMPOSTING SYSTEMS

At this point it is appropriate to consider the characteristics of certain earthworms which make them suitable for utilisation in vermicomposting. Studies of habitats containing high concentrations of decomposing organic matter, such as compost and manure heaps and sewage sludge drying beds, showed that one earthworm was frequently dominant in numbers in such substrates. The life cycle and ecology of this species, the 'brandling' or 'tiger worm' *Eisenia fetida*, have therefore been the subject of intensive research. The information which has been obtained illustrates the general attributes which are desirable in considering a species for its vermicomposting potential.

The life cycle of *E. fetida* is summarised in Figure 1. Reproduction is by the production of cocoons. As this species is hermaphrodite (i.e. adults contain both male and female sexual organs), two cocoons are produced at each mating. The cocoons are lemon-shaped, approximately 4 mm by 3 mm in size. Up to 20 ova are present in each cocoon, although, of these, a maximum of 8 usually develop, with most cocoons producing two hatchlings.

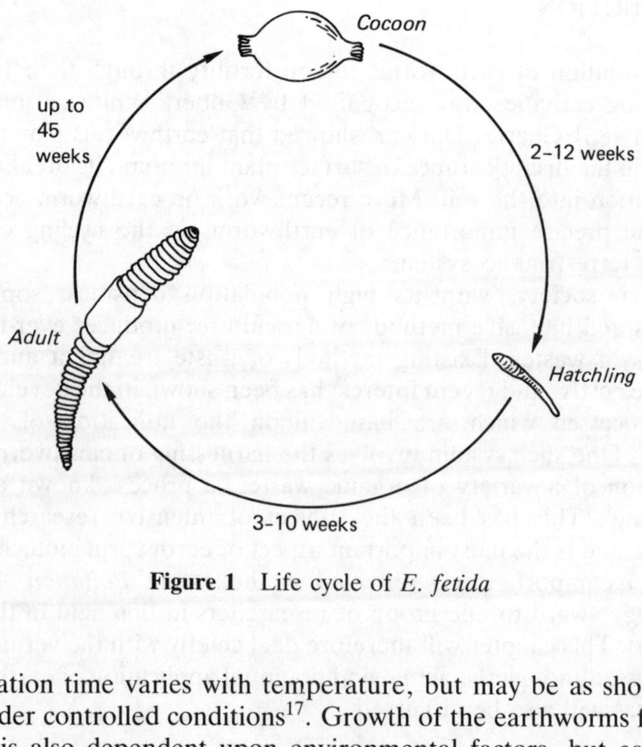

Figure 1 Life cycle of *E. fetida*

Incubation time varies with temperature, but may be as short as 2–3 weeks under controlled conditions[17]. Growth of the earthworms following hatching is also dependent upon environmental factors, but earthworm weights of between 700 and 1400 mg are usually achieved after 10 weeks at 25°[14]. Wide temperature fluctuations are tolerated by this species: growth occurs between 5 and 29°, although the optimum range is between 20 and 25°. The onset of sexual maturity occurs between 3 and 10 weeks after hatching and, when food is not limiting, reproduction is continuous for an extended period. Reproduction rates of up to three cocoons per worm per week over a 22 week period have been reported[8], and earthworms have been kept under laboratory culture conditions for up to $4\frac{1}{2}$ years[3].

These figures give some idea of the high fecundity of *E. fetida*. These growth and reproduction rates are also associated with intensive feeding activity; it has been shown that material passes through the gut of the earthworm in approximately 4 h, and that individuals may consume approximately their own body weight of substrate each day.

Unfortunately, little is known of the suitability of other earthworm species for vermicomposting. The few species which have been tested have been found to be inferior to *E. fetida* in terms of growth and reproduction rates[11]. However, more than a thousand species are recognised world-wide, many of which have been inadequately documented, and it may prove possible in the near future to select other earthworms which are more efficient at vermicomposting under certain conditions.

Many types of organic waste are suitable for stabilisation by earthworms, and different systems have been developed to suit the particular characteristics of each. The basic requirements of a successful system, however, remain the same: (1) provision of a solid or semi-solid 'bedding material' which physically supports the earthworm population; (2) the addition of a nutrient source for earthworm growth and reproduction (the waste); and (3) separation of the worm-worked material from the earthworms after a suitable time interval. The following examples illustrate how these principles have been applied to the vermicomposting of large-volume waste materials from sewage treatment, agriculture and industry.

SEWAGE SLUDGES

The population of England and Wales produces 1.2 million dry tonnes of sewage sludge per annum, and the treatment and disposal of this sludge costs £85 million per year. In the United States 7 million dry tonnes of sludge is produced each year. The low dry matter content and, hence, large volume of this material is a major factor in the cost of disposal.

Earthworms are common inhabitants of trickling filter beds at sewage works, and *E. fetida* in particular is found in sludge drying beds[9]. Both aerobically and anaerobically produced sludges are suitable substrates for earthworms, although some fresh anaerobic sludges have been found to cause earthworm mortality and must be stored for several days under aerobic conditions before earthworms can be introduced.

A recently completed project in the United States[7] has shown that even primary sludges, with a solids content as low as 0.6 per cent, can be vermicomposted, dewatering and stabilisation being achieved in the same process. The capital and annual costs of this process are stated to be competitive with those for other sludge management systems.

The presence of earthworms in a sludge produces increased oxygen consumption, decreased anaerobic decomposition and increased mineralisation[13]. Many of these effects have been attributed to the stimulation of microbial activity by the tunnelling and casting of the earthworms, which increases the aeration and surface area of the sludge[6].

Extrapolation of results obtained from laboratory experiments has indicated that large-scale vermicomposting of sewage sludges is feasible, and several pilot-scale treatment plants have been set up and operated successfully for short periods, notably that at Lufkin, Texas[4]. More recently the town of Toulouse in France has invested 300 000 francs in a vermicomposting system to treat 240 tonnes of sludge and waste fats per month[12].

AGRICULTURAL WASTES

Approximately 170 million tonnes of livestock excreta are produced in the United Kingdom each year, of which 60 million tonnes are voided indoors[16]. Most of this intensive farm waste is in the form of slurry, and problems arise because insufficient land is available for safe disposal.

Research has shown that a wide range of animal wastes are suitable for vermicomposting, including chicken, duck, turkey, pig, cattle and horse manures. Several 'on-farm' vermicomposting systems have in fact been developed and operated successfully, with waste to earthworm biomass conversion rates of up to 10 per cent. Systems are tailored to the properties of individual waste for optimum performance: for example, slurries may be separated, with the fibrous portion used as a bedding material and the liquid fraction irrigated onto the surface of the earthworm beds. Alternatively, whole slurries can be pumped directly from a holding tank onto the surface of earthworm beds containing a suitable solid substrate.

A recent interesting development in this field is the suggestion that anaerobically digested animal wastes could be utilised as a substrate for earthworm growth. The end result is an upgrading of the fertiliser value of the digester residues. A combined anaerobic digestion/earthworm culture system may therefore become economically attractive on small-scale livestock units.

A range of other agricultural wastes, such as dairy wastes, spent mushroom composts, plant residues and food processing wastes, may also be used as substrates for earthworm growth, with complete bioconversion of the waste being effected several times more rapidly than by conventional methods.

INDUSTRIAL WASTES

The recent discovery of specific feeding interactions between *E. fetida* and micro-organisms from its substrate has led to the suggestion that certain types of industrial waste, which were hitherto regarded as unsuitable for disposal by earthworms, may be seeded with appropriate microbes to render them amenable to vermicomposting.

This method has been successfully employed on an experimental scale in the treatment of solid wastes from the manufacture of tissue paper. Using a system based upon that employed in sewage treatment trickling filters (Figure 2), the rate of breakdown of the waste was significantly enhanced when cultures of a food bacterium were added at regular intervals[5].

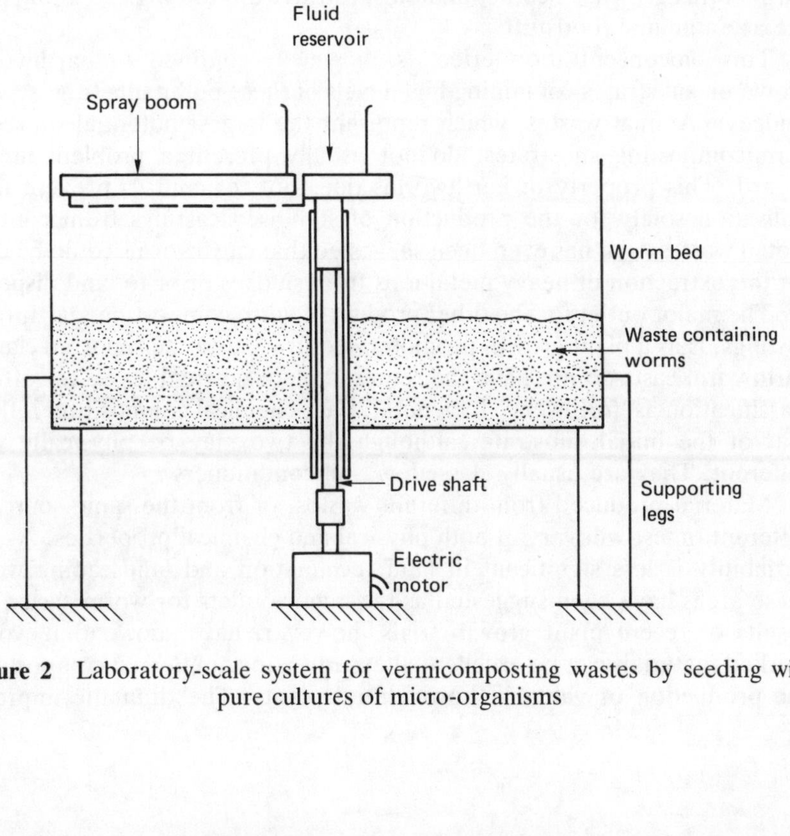

Figure 2 Laboratory-scale system for vermicomposting wastes by seeding with pure cultures of micro-organisms

PRODUCTS OF VERMICOMPOSTING

The products of vermicomposting are earthworm biomass and wormcast substrate. Potential markets exist for both of these products, although little research has been carried out on the size of these markets or on the value of the products as commodities.

The sale of the earthworms to fishermen for bait is an established practice, but demand tends to be localised and is already catered for without the need for widespread intensive vermiculture.

The use of earthworms in bulk as food for animals (or humans) has been proposed. Earthworms contain 70–80 per cent protein on a dry weight basis and this protein is of a high quality. Analyses have shown that it has a good balance of essential amino acids and is especially rich in lysine. It has been suggested that earthworm protein could be used to supplement animal grain rations, which are lysine-poor. Several feeding trials have shown that earthworm meal is nutritionally equivalent to fishmeal in the diet of rats, poultry and pigs.

Despite these promising results, the indiscriminate use of earthworms as animal food is not at present recommended by regulatory bodies. It has been found that earthworms have the unfortunate ability to greatly concentrate heavy metals within their tissues. The levels present in earthworms grown on contaminated substrates are too high to allow their use as an animal foodstuff.

This bioconcentration effect is, however, confined to earthworms grown on substrates containing high levels of these pollutants (e.g. sewage sludges). Animal wastes, which represent the largest potential source of vermicomposting substrates, do not usually present a problem in this regard. This property of earthworms does not, of course, prevent their utilisation solely for the production of stabilised castings from contaminated wastes — it has even been suggested that earthworms could be used for the extraction of heavy metal ions from sludges prior to land disposal.

The major outlet for the other product of vermicomposting, earthworm castings, is as a plant growth promoter. Contrary to some reported claims, earthworm casts do not have the required plant nutrient levels for official classification as 'fertilisers'. In general, their chemical composition reflects that of the initial substrate, although the two may be physically very different. They are usually classed as 'soil conditioners'.

Material produced from different wastes, or from the same source at different times, will vary in both physical and chemical properties. As this variability is less significant in land reclamation and landscaping work, these areas have been suggested as the major outlets for worm casts. The results of recent plant growth trials, however, have shown that worm-worked materials can be excellent alternatives to commercial composts for the production of valuable horticultural plants. The dramatic improvements in plant growth which have been obtained using vermicomposts suggest that further work on quality control and product variability may be worth-while, in view of the greater financial returns to be expected from sales to the horticultural market.

THE ECONOMICS OF VERMICOMPOSTING

An economic assessment of the use of earthworms in waste treatment is difficult, as many of the data on earthworm growth rates and substrate turnover are based upon relatively small-scale laboratory studies. Also, as previously indicated, the precise value of earthworm biomass and castings, and the size of the markets for these products, are difficult to define at present.

However, a detailed study of one full-scale operational vermicomposting set-up at Lufkin, Texas, has been performed[15]. At this site, earthworms were used to treat a proportion of the primary sludge produced by a sewage treatment plant. The annual cost of such a plant to treat 1 dry tonne of sludge per day (equivalent to a plant serving a community of 10 000–15 000 people) was calculated at $38 000–86 000. The cost per dry tonne of sludge processed was $105–235, which compared well with alternative treatment and disposal methods, such as landspreading or dewatering and composting, at $150–250 per tonne. These figures do not include any possible revenue from the products of the process, the operation being regarded purely as one of the treatment of a potentially polluting waste. Any such revenue would obviously have a further favourable effect upon the economics of the system.

OTHER ASPECTS OF EARTHWORM BIOTECHNOLOGY

In addition to the utilisation of earthworms in waste treatment, several other applications of earthworm biotechnology have been proposed which show considerable promise for further development.

Soil Reclamation and Amelioration

In many areas of the world, agricultural productivity is limited by low soil fertility or poor physical conditions. Elsewhere, previously uncultivated land is being reclaimed for agricultural use. In many areas the introduction of suitable earthworm species has been shown to have a significant positive effect upon nutrient cycling and physical structure within the soil.

In Eire earthworms have been used experimentally to increase the rate of reclamation of worked-out peat bogs. Planted grasslands produced 25 per cent more herbage when earthworms were introduced, compared with sites where worms were excluded. In New Zealand similar increases in productivity have been reported for pastures where earthworms have been introduced. Earthworms are deliberately added to land which has been reclaimed from the sea in Holland, in order to reduce the time taken to bring such land under cultivation.

The beneficial effects of earthworms on soil structure and aggregate stability are crucial in the rehabilitation of open-cast mining sites and industrial spoil heaps. The presence of earthworms rapidly improves water balance, organic matter availability and microbial activity, and, hence, facilitates the establishment and growth of vegetation.

Ecotoxicological Testing Using Earthworms

Ecotoxicology is the study of the ecological effects of toxic substances or pollutants. As soil organisms, earthworms are susceptible to exposure to anything buried, spilled or applied to the soil. Conventional chemical methods of analysis of contaminants are costly and often give no indication of the bioavailability of materials present in the environment. Alternative methods of assessment using earthworms have been proposed by the EEC and OECD for adoption as standardised tests. Results indicate that earthworms may be extremely useful in assessing the environmental impact of heavy metals, pesticides, radioactive wastes and other toxic chemicals, and in the screening of novel agricultural chemicals for their toxicity to the soil fauna.

FUTURE PROSPECTS

This chapter has provided an assessment of the current position of earthworm biotechnology. At present the vermicomposting of waste materials is at the forefront of this field. Increasing environmental awareness in the population at large, and the ever-rising costs of energy and raw materials, mean that waste treatment and recycling will become even more politically and economically desirable in the future. As evidence accumulates that the large-scale harnessing of natural decomposition pathways involving earthworms can be achieved at a cost comparable to that of other waste treatment alternatives, the future of vermicomposting seems assured.

Earthworms as a high-quality, low-cost source of protein for animals and humans may have a more widespread application in the future, especially in the developing countries. The production of earthworm biomass is well suited to a small-scale, low-technology approach, using locally available labour and raw materials. The use of earthworms for soil amelioration and land reclamation (directly or through the use of vermicomposts) is also likely to increase in both developed and underdeveloped countries, as population and land-use pressures become more acute.

Moving into the more distant future, other aspects of earthworm biotechnology may become important: folklore contains several references to the medicinal properties of earthworms, and several therapeutic compounds have been isolated from earthworms, though at very low concentrations. The ability of some species to regenerate segments following injury has also been the subject of much research, which may one day find an application in human medicine.

These, and other future developments in earthworm biotechnology, are largely dependent upon continued investment. In common with many areas of biotechnology, such investment is of the high-risk category, but this field, unlike some others, does have the advantage of a solid scientific basis from which to evolve.

REFERENCES

1. Anon. (1984). Worms combat pollution. *New Scientist,* **101**, No. 1401, 23
2. Darwin, C. E. (1881). *The Formation of Vegetable Mould through the Action of Worms*. John Murray, London
3. Edwards, C. A. and Lofty, J. R. (1972). *Biology of Earthworms*. Chapman and Hall, London
4. Green, E. and Penton, S. (1981). Full-scale vermicomposting at the Lufkin water pollution control plant. In *Workshop on the Role of Earthworms in the Stabilization of Organic Residues* (Ed. Applehof, M.). Beech Leaf Press, Kalamazoo, Michigan, pp. 229–231
5. Hand, P. and Hayes, W. A. (1986). The composting of slurries by earthworms. In *Proceedings, Conference on the Composting of Solid Wastes and Slurries*, University of Leeds, England, September 1983
6. Hartenstein, R. and Hartenstein, F. (1981). Physicochemical changes effected in activated sludge by the earthworm *Eisenia fetida. J. Environ. Qual.,* **10**, 377–382
7. Hartenstein, R., Kaplan, D. L. and Neuhauser, E. F. (1984). Earthworms and trickling filters for waste water treatment. *J. Wat. Polln Control Fedn,* **56** (3), Part 1, 294–298
8. Hartenstein, R., Neuhauser, E. F. and Kaplan, D. L. (1979). Reproductive potential of the earthworm *Eisenia fetida. Oecologia,* **43**, 329–340
9. Hawkes, H. A. (1963). *The Ecology of Waste Water Treatment*. Pergamon Press, Oxford
10. Joint working party of the Advisory Council for Applied Research and Development, The Advisory Board for Research Councils and the Royal

Society (1980). *Biotechnology: Report of the Joint Working Party*. HMSO, London

11. Kaplan, D. L., Hartenstein, R., Malecki, M. R. and Neuhauser, E. F. (1980). Role of the pheretemoid worms *Amynthas hawayana* and *A. rodericensis* in soils and recycling. *Rev. Ecol. Biol. Sol.*, **17**, 165–171
12. Milhau, C. and Vignoles, M. (in press). Treatment of fats and other wastes. One solution — vermicomposting. In *Proceedings, International Conference on Earthworms in Waste and Environmental Management*, Cambridge, England, July 1984
13. Mitchell, M. J., Hornor, S. G. and Abrams, B. I. (1980). Decomposition of sewage sludge in drying beds and the potential role of the earthworm *Eisenia fetida*. *J. Eviron. Qual.*, **9**, 373–378
14. Neuhauser, E. F., Kaplan, D. L., Malecki, M. R. and Hartenstein, R. (1980). Materials supporting weight gain by the earthworm *Eisenia fetida* in waste conversion systems. *Agric. Wastes,* **2**, 43–60
15. Pincince, A. B., Donovan, J. F. and Bates, J. E. (1980). Vermicomposting municipal sludge: an economical stabilisation alternative. *Sludge,* **3** (5), 26–30
16. Royal Commission on Environmental Pollution (1979). *Seventh Report: Agriculture and Pollution*. HMSO, London, pp. 131–132
17. Stephenson, J. (1930). *The Oligochaeta*. Oxford University Press
18. White, G. (1789). *The Natural History of Antiquities of Selbourne in the County of Southampton*

5

BRAVE NEW VEGETABLES —A CHALLENGE TO BIOTECHNOLOGY

Tudor H. Thomas

INTRODUCTION

Although cereals feature as the main world crops (1638 million tonnes — 1983 FAO figures), field vegetables (372 million tonnes), roots and tubers (556 million tonnes) and pulses (44 million tonnes) form the other major components of food crop production (see Table 1 for UK vegetable statistics). Thus, in most parts of the world, vegetables comprise a large part of the human diet, providing a high proportion of the nutrients and vitamins essential for normal body metabolism. Indeed, in some countries they constitute the main food intake, particularly where the protein levels of some indigenous vegetables is high and in areas where availability of other foods is very limited.

Vegetable crops are grown on a considerable proportion of the world's potential arable land, at least 90 per cent of which is already under cultivation. There is little prospect of new land being made available for cultivation, since much of the non-cropping land lies in climatic zones that present serious problems to agricultural development. The necessity for improving and increasing crop production in existing areas is indisputable, since it has been estimated that on average the world food demand during the next 20 years will rise by about 3 per cent per annum.

Further, the UN world food conference assessment of the world food situation has estimated that about 25 per cent of the population in developing market-economy countries, which are highly dependent on vegetables as food, has an inadequate daily intake of dietary energy, as compared with 3 per cent in industrial areas.

The demand for increased crop production has previously been met by a combination of agronomic techniques, improved plant protection, the development of new varieties and the introduction of sophisticated agricultural machinery. As far as the future is concerned, considerable advances can yet be envisaged using these conventional approaches, but novel technologies, including advanced fermentation, *in vitro* cloning, genetic manipulation, biological control and the use of plant growth regulators, are already beginning to provide the basis for a new impetus towards sufficiency of food supplies.

Table 1 Statistics of major vegetables produced in the United Kingdom (1985). Source: Basic horticultural statistics for the UK

Vegetable crop	*Cropped area* (ha×10^3)	*Output* *Marketed* (t×10^3)	*Output* *Value* (t×10^3)	*Imports* *Marketed* (t×10^3)	*Imports* *Value* (t×10^3)
Potatoes	192.3	6,404	323,060	299	45,233
Potatoes (processed)	—	—	—	345	53,873
Peas (green)	47.3	197	53,425	1	1,150
Peas (dry)	28.0	35	10,384	27	10,417
Cabbage	21.8	621	65,599	48	6,580
Brussels sprouts	11.4	142	33,653	3	1,171
Cauliflower	17.4	344	55,437	41	17,760
Carrots	14.6	523	53,493	56	17,075
Onions (dry bulb)	7.2	213	21,385	228	22,296
Beans (runner, broad and French)	11.6	81	22,799	6	5,042
Lettuce	6.4	158	42,252	34	24,486
Turnips and swedes	4.6	137	11,188	—	—
Parsnips	2.7	56	8,959	—	—
Celery	1.0	51	12,721	25	7,637
Tomatoes	—	120	53,081	253	115,739
Beans (dried)	—	—	—	117	54,011

Total value of output of outdoor and protected vegetables in the UK, £1 027 031 000.

Over the past 30 years there has been considerable research into molecular biology and related disciplines, to the extent that conventional plant breeding and production techniques could be changed dramatically by the commercial exploitation of a variety of procedures and processes collectively termed 'biotechnology'. However, the shining promise of biotechnology must be coupled with some caution and concern. It has been pointed out that while biotechnologies promise to lead to major improvements in agriculture and industry in both high- and low-income countries, there are major reasons for concern about the impact of these new technologies, particularly on developing nations[8]. It may be that biotechnology will revolutionise agriculture almost into a non-farm industry in which development will be influenced mainly by multinational corporations. A move in this direction is evident from the steady acquisition of seed companies by many of the major multinational agrochemical companies. As yet, biotechnology is an unknown quantity, the outcome of each research line being highly uncertain. It is perhaps premature to assume that the use of new techniques in genetics and cell biology will completely revolutionise plant breeding, although obviously they will have an important role in many aspects of the applied biological sciences.

BIOTECHNOLOGY AND PLANT BREEDING

The extent to which conventional plant breeding has influenced the production of food crops is well illustrated by the creation of high-yielding, fertiliser-responsive dwarf cultivars of cereals. This, together with the increased use of technology in developing countries, resulted in what was commonly termed the 'green revolution' and advanced a number of tropical countries some way towards self-sufficiency in foodstuffs. There have been no advances of the same magnitude in field vegetable breeding; nevertheless the plant breeder may hold the key to a number of doors leading towards improved vegetable crop production.

The breeding of field vegetables has been reviewed by Innes[7], who commented on the fact that among vegetables there is a wide range of natural breeding systems and that a knowledge of these is essential before appropriate breeding strategies can be devised to develop improved varieties. Further, there is a need for such new varieties to be adapted not only to the environment for which they were designed, but preferably also to a wide range of environments. The plant breeder not only is interested in increasing yield and improving uniformity, but also aspires to introducing resistance to pests, diseases and physiological disorders, tolerance of or responsiveness to agrochemicals, and the capacity to produce economic yields in the presence of stress factors such as heat, cold, drought, saline soil and pollution. All these features can have a major influence on the type, quality and quantity of vegetables produced in the UK and abroad.

Breeding for resistance to pests, diseases and disorders has been and continues to be a lengthy and elaborate operation in which germplasm of foreign origin, often of related wild species, has been of considerable importance. Complications can arise, since genetic variability occurs not only in host resistance, but also in pathogen or pest pathogenicity. Despite such problems, breeding of vegetables for resistance to diseases has been fairly successful, although pest resistance has proved to be more difficult to introduce. The rewards for success are considerable, as is illustrated by the fact that despite over £20 million expenditure each year on herbicides, insecticides and fungicides in the UK alone, the possible crop losses due to weeds, pests, diseases and disorders are still of the order of about 20 per cent. Obviously, introduction of even partial resistance to attack from pests and pathogens could result in a much lower input into agrochemical treatment of many vegetable crops, with the additional possibility of considerably lower crop losses. The development of biotechnology as a means of introducing resistance genes into crop plants has itself raised political problems, with increasing conflict between industrial and less-developed countries over access to and control over germplasm resources.

The concept of breeding crops for tolerance, resistance or even responsiveness to agrochemicals is intriguing. It may not be widely recognised that, owing to the wide range of species involved, at least 70 herbicides are used alone or as mixtures on vegetable crops in the UK. Early weed control in vegetable crops is essential in order to obtain good stands of uniform plants. The conventional production of new herbicides is long and costly, so that any method which would reduce the necessity for producing new herbicides would be an attractive proposition. It may be possible to use existing cost-effective herbicides which would normally be destructive to both crops and weed species but to select for tolerance in the crop. This has already been exploited in some crops — for example, paraquat-resistant ryegrass and simazine-resistant oilseed rape have been developed by traditional selection techniques. Isolation of resistant mutants *in vitro* using cultured plant cells has already been considered as a potential method for producing herbicide-resistant lines and the possibility of producing lines with positive responses to plant growth regulators seems equally feasible. As an example, an important development could be the selection of sugar-beet cultures which produce higher concentrations of sucrose in response to ripening chemicals such as glyphosphate. Increased response to plant growth regulators in terms of developmental changes, yield, ripening and senescence could all be of major importance in vegetable production. Protoplast techniques have already been used to introduce disease resistance, as exemplified by maize, where protoplasts were screened for their resistance against toxins from *Helminthosporium* (leaf blight) and resistant types regenerated. Of course, success depends to a large extent on the selection pressures inducing the same response in tissue culture as in the whole plant, rapid identification of the desired variants and good regeneration capacity of the variants.

In addition to pests and diseases, environmental factors such as salinity, extremes of temperature, day length and lack of water often affect crop quality and yield. In the UK alone, climatic conditions impose considerable restraints on the type of vegetables grown and so influence the extent to which perishable foodstuffs have to be imported from more favourable environments. Temperatures in early spring and autumn are generally too low to ensure consistently rapid growth and early ripening of some crops of tropical or subtropical centres of origin, but breeding of low-temperature tolerance into such crops could make them more suitable to the UK climate. As an example, the introduction of a cold-tolerant *Phaseolus* dry bean that would produce economic yields in the UK could save import costs of about £54 million for baked bean production (Table 1) (market value of *c*. £200 million). Although some cold-tolerant *Phaseolus* lines have been identified in both growth-cabinet and field trials, at the National Vegetable Research Station, UK, the difficulties in devising suitable large-scale screening techniques needed by the plant breeder have prevented complete success[6]. Similarly, it has been calculated that the development of outdoor tomato varieties with an increased cold tolerance of 1°C could save UK import costs of up to £45 million on tomatoes for fresh market and processed products[15]. Interestingly, the introduction of heat tolerance into tomato lines is equally important, particularly as an adaptation to crops grown under tropical conditions, since fruit set is very much reduced at night temperatures in excess of 20°C.

Photomorphogenic responses to light and temperature change may often be important in relation to vegetative organ and flower formation, and present an interesting problem to the plant breeders in adapting crops to new environments. In onions the availability of day-neutral plants, which produce bulbs irrespective of the day length, may significantly increase the world-wide production of this crop. Already the introduction of short-day-responsive onions into the UK as an autumn-sown crop has saved import costs in excess of £4 million, there now being over 2000 ha sown each year[13].

Having identified the importance of plant breeding in vegetable crop production, it is appropriate to consider the ways in which biotechnological methods may be used to complement or even replace conventional plant breeding. When one considers that each plant cell contains anything from 10 000 to 100 000 genes, it seems quite surprising that plant breeding has met with so much success. The plant breeder is faced with the problem of extricating from this tangle of genetic potential an optimal assortment of genes in order to introduce useful characteristics into crop plants. The disadvantages of conventional plant breeding are quite clear: it is a random process of long duration, genes can only be transferred between compatible plants, and transfer is limited to existing plant genes.

The UK Agricultural and Food Research Council has defined the main aim of genetic manipulation of crop plants as 'to improve crop plants by the transfer and manipulation of genes in ways which extend or circumvent the normal genetic exchange of sexual crossing'[1]. The ability to multiply rapidly the products of genetic manipulation has already been demonstrated in many crop plants by the use of *in vitro* cloning techniques. It is already apparent that genetic manipulation can be controlled at either the whole tissue, plant cell or even gene level by techniques such as tissue culture, haploid plant formation, protoplast fusion, somacloning and genetic engineering of DNA.

HYBRIDISATION

The improved characteristics of many modern vegetable varieties as compared with their predecessors is largely due to the introduction of F_1 hybrids. Successful production of such material will normally take up to at least 7 years, for biennial vegetables. The use of anther culture has the advantage of producing instant homozygotes from haploid plants, after chromosome doubling with colchicine, thus possibly saving many years as compared with conventional plant breeding. Haploids are readily obtainable within the family *Solanaceae*, which contains a number of important economic species, but most other vegetable species are less responsive. However, haploid plants have been obtained from asparagus, broccoli, cucumber, pepper, potato, sweet potato and tomato. Anther culture has also been successfully used in rape (*Brassica rapus* L.) breeding and is showing promise in horticultural brassicas, but so far on average only 12–120 embryos have been produced for each 100 anthers cultured[9]. Numerous factors affect the degree of success in anther culture, such as the culture medium composition, the growth and age of donor plant and the optimum developmental stage of the anthers. In addition, pretreatment of the anthers with cold or heat shocks following removal from the plant may be beneficial or even essential. Obviously, the potential benefits of this approach fully justify a major input into determining the conditions necessary for successful anther culture of major vegetable crops.

The introduction of pest and disease resistance into vegetables may pose special problems, since it can involve a complicated process of resistance-gene transfer from related wild species. Often the wild and cultivated species may be incompatible or sterile crosses may result due to failure of endosperm development in the fertilised ovule. The sterile culture of immature embryos in the absence of endosperm could provide a solution to these problems, and already introduction of resistance to *Fulvia* into cultivated tomato from the wild *Lycopersicon peruvianum* has been achieved in this way. Another potentially successful method of transferring useful resistance genes, which is unlikely to necessitate recurrent back-crossing, is the use of irradiated pollen to transfer a limited number of genes (instead of one haploid set of chromosomes) from a paternal source into an *unrelated* maternal genotype[11]. Again, production of a fully developed progeny could necessitate the use of embryo culture. Attempts are being made to transfer genes for resistance to clubroot from turnip into horticultural brassicas such as cauliflower at the National Vegetable Research Station using pollen irradiation techniques.

PROTOPLAST TECHNOLOGY

Protoplast isolation techniques are now common, with the possibility of fusing protoplasts to obtain hitherto unattainable interspecific and intergeneric crosses or of introducing foreign genetic material into isolated protoplasts. So far, success has been mainly limited to somatic hybridisation within the genera *Petunia* and *Nicotiana*[12], but these investigations have provided material for analysing nuclear and cytoplasmic interactions and breeding characteristics. Meanwhile, protoplast fusion techniques are being used in an attempt to introduce disease and pest resistance into agronomically valuable species such as lettuce (resistance to *Bremia*), tomato (*Fusarium* wilt resistance) and peanut (leaf-spot and mite resistance). Cell fusion may also be a potentially important way to mediate the transfer of cytoplasmic male sterility as a step towards the production of F_1 hybrid cultivars, as are already used in onion and sugar-beet.

Protoplast technology not only offers advances through the production of interspecific hybrids by fusion of genetic manipulation, but also may provide an untapped source of plant (somaclonal) variation[5, 14]. In 1979, field evaluation in the USA of 60 clones produced from leaf mesophyll protoplasts of Russet Burbank potatoes showed a range of potentially useful variation which was stable over three tuber generations; in the UK similar results have been obtained with wheat and potato at Rothamsted Experimental Station. The variations obtained in potatoes included differences in habit, yield, earliness by up to 4 weeks, photoperiodic response and disease resistance. Although extrapolation to other species may be premature, the potential for improving existing vegetable crops and introducing new ones to the UK is considerable. As already indicated, cold tolerance and earliness in crops such as *Phaseolus* beans and outdoor tomatoes would almost certainly guarantee an economically viable crop in most if not all years. Variation in photoperiodic response could lead to the selection of non-bolting strains of biennial vegetables for overwintered production. The potential for such crops in the UK is clear, particularly for storage root vegetables, where yield depends heavily on the early development of a leaf canopy. It assumes even more economic importance if sugar-beet is considered, where it has been estimated that even the earliest-sown crops fail to intercept some 40 per cent of the incident radiation between sowing and harvest.

INSERTION OF NEW GENES BY GENETIC ENGINEERING

The insertion of new genes into the chromosomes of plants with complete stability of inheritance has now become a realistic concept. It is clear that the recognition of desirable genes will increase and that ‘transformation’ of the insertion of these genes will lead towards considerable improvements in crop plants. Since purified DNA can be susceptible to digestion by host cell nucleases, transfer may only be possible through the use of gene vectors which actively incorporate into the plant genome. Many potential vectors for plant transformation have been proposed, but most attention has been paid to the use of plasmids from *Agrobacterium tumefasciens* possibly through viral DNA. Following infection by the vector, part of the Ti-plasmid integrates into the host cell nuclear DNA, thus offering a natural method for stable introduction of foreign DNA into the plant cells. Transformation has normally been achieved using cultured plant cells and protoplasts, and although these must then be cloned, redifferentiated into non-tumerous cells and finally regenerated into complete plants, technologies for this are available and are improving rapidly. However, one major problem which has arisen is that no vectors have been found for monocotyledonous plants.

Nevertheless, the future for genetic engineering of crop plants is considerable, with the possibility of introducing disease resistance, enhancing nitrogen-fixing capacity, increasing stress tolerance, improving nutritional quality, improving yield by increasing enzyme efficiency and introducing resistance to pesticides. Some of these improvements may be possible through gene amplification, i.e. introduction of more copies of important genes per plant in order to amplify physiological and biochemical processes. Even more futuristic possibilities include the introduction of non-plant genes (e.g. genes which would produce bacterial toxins as defensive mechanisms against pathogen attack) and the introduction of newly constructed complete artificial genes. Although the potential application of genetic engineering to agricultural crops is truly astonishing, there will undoubtedly be limitations to these applications. For example, it now seems less likely that the transfer of nitrogen-fixing (*n.f.*) genes from *Klebsiella* to *E. coli*[3] will lead to their introduction into crop plants to produce an autonomous nitrogen fixer, because of the numerous genes involved and the complexity of the control mechanisms. More immediate success is likely to be achieved in leguminous crops such as peas and beans, which can already fix nitrogen via a symbiosis by generating strains of *Rhizobium* with improved symbiotic performance.

It is clear that more critical studies on the regulation of gene expression are essential both to understand how gene activity is controlled during normal plant development and to devise ways in which foreign genes can be inserted to provide controlled plant modification. Genes from different species are controlled differently, and it is often necessary to reconstruct them, before insertion, to ensure that they are appropriately activated and regulated by the new host. A major advantage of constructing new Ti-plasmids is that construction can be accurately controlled to ensure expression of the inserted genes.

GENETIC ENGINEERING, HORMONAL CONTROL AND PLANT GROWTH REGULATORS

As mentioned previously, the use of plant growth regulators has not been exploited fully in agriculture, but there is now the possibility of a most interesting interface with genetic manipulation. It has already been demonstrated that two different Ti-plasmid mutants derived from *Agrobacterium tumefasciens*, when introduced into tobacco, induced tumours that gave rise specifically to either roots or shoots and that this was probably related to the ratio between auxin and cytokinin synthesised by the tumours[10]. Coinfection with equal amounts of both types of mutants resulted in the formation of a tumour with a normal wild-type morphology. Thus, it seems possible to mutate T-DNA genes involved in hormonal activity and to insert these into crop plants in order to control hormone synthesis or activity.

Genetic activation by plant hormones has already been studied in important crops such as cereals and legumes, and such work also introduces the possibility of introducing foreign genes which could be triggered by growth regulator application. In wheat, genes activated by gibberellic acid have been isolated and clones representative of two sets of wheat α-amylase have been characterised and sequenced. In kidney beans a translocation product for cellulase synthesis has been detected in response to ethylene treatment, and this is suppressed by auxin treatment. This work has led towards the cloning of abscission-zone cellulase DNA. Experts agree that the future of plant growth regulator research must be in multidisciplinary approach, including an understanding of the basic biological processes and how they regulate the productivity of crops. With increased knowledge of the hormonal control of gene regulation, the possibilities of discretely controlling specific events such as flowering, fruiting, abscission, sex expression, rooting, senescence and assimilate partitioning by inserting genes which either alter endogenous hormone levels or those which respond to plant growth regulator treatment become more realistic.

IN VITRO CLONING FOR PROPAGATION OF FOOD CROPS

Many plant species can be propagated *in vitro*, and this is regarded as an ideal method, since large numbers of fairly identical, clonal offspring can be produced from a very limited amount of parental material. Indeed, the future success of genetic manipulation techniques may depend considerably on the ability to clone *in vitro* the genetic products obtained, particularly if these do not give rise to fertile progeny from which seeds can be obtained. The use of this technique for asexual propagation is the most advanced area of plant tissue culture. By 1978 there were over 100 concerns engaged in commercial propagation of a variety of plants through tissue culture, and this number has increased steadily. There are obvious advantages of tissue culture methods as compared with traditional propagation, including space-saving, provision of aseptic and artificial environment and *very* substantially faster plant multiplication.

Some uses for *in vitro* propagation are as follows:

(1) Rapid multiplication of new hybrid cultivars which arise as single plants for testing and eventual commercial production. A good example of this is the development of clonal oil palms by the Unilever Research Laboratories.

(2) Elimination of virus diseases from infected stock.

(3) Vegetative propagation of species which are difficult to propagate by sexual means (e.g. orchids).

(4) Propagation independent of climatic and environmental problems.

(5) Production of genetically uniform parental types in large numbers for hybrid seed production.

(6) Storage of unique genetic material.

(7) Induction of mutations for breeding programmes.

Rapid asexual multiplication can be achieved in a number of ways, which include meristem culture, adventitious shoot development on callus and somatic cell embryogenesis in liquid cell culture. Although meristem culture is one of the most successful methods of cloning, particularly for high-value ornamental species, there are a number of reasons why it is unlikely to be so successfully employed for food crops. Principally, meristem culture methods are labour-intensive, are costly and may vary widely in the potential multiplication rate in different species. Propagation of high-density agricultural crops introduces a number of specific problems, not least of which are the large numbers of propagules required and the need to introduce less costly, automated, bulk-handling techniques. This must inevitably lead to the utilisation of somatic cell embryogenesis, using cell suspensions in liquid culture. Obviously, embryoids produced in cell suspension cultures could provide the most convenient starting material in terms of both propagule number and ease of dispensing the cultured plants. However, the problem of inducing embryogenesis consistently in cell suspension cultures, as can be easily done with carrots, is one which will not be easily resolved. Another major difficulty is the lack of methods for bulk-handling millions of small plantlets, so avoiding the need for individual handling. Research on the post-micropropagation stages of tissue regeneration is rare, particularly for low-value crops, and more work is required in extending and modifying systems which have already been developed for sowing pregerminated seeds or dispensing transplants.

These modified systems could provide potential microenvironments in which to introduce agrochemicals or to apply environmental variables. Incorporation of pesticides, growth regulators or even nutrient-fixing bacteria and mycorrhizae into the medium could result in more efficient plant protection, decreased pesticide use, improved growth regulation and more rapid growth. The attractions of chemical incorporation into these systems has already been demonstrated with some vegetable crops, but the potential advantages and disadvantages of combining these techniques with *in vitro* propagation should be carefully assessed.

A further problem arises in that plants produced from culture are not always capable of normal biochemical and physiological performance. Plantlets may possess seemingly normal shoots and roots, but may have poor vascular connections, little epicuticular wax and very low rates of photosynthesis[4]. Consequently, a period of appropriate hardening treatment is often necessary for plantlets to make the transition to normal morphology and physiology under either glasshouse or field conditions. This problem may be accentuated when small plantlets obtained by somatic embryogenesis are used. Very often embryoids must reach a relatively advanced stage of development before they become fully autotrophic. Recently photoautotrophy has been achieved with spinach cells grown in continuous culture using a 'sugar-famine' technique[2], but considerably more research into this problem is required. It is clear that there are many problems to be solved before *in vitro* cloning becomes a common practice for vegetable and other agricultural crops.

THE FUTURE FOR AGRICULTURAL BIOTECHNOLOGY

The future production of agricultural and horticultural crops may well depend on the selection and breeding of suitable plant material by biotechnological methods, *in vitro* cloning for multiplication, and the use of novel chemical treatments during late culture and early establishment phases. In this context, there is an increasing demand for chemicals which can affect photosynthesis, assimilate partitioning, nitrogen fixation and leaf senescence. The problems already encountered in devising methods of screening chemicals for biological activity may well be increased as biotechnology advances. Conversely, biotechnology could provide new techniques for monitoring the effects of plant growth regulators on fundamental plant processes, thus speeding up the release of sophisticated new agrochemicals for improving crop production.

It would seem that biotechnology will initially be best applied to high-cost, high-input crops where the methods introduced will be financially viable. However, with increasing experience there should be a gradual broadening-out of applications towards lower-cost, intensively cultivated crops as the costs of biotechnology decrease. It also seems likely that, in view of the high costs of biotechnological development and agrochemical production, the initial benefits may ably be felt by the developed, industrial countries. It is to be hoped that the poorer Third World countries will not be overlooked in the tide of agricultural biotechnology.

REFERENCES

1. Anon. (1984). *Genetic Manipulation of Crop Plants; Five Years on*. Agricultural and Food Research Council pamphlet, London
2. Dalton, C. C. (1980). Photoautotrophy of spinach cells in continuous culture. *Journal of Experimental Botany,* **31**, 791–804
3. Dixon, R. A. and Postgate, J. R. (1972). Genetic transfer of nitrogen fixation from *Klebsiella pneumoniae* to *Escherichia coli. Nature, Lond.,* **237**, 102–103
4. Grout, B. W. W. and Crisp, P. (1977). Practical aspects of the propagation of cauliflower of meristem culture. *Acta Horticulturae,* **78**, 289–296
5. Gunn, R. E. (1983). Breeding new potato varieties from protoplasts. In *Proceedings of the Eighth Long Ashton Symposium on Improvement of Vegetatively Propagated Crops, Bristol*, 12–15 September 1982. Academic Press, London
6. Hardwick, R. C. (1983). Genetic variation in yield processes of grain legumes and their responses to temperature. In *The Physiology, Genetics and Nodulation of Temperate Legumes* (Ed. Davis, D. R.). Pitman, London, pp. 55–76
7. Innes, N. L. (1983). *Breeding Field Vegetables*. 10th Anniversary Monograph Series. Asian Vegetable Research and Development Center, Taiwan
8. Kenney, M., Kloppenburg, J. Jr, Buttel, F. H. and Cowan, J. T. (1983). Genetic engineering and agriculture: socioeconomic aspects of biotechnology R&D in developed and developing countries. *Biotech* 83, Online Publications, Pinner, 475–489
9. Ockendon, D. J., Currah, L. and Robinson, H. (1983). Anther culture. In *Annual Report of the National Vegetable Research Station for 1982*. Long Ashton Research Station, Bristol, pp. 51–52
10. Ooms, G., Hooykaas, P. J. J., Moolenaar, G. and Schilperoort, R. A. (1981). Crown gall plant tumors of abnormal morphology, induced by *Agrobacterium tumefasciens* carrying mutated octopine Ti-plasmids; analysis of T-DNA functions. *Gene,* **78**, 33–50
11. Pandey, K. K. (1975). Sexual transfer of specific genes without gametic fusion. *Nature, Lond.,* **256**, 310–313
12. Power, J. B., Berry, S. F., Chapman, J. V. and Cocking, E. C. (1980). Somatic hybridization of sexually incompatible petunias. *Theoretical and Applied Genetics,* **57**, 1–4
13. Salter, P. J. and James, J. M. (1975). The performance of Japanese and European cultivars of onions from autumn sowing for early production. *Journal of the National Institute of Agricultural Botany,* **13**, 367–379
14. Secor, G. A. and Shepard, J. F. (1981). Variability of protoplast-derived potato clones. *Crop Science,* **21**, 102–105
15. Thomas, T. H., Gray, D. and Drew, R. L. K. (1977). Potential for outdoor tomato production and storage in the U.K. *Acta Horticulturae,* **62**, 101–108

6
FUNGAL CULTURE COLLECTIONS FOR THE BIOTECHNOLOGY INDUSTRY

Dennis Allsopp

INTRODUCTION

About 130 years ago, with the advents of sterile techniques and solid media, the first efforts were made to preserve living isolates of fungi. The first independent centre to preserve and supply a wide range of fungus cultures was the Centraal-bureau voor Schimmelcultures (CBS, Baarn, The Netherlands), established in 1904.

This and other early collections served mainly agriculture, brewing and medicine, but with the discovery of penicillin in the 1930s, the importance of such collections was dramatically increased, and today the pharmaceutical industry continues to find that the fungi are a rich source of biologically active novel compounds. The 1980s have witnessed a further development of culture collections in response to the needs of biotechnology-based industries. The organisms held in culture collections constitute a vital but often untapped resource for the new technologies.

ORGANISATION

International

The World Federation for Culture Collections (WFCC), founded in Paris in 1966 under the auspices of UNESCO, provides a forum to culture collections to discuss matters of interest. International conferences are held every 4 years, and a newsletter is issued. The holdings of the collections on a world basis is the responsibility of the World Data Centre (WDC), whose activities are considered further below.

In Europe the European Culture Collection Curator's Organisation (ECCCO), founded in 1982, provides a similar forum to the WFCC.

National

In the UK all collections and users can join the UK Federation for Culture Collections (UKFCC). The USA, Japan and some other countries have similar national organisations.

The total number of culture collections in the world is unknown. The WDC catalogue[3] at present lists 566 collections, of which at least 200 hold some strains of fungi. Many private, undocumented and commercial collections, however, are not included in these lists. Further information on the groups held, preservation methods used and ancillary services provided by over 70 of the large fungus culture collections is provided by Hawksworth[1], whose paper may be consulted for further information on many of the topics considered in the present contribution.

RESOURCES AND SERVICES

Conservation

It has been estimated that there could be over 250 000 species of fungi (64 500 accepted at present[2]; over 1500 new species are currently described each year). Living isolates of fungi (including yeasts) in the world's collections amount to around 170 000, representing some 7000 species. The potential for expansion and investigation is therefore very great. The strains held provide a tremendous genetic resource, and are an obvious starting point for cost-effective screening programmes.

In order to maintain a broad spectrum of isolates in a stable condition over long periods, any collection must have the equipment, personnel and financial resources to operate a variety of preservation methods in parallel. For fungi, these include frequent transfer; cool storage; deep freezing; storage under mineral oil, in soil, in water, in silica gel; freeze drying (lyophilisation); and storage in liquid nitrogen[4]. Most major culture collections endeavour to preserve each strain by two or more methods.

Research projects involving micro-organisms should always be backed up by deposits in a culture collection as an insurance against loss and to ensure that the strain will be available for any later work. Some collections have safe-deposit facilities for strains which are commercially sensitive or part of current research programmes; these are not listed in published catalogues.

Supply of Strains

The major service collections all supply strains, either for a fee or in exchange for new isolates. The history of culture collections involved subsidies by agencies of various kinds; consequently, cultures are sold at well below their true economic cost. Health and quarantine requirements can limit the supply and movement of some strains, and the collections can advise on this aspect. In the future, on-line ordering facilities can be envisaged as on-line catalogues become more generally accessible.

Information

The extent of information on the holdings of collections varies very greatly. Some collections do not publish a catalogue, but there are welcome moves to improve this situation through the production of national and international on-line databases. Such moves will also facilitate access to metabolic and physiological data being gathered by major collections, and such data will in future be included, as they become available, into such databases.

Identification

Where culture collections have experienced taxonomists on their staff or access to them, identification services can also be provided, usually for a fee. Correct identification or the verification of the identity of cultures is vital before publications are produced.

Much physiological and biochemical work currently published is marred and its value diminished by insufficiently or inaccurately identified fungi, a situation which is even more serious if the strains used have not been deposited in a culture collection.

Research

Culture collections are obvious centres for research into taxonomy and the long-term preservation and maintenance of organisms, but they are also ideally placed to undertake other studies of biotechnological importance.

Such work can include: (1) screening of strains for desirable biotechnological properties; (2) comparative studies of strains related to known enzyme or metabolite producers; (3) development of screening and assay techniques; (4) studies of growth requirements of related groups of micro-organisms; and (5) testing of fermentation equipment with a wide range of fungi.

Consultancy

Using internal data and personal knowledge, the major culture collections can provide confidential services to organisations requiring details of the activities of potentially useful strains. In future, with the development of on-line services, outside organisations will be able to obtain data with a high level of security, in that the aims of their searches will be known only to them. Such facilities are also well established in regard to scientific literature, and in culture collections which are also involved in on-line database production, such as the Commonwealth Mycological Institute (CMI).

Consultancy activities also extend much further than those concerned with cultures held. CMI and CBS, for example, are able to offer investigative consultancy work and testing facilities, not only in mycology, but also in applied biology generally, especially in the fields of the biodeterioration of materials and plant pathology.

Patents

Several culture collections around the world are now recognised as International Depository Authorities (IDAs), under the Budapest Treaty on the International Recognition of the Deposit of Micro-organisms for the Purposes of Patent Procedure.

To be recognised as IDAs, collections need to comply with a complex list of provisions, including the undertaking to safeguard the strain for 30 years. The fees involved in patent deposits may seem high (CMI fees are around £400) but are to allow for their long-term commitment; in comparison with other costs incurred in the filing of patents, they are rather modest.

Training

Some of the major culture collections are able to offer training of immediate relevance to biotechnologists, from the handling of micro-organisms, and their preservation and maintenance, through to state-of-the-art courses and seminars on specialist topics. Some collections also have strong links with universities and can assist with postgraduate training leading to higher degrees.

DEVELOPMENTS

If the potential afforded by the microbiological resource maintained in the world's culture collections is to be realised, developments are needed to strengthen the resource and increase the effectiveness with which it can be used.

Acquisitions

To expand beyond the traditional areas mentioned above, there is a need for collections to develop and implement vigorous acquisition policies aimed at increasing their holdings of strains of actual or potential biotechnological importance. Literature scanning and exchanges by collections can go some way to achieving this, but collections which provide or are associated with world-wide identification services are assured of a continuous flow of fresh isolates for a wide variety of substrates. Collections can also be enriched by encouraging research workers to lodge strains which are cited in published work and references, and editors of books and papers can play a key role by insisting that key strains mentioned are preserved in a culture collection prior to acceptance of material for publication.

Strain Data

The value of strains held will be increased according to the amount of information which is available on them. Collections are now recognising that, in the future, biotechnologists will wish to locate organisms with particular culture, physiological or metabolic properties and not only by their scientific name. Culture collections are in a unique position to check on such properties and build up strain data on the organisms held. For example, the UK Department of Trade and Industry is now supporting CMI in this area.

Information Availability

Information on the names of isolates held in various collections is already available through the *World Directory*[3], and both hard-copy and on-line catalogues of individual collections. International and national moves are now being made with a view to making detailed strain data available on-line — for example, through the UK Biotechnology Database being organised by the Laboratory of the Government Chemist. A major constraint on the utilisation of the existing resource will therefore be removed.

Service Development

In theory, many culture collections could offer a range of ancillary services, such as consultancy, industrial investigations, laboratory studies and microbial testing of materials. In practice, rather fewer are able to provide such services, owing to a lack of resources and appropriate personnel. Government departments concerned with the development of biotechnologically based industries need to be appraised of this situation, so that support of the type currently being given to CMI by the UK Department of Trade and Industry can be provided. The costs are modest in comparison with the potential, and such services can be expected to make a substantial contribution to their real costs or to be made self-financing.

Finance

Culture collections are a long-term resource. Stability and continuity of funding are therefore essential if the resource is to be maintained and developed. The value of the resource needs to be recognised by governments and other funding bodies wishing to promote the development of biotechnology. Many small collections, especially those in universities and colleges and run by individuals, are particularly vulnerable to financial cuts and rapid changes in policy. Research grant costings rarely include provision for culture collection support and the services of outside collections. The failure to lodge strains or have them authoritatively identified can negate research results which have often been obtained at a not inconsiderable cost.

The sale of cultures is most unlikely to meet the running costs of any collection, and core support will continue to be needed, even though new ancillary services being built up may reduce the amount required.

It is an exciting time for culture collections. Major changes are taking place within them, and enlightened biotechnological companies are increasingly utilising the resources and services offered. Demand will ensure the continuity of these valuable resources and also the way they develop in the years ahead.

ACKNOWLEDGEMENT

I am grateful to my colleague Professor D. L. Hawksworth for making available the manuscript of his review of culture collections, on which this chapter is partially based, prior to its publication.

REFERENCES

1. Hawksworth, D. L. (1985). Fungus culture collections as a biotechnological resource. In *Biotechnology and Genetic Engineering Reviews*, Vol. 3
2. Hawksworth, D. L., Sutton, B. C. and Ainsworth, G. C. (1983). *Ainsworth and Bisby's Dictionary of the Fungi*. Seventh edition. Commonwealth Mycological Institute, Kew
3. McGowan, V. F. and Skerman, V. B. D. (1982). *World Directory of Collections of Cultures of Microorganisms*. Second edition. World Data Centre, University of Queensland, St. Lucia
4. Smith, D. and Onions, A. H. S. (1983). *The Preservation and Maintenance of Living Fungi*. Commonwealth Mycological Institute, Kew

7

A NEW RESOURCE FOR BIOTECHNOLOGY — THE EUROPEAN COLLECTION OF ANIMAL CELL CULTURES

Alan Doyle

The UK Centre for Applied Microbiology and Research (CAMR; previously the Microbiological Research Establishment) has a distinguished history in biotechnological innovation which dates back to the 1950s and 1960s, with pioneer work in fermentation vessel and process design and continuous-flow culture. This expertise was retained when the establishment became part of the Public Health Laboratory Service, and this tradition has been recognised by the award of substantial grants by the Department of Trade and Industry (DTI) to fund research and development in fermentation technology, downstream processing and biosensors. In addition, the DTI has supported the establishment of a National Collection of Animal Cell Cultures, which was opened in July 1984 by the Government Chemist, Dr R. F. Coleman. The initial success of the collection has resulted in its becoming known as the 'European Collection' as from November 1985 and it now also gains financial support from the EEC under the Research Action Programme in Biotechnology.

Culture collections provide important support services for all biotechnology programmes. They supply reliable, authenticated cultures and offer informative catalogues of the world's genetic resources. The UK has a decentralised network of culture collections in which each type of organism is kept in an institute with the appropriate expertise and back-up facilities. The UK network has been developed to provide collections of yeasts, bacteria, fungi, algae, protozoa and, most recently, this new one for animal cells.

The need for an animal cell culture collection within the European Community has been recognised for a long time. The only animal cell facility available up to now has been that of the American Type Culture Collection (ATCC); and although this organisation provides an excellent service, it is inconvenient and expensive for general European use, especially in terms of freight costs. European scientists are world leaders in many areas of animal cell biotechnology, particularly the use of large-scale cell culture (up to 8000 litres) for the manufacture of biological products. It is unfortunate that a European cell culture collection has not been available to support this work. Two factors are responsible for the change in attitude within Europe as regards an animal cell bank. First, there is the importance of depositing a cell line involved in a patented process in an internationally recognised patent depository. Second, there are restrictions on the import into the USA of cell lines that have been in contact with bovine material (many cell lines contain calf serum in their growth medium), for fear of the transfer of foot and mouth disease virus. This has isolated the ATCC as a depository for potential users outside the USA.

The exploitation of animal cell cultures in biotechnological manufacturing has always lagged behind the use of micro-organisms (bacteria, fungi, etc.), because they are more fastidious, requiring complex and expensive growth media and culture systems; grow slowly (animal cells have a doubling time of 24 h, compared with only 20–200 min for many bacteria); and are not as productive. Until recently their main industrial role was as a substrate for virus production in viral vaccine manufacture (e.g. polio, measles, rubella and rabies vaccines). However, recent technical developments now make them an important growth area in biotechnology. One of these developments is the ability to hybridise (fuse) non-dividing antibody-producing cells with cells which grow in culture indefinitely to produce a cell type that will both continually replicate and secrete antibodies, all of which are identical (monoclonal antibodies).

The potential use of monoclonal antibodies for diagnostic, therapeutic and biochemical purification purposes is enormous and is the underlying rationale of a large number of biotechnology companies which have been established recently. As it is proving difficult to get genetically engineered bacteria to secrete some viral and mammalian proteins in an active form, more attention is now being paid to the engineering of animal cells for the production of vaccines (e.g. hepatitis B, herpes simplex) and therapeutic products (e.g. interferon, tissue plasminogen activator and hormones). The potential of cell lines for producing health care products, especially in the area of immunoregulation (interferon, interleukins, etc.), is still in a very early stage and means the search is on for suitable cell substrates.

How does the setting up of an animal cell culture collection at CAMR support these programmes and provide a much needed role in the infrastructure of biotechnology? One function is to provide the starting material for both academic research workers and industrial users. Laboratories all over the world are developing cell cultures for many and varied reasons. If you have one particular purpose for a cell line in mind, then you could screen many existing cultures to see whether they are suitable rather than returning to square one and establishing them for yourself. Without a culture collection this has always tended to be rather a hit and miss procedure, with an 'old boy' network establishing itself for the distribution of suitable cultures. This is a far from satisfactory state of affairs, in that an individual has very limited access to what is available — what is needed is a central store of information on all potentially useful cultures, and this is one of the major roles of any culture collection. In addition, it is up to the collection to ensure not only that cell lines which it distributes have the correct properties, but also that strict quality control procedures are adhered to.

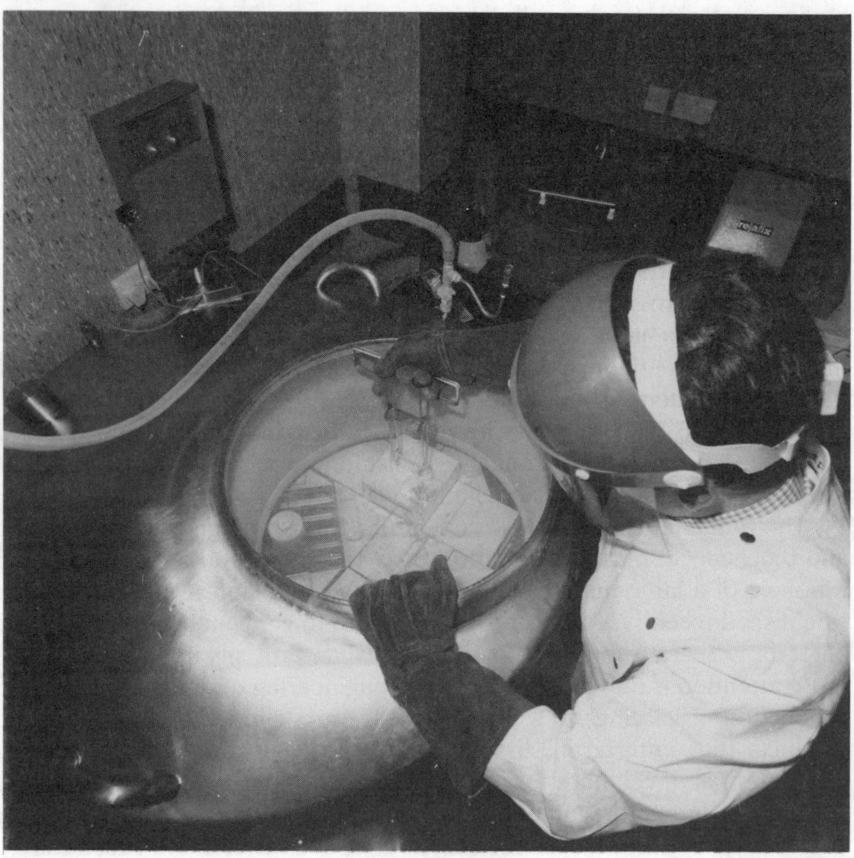

This latter function is vital, in that one of the major problems in all tissue culture work is contamination by other cells (there was one notorious instance when some supposedly mouse cell cultures were found to possess all the characteristics of a well-known human cell line) or by micro-organisms. Specialist techniques are needed to carry out these tests, and culture collections are often at the forefront of developing new technology to increase both sensitivity of detection and the speed at which such tests are carried out. The distribution of cell lines between colleagues can be an unfortunate means of disseminating micro-organisms, such as mycoplasmas, which may then infect all the cell lines in the receiving laboratory.

Another area in which culture collections are very active is that of preservation techniques. Animal cells are kept in suspended animation at a temperature of −196°C by the use of liquid nitrogen. Special conditions are needed to ensure that the cells remain viable during the freezing process — in terms of both the chemical composition of the medium and the mechanics of the cooling process. However, little work has been done to ensure that the specialist functions which are carried out by animal cells important in biotechnology can survive long periods of storage from the standpoint of genetic stability and retention of these functions. Culture collections are in a unique position to carry out such long-term prospective studies. The fact that cells need to be stored at very low temperatures (in liquid nitrogen) poses yet another danger to the long-term preservation of genetic material. Liquid nitrogen refrigerators have to be regularly maintained, and instances of material being permanently lost, because of failure to top up with liquid nitrogen before a weekend, or forgetting to replace the lid, are commonplace. Culture collections thus provide a safe storage facility which acts as a back-up to that in the laboratory and also safeguards the confidentiality of the clients' deposits.

In addition, many laboratories which initiate cell lines do not have facilities to freeze or preserve cells in their maintenance, and these cell lines thus suffer all the dangers of continuous propagation, with the associated contamination risks. A culture collection preserves in perpetuity many valuable cell lines which might otherwise be lost. Many types of research work and all production processes require a seed stock which can be referred back to continually for comparison of properties in case of loss or for consistency in production runs. This is what is meant by a culture collection playing a supporting role in biotechnology development.

Hitherto, the funding of such collections has been a haphazard affair. Funds have come from various government agencies but the sources are as diverse as the collections — the Health Service, the Research Councils and the universities. Now the DTI has taken the initiative in support of this new cell culture collection, and it is hoped that the future of all UK culture collections is secure, as their value is recognised as part of the essential background framework to a thriving biotechnology industry.

8
CHEMILUMINESCENCE — AN ULTRASENSITIVE ANALYTICAL TOOL IN MEDICINE AND BIOTECHNOLOGY

Anthony K. Campbell

INTRODUCTION

Walk out in one of the quiet Welsh valleys on a warm summer's night and you might be lucky enough to see the bright green glow of a female beetle. This is the European glow-worm (Figure 1) trying to attract a mate. Such luminous phenomena are rare on land, but are much more common in the sea. In fact, animal light plays a major role in the ecology of the deep oceans, where as many as nine out of ten of all the fish may be luminous[7]. Here we find the extraordinary angler fish with its luminous lure containing a culture of bacteria, and the remarkable *Malacosteus* with its two pairs of light organs, one emitting blue, the other red, light (Figure 2). There are also many invertebrate examples of living light, including starfish, shrimps, tunicates and jellyfish. The sources of this light are chemical reactions within specialised cells in the animal. That synthetic chemical reactions can also produce light has been known since the last century, but it is only during the last 20 years or so that the secrets of these two sets of reactions have been uncovered, to reveal a chemistry which not only can provide the research and clinical scientist with a uniquely sensitive analytical tool for

Figure 1 The European glowworm, *Lampyris noctiluca*: (a) female beetle on a frond of grass; (b) female beetle; (c) upside down, showing light organs; (d) photographed by its own light

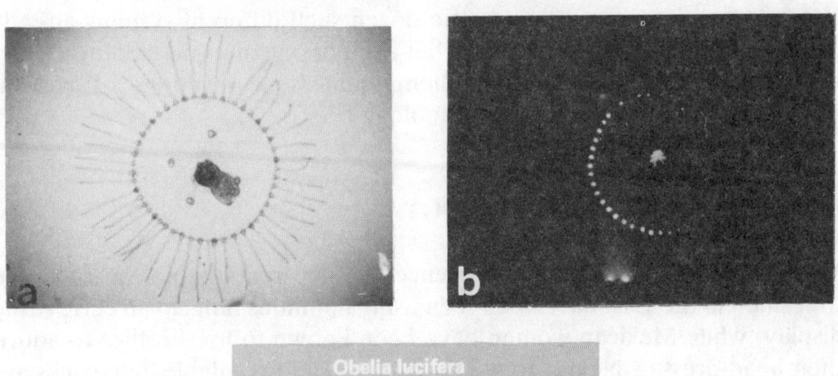

Figure 2 Some luminous marine organisms: (a) ambient light; (b) fluorescence

measuring many substances of biological interest, but also has an exciting potential in the new biotechnologies[2, 4].

The phenomenon is known as chemiluminescence.

WHAT IS CHEMILUMINESCENCE?

Most people expect chemical reactions to produce heat, but in chemiluminescence the energy from the reaction is released as light. In 1888 the German physicist Eilhardt Wiedemann first defined the term 'luminescence' to describe the phenomena producing such 'cold' light, distinguishing these from incandescence, where the energy for light emission arises from molecular motion induced by heat. He then subdivided the various types of luminescence, using a prefix to identify the source of energy for light emission. Thus, in chemiluminescence the energy arises from a chemical reaction, but in photoluminescence, which is the correct term to embrace what is usually referred to as fluorescence and phosphorescence, the energy comes from the absorption of electromagnetic radiation, usually ultraviolet or visible light. The modern definitions of these phenomena incorporate the concept of the electronically excited state. Chemiluminescence is thus the emission of light as the result of an exergonic chemical reaction when the vibronically excited product relaxes to ground state. The product of the reactions is, therefore, the actual light emitter, which is structurally different from the initial substrate, another difference from photoluminescence. Bioluminescence is chemiluminescence occurring in visibly luminous organisms. Here the substrate producing the excited state is usually called the luciferin, the reaction being catalysed by the enzyme luciferase, except when the small organic moiety is isolated with the protein as a fixed complex, when it is known as a photoprotein. Despite this genetic nomenclature, each luciferin and luciferase is distinct for each class of organism, more than half a dozen such different systems already having been identified. Oxygen, either itself or one of its metabolites, plays a central role in virtually all chemiluminescent reactions of current significance to biology and biotechnology (Figure 3).

POTENTIAL IN BIOTECHNOLOGY

Man has exploited chemiluminescence in some remarkable ways. Native tribesmen in the East have been seen using luminous fungi in an eerie ritual display, while Mexican women have been known to use fireflies to adorn their head-dresses. Nearer to home, commercially available light sticks are used by sailors as 'safe' lights and are familiar features in nightclubs on waitresses or exotic dancers. More practical uses include the attachment of

Obelin for intracellular calcium

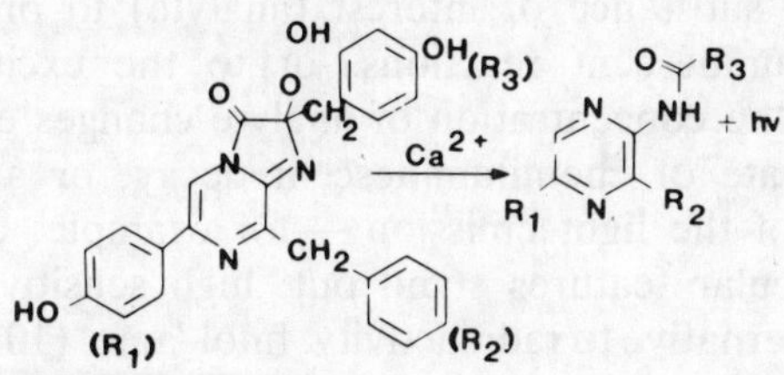

Phthalazine diones for reactive oxygen metabolites

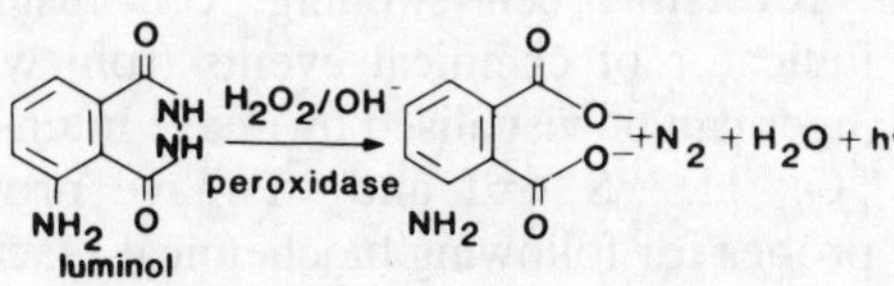

Acridinium esters as antibody labels

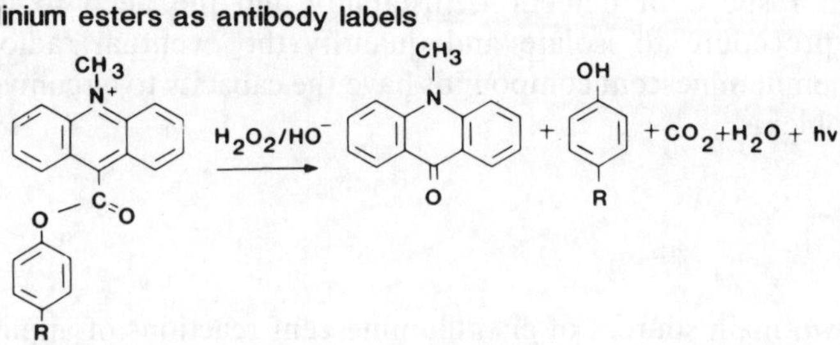

OXALATE ESTER

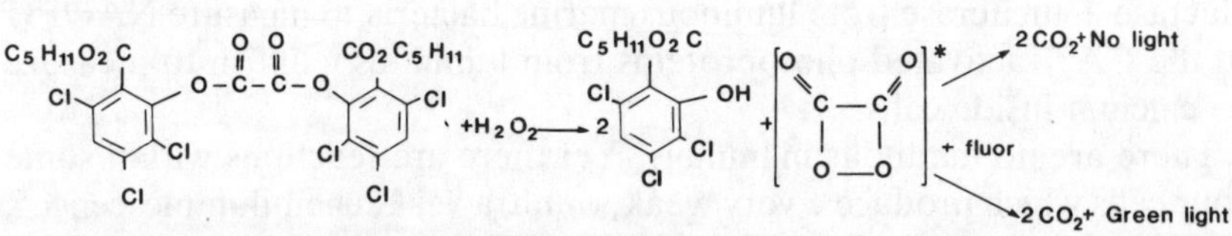

Figure 3 Some chemiluminescent reactions

bits of luminous fish or squid to fishing lines as a lure and the allocation of dried sea fireflies to Japanese soldiers during the war as a 'torch' when moistened. Attempts have even been made to detect large shoals of fish or nuclear submarines from the air by their ability to spark off luminous organisms which they may hit. However, the unique potential of chemiluminescence as an analytical tool in cell biology, medicine and biotechnology is only realised when the light emission is undetectable by the naked eye (<30 000 photons s^{-1} cm^{-2}), but can be detected by a modern photomultiplier tube capable of detecting just a few photons per second. It depends on coupling the substance of interest (analyte) to one of the reactants of the chemiluminescent reactions, or to the excited state product, so that altering the concentration of analyte changes either the total light yield or the rate of chemiluminescent decay, or a physical parameter characteristic of the light emission — for example, colour or polarisation. Three particular features stand out: high sensitivity, non-invasiveness and it is an alternative to radioactivity. fmol–amol (10^{-15}–10^{-18} mol) quantities are easily detected by chemiluminescence and potentially 10^{-21} mol (tipomol, from *tipyn*; Welsh = small). Dare I suggest an impossomol for 10^{-24} mol? Incorporation of chemiluminescent probes into living cells by microinjection, reversible cell swelling, cell fusion or micropinocytosis provides an indicator of chemical events from within. Furthermore, distributional changes can be visualised by image intensification. Radioisotopes such as ^{14}C, ^{3}H, ^{35}S, ^{125}I and ^{32}P have provided biologists with highly sensitive probes for following biochemical reactions. However, they suffer from problems of hazard, poor sensitivity of some isotopes, poor speed of detection, instability and the need to have a separation procedure to isolate and quantify the eventual radioactive product. Chemiluminescent compounds have the capacity to circumvent all of these problems.

SOURCES

There are two main sources of chemiluminescent reactions of application to biology and biotechnology: synthetic and those from living organisms[4]. The latter include firefly luciferin-luciferase to measure ATP, the oxido-reductase + luciferase from luminous marine bacteria to measure NADPH and the CA^{2+}-activated photoproteins from luminous jellyfish to measure free calcium inside cells.

There are no luminous mammals. Yet there are reactions within some of our cells which produce a very weak, or ultraweak, chemiluminescence[1], often less than 1 photon per cell per second. These cells include phagocytes responsible for killing invading micro-organisms, liver and heart cells, dividing yeast cells and plants. The source of the chemiluminescence is the reaction of toxic metabolites of oxygen O_2, OH, H_2O_2, OCl and $^{1}O_2$ with endogenous substrates to produce the key intermediate in many chemiluminescent reactions, the dioxetan. This four-membered ring, consisting of two carbons bridged by two oxygens, breaks down spontaneously to form excited carbonyls which emit the light. The overall quantum yield of this process is usually very low, but the light intensity can be increased some ten thousand times by adding a compound such as luminol or lucigenin capable of reacting with one or more of the oxygen metabolites.

The significance of this in industry is that oxidative processes in foods, textiles and other materials often involve these metabolites. They also play a key role in the action of certain herbicides. On the other hand, for a patient with rheumatoid arthritis or heart disease, these toxic derivatives of oxygen may be involved in the pathogenesis of the disease. The investigation of these problems requires methods for measuring the oxygen metabolites.

One final fascinating feature of some sources of chemiluminescence is the phenomenon of energy transfer. Certain jellyfish and fish (Figure 2)[5], as well as several other luminous organisms, can change the colour of the light they emit by transferring the energy from the product of the initial reaction to a fluor within the same cells. This occurs without the direct transfer of a photon by dipole–dipole resonance and requires that donor and acceptor are within about 80 Å. By coupling the chemiluminescent donor to an antigen and a fluorescent acceptor to an antibody the concentration of antibody–antigen complex can be determined by detecting the resulting change in colour. The method is 'homogeneous', not requiring separation of bound and free antigen, and has important potential in an assay system where ligand–ligand interactions require quantification[6, 13].

Another type of energy transfer is found in commercial light sticks. Here transfer occurs via electrons; without the fluor there is virtually no light detectable.

MAJOR APPLICATIONS IN BIOTECHNOLOGY

There are four main areas of analysis where chemiluminescence has provided the research or clinical scientist with a practical assay system (Table 1):

(1) Measurement of enzymes, metabolites and trace substances in small samples or at very low concentrations (nM–fM)[2, 4].
(2) Measurement of intracellular free Ca^{2+} [3].
(3) Measurement of oxygen metabolites[1, 9].
(4) Replacement of radioisotopes, particularly in immunoassay[10–13].

While the second of these has provided important information about the role of Ca^{2+} in the physiology and pathology of cells, the other three areas are attracting wide attention from the chemical, brewing, textile and food industries, and from companies responsible for supplying reagents and equipment to research and clinical laboratories in human, forensic and veterinary medicine. The detection of trace chemical and microbiological contaminants using chemiluminescence also has potential application in

Table 1 Major biomedical applications of chemiluminescence

Source	*System*	*Analyte*
A. Synthetic		
phthalazine	luminol	reactive oxygen metabolites
diones	aminobutylisoluminol (ABEI)	immunoassay label
	ABEI + fluor-antibody	homogeneous immunoassay
	luminol + H_2O_2	trace metals
	luminol + H_2O_2	peroxidase label
acridinium	lucigenin	reactive oxygen metabolites
salts	acridinium esters	immunoassay label
oxalate esters	+ H_2O_2	fluor
	+ H_2O_2 + fluor	light stick
B. Biological		
firefly marine	luciferin + luciferase	ATP and related metabolites
bacteria	oxidoreductase + luciferase	FMN, NAD(P)H and related metabolites
hydrozoa	aequorin, obelin	intracellular free Ca^{2+}
anthozoa	*Renilla* luciferin + luciferase	phosphoadenosine phosphosulphate
ultraweak	blue	reactive oxygen metabolites
	red	singlet oxygen

water purification, in sewage plants and in studying fresh-water and marine pollution. The military have also been attracted to the techniques as a potential means of detecting minute quantities of virulent organisms, or for detecting nuclear submarines as they spark off 'living light' in the deep oceans.

Two good examples of considerable commercial potential are the measurement of living organic material by ATP-triggered firefly luciferin-luciferase and the replacement of radioisotopes in immunoassay by chemiluminescent compounds[10–13]. ATP is essential to the survival of all prokaryotic and eukaryotic cells. Its measurement therefore provides a biochemical criterion of biomass and cell viability, provided that it can be extracted quantitatively from the sample to be analysed. The firefly assay can detect down to about 0.1 fmol ATP, sufficiently sensitive, in principle, to detect a single mammalian cell with an ATP content of 1–10 fmol, but not quite sensitive enough to detect a single bacterium with an ATP content of 10 amol. It has been used to assess biomass in sewage and water purification plants and even to search for life on Mars.

Immunoassay, invented in the 1960s, has become one of the most significant analytical techniques in biology in recent years. The availability of monoclonal antibodies has widened the scope of this technique, which can measure, specifically, many substances to which an antibody can be raised down to the pM region (fmol–amol in sample to be analysed). Chemiluminescent substances attached to either the antigen or the antibody in competition or immunometric assays have equivalent sensitivity to ^{125}I yet circumvent all of the problems associated with radioactivity. Phthalazine diones, acridinium esters and peroxidase detected by 6-hydroxybenzothiazole-enhanced luminol chemiluminescence have enabled a wide range of clinically viable immunoassays for drugs, hormones and other substances to be established. One for thyroid-stimulating hormone even has superior performance over its radioactive counterpart. Several of these are about to be launched commercially as 'kits', the potential market exceeding billions of dollars per year on a world-wide basis. Automated chemiluminometers will enable large numbers of samples to be processed, whereas the detection of chemiluminescence by very sensitive Polaroid film may have valuable application in the field or in the Third World.

FUTURE PROSPECTS

The increasing number of research, clinical and industrial applications of chemiluminescence is likely to be enhanced through the ingenuity of the chemist and by the possibility of cloning the genes responsible for bioluminescent proteins. Three have already been cloned. The full potential of chemiluminescence energy transfer has yet to be realised, while the widespread use of recombinant DNA technology in medicine and industry is likely to be restricted until a satisfactory alternative to ^{32}P is found. Chemiluminescence offers this opportunity[4, 8].

I have spent many hours on the beach, on research ships or in the mountains collecting luminous animals at night. The question now is whether this enthusiasm for the beauties of 'living light' can be harnessed not only to provide new insights into the molecular biology of living cells, but also to enable the community to benefit fully from advances in science and technology.

ACKNOWLEDGEMENTS

I thank my research group, the MRC, SERC, ARC and DHSS for financial support, and the Director and staff of the Marine Biological Association Laboratory, Plymouth, Dr Peter Herring, RRS *Discovery*, and Professor Frank McCapra.

REFERENCES

1. Boveris, A., Cadenas, E. and Chance, B. (1981). Ultraweak chemiluminescence: A sensitive assay for oxidated radical reactions. *Fed. Proc.*, **40**, 195–198

2. Campbell, A. K. (1986). Living light: chemiluminescence in the research and clinical laboratory. *TIBS*, **11**, 104–108
3. Campbell, A. K. (1987). Intracellular calcium: friend or foe? *Clinical Science*, **72**
4. Campbell, A. K., Hallett, M. B. and Weeks, I. (1985). Chemiluminescence as an analytical tool in cell biology and medicine. *Methods of Biochemical Analysis*, **31**, 317–416
5. Campbell, A. K. and Herring, C. J. (1986). A novel red fluorescent protein from the deep sea luminous fish *Malacosteus niger*. *Comparative Biochemistry and Physiology*
6. Campbell, A. K., Roberts, P. A. and Patel, A. (1985). Chemiluminescence energy transfer: a technique for homogeneous immunoassay. In Collins, W. P. (Ed.), *Alternative Immunoassays*, pp. 153–183. John Wiley, Chichester
7. Herring, P. J. (Ed.) (1978). *Bioluminescence in Action*. Academic Press, London
8. Matthews, J. A., Batki, A., Hynds, C. and Kricka, L. J. (1985). Enhanced chemiluminescence method for the detection of DNA dot-hybridization assays. *Analytical Biochemistry*, **151**, 205–209
9. Roberts, C. A. and Campbell, A. K. (1987). Pholasin — a bioluminescent indicator for detecting activation of single neutrophils. *Analytical Biochemistry*
10. Thorpe, G. H., Kricka, L. J., Gillespie, E., Moseley, S., Amess, R., Baggett, N. and Whitehead, T. P. (1985). Enhancement of the horseradish peroxidase-catalysed chemiluminescent oxidation of cyclic diacylhydrazides by 6-hydroxybenzothiazoles. *Clinical Chemistry*, **31**, 1335–1341
11. Weeks, I., Beleshti, I., McCapra, F., Campbell, A. K. and Woodhead, J. S. (1983). Acridinium esters as high-specific activity label in immunoassay. *Clinical Chemistry*, **29**, 1474–1479
12. Whitehead, T. C., Thorpe, G. H. G., Carter, T. J. N., Groucutt, C. and Kricka, L. J. (1983). Enhanced luminescence procedure for sensitive determination of peroxidase-label conjugates in immunoassay. *Nature, Lond.*, **305**, 158–159
13. Williams, E. J. and Campbell, A. K. (1986). A homogeneous assay for biotin based on chemiluminescence energy transfer. *Analytical Biochemistry*, **155**, 249–255

9
THE LEACHING OF MINERAL ORES USING BACTERIA

D. Barrie Johnson

INTRODUCTION

The exploitation of micro-organisms by *Homo sapiens* to obtain desirable products — one definition of biotechnology — is not exclusively a twentieth century innovation. Wine and beer production are frequently quoted examples of long-established biotechnological processes, and another, though less readily recognised, is that of 'microbial leaching', which refers to the use of bacteria to extract metals from mineral ores. It is thought that the Phoenician and Roman civilisations used microbes in mining operations some 2000 years ago, and in more recent history there is a documented account of the process occurring in Spain in the eighteenth century. However, it is only within the last 30 years that the central involvement of bacteria in the process has been recognised, and during this time the economic significance of microbial leaching has greatly increased, primarily as a result of the increased need to utilise lower-grade ores with the decline of higher-grade, conventionally processed ores. Two non-ferrous metals, copper and uranium, are currently recovered by using bacteria, but their potential to extract numerous other metals will undoubtedly be realised in the future. Microbial leaching is also being developed as a means to reduce the sulphur contents of coal; here pollution control is the aim, rather than metal recovery.

The principles of, and possible future developments of, microbial leaching are the subject of this chapter.

TARGET ORE BODIES

Ore bodies which are sulphidic in nature are susceptible to microbial leaching. Metal sulphide minerals come in many varieties, and some, such as copper, lead and zinc, represent the major source of that metal in the lithosphere; examples are given in Table 1.

Table 1 Some examples of sulphidic minerals

Iron sulphides	
Pyrite	FeS_2 ('fools' gold')
Marcasite	FeS_2
Pyrrhotite	Fe_7S_8–FeS
Hydrotroilite	$FeS \cdot nH_2O$
Mixed iron sulphides	
Chalcopyrite	$CuFeS_2$
Bornite	Cu_5FeS_4
Stannite	Cu_2FeSnS_4
Pentlandite	$(Fe,Ni)_9S_8$
Arsenopyrite	$FeAsS$
Cobaltite	$(Co,Fe)AsS$
Tetrahedrite	$(Cu,Fe)_{12}Sb_4S_{13}$
Other metallic sulphides	
Chalcocite	Cu_2S
Covellite	CuS
Galena	PbS
Sphalerite	ZnS
Millerite	NiS
Realgar	AsS
Cinnabar	HgS
Stibnite	Sb_2S_3
Molybdenite	MoS_2
Argentite	Ag_2S
(Associated ores)	
Uranite	UO_2
Coffinite	$USiO_4 \cdot nH_2O$
Chromite	$FeO \cdot C_2O_3$

The most abundant and familiar sulphide is pyrite (iron pyrites or 'fools' gold'), a mineral which is frequently associated with other sulphidic minerals and also non-sulphidic minerals in ore bodies. All sulphidic minerals are formed under reducing conditions, and when exposed to moist, aerobic conditions are unstable and will tend to oxidise. The process can be entirely chemical, but proceeds one million times faster when biologically catalysed (Figure 1).

$$FeS_2 + Fe_2(SO_4)_3 \rightarrow 3FeSO_4 + 2S \quad (1)$$

$$Fe^{2+} \rightarrow Fe^{3+} + e^- \quad (2)$$

$$2S + 3O_2 + 2H_2O \rightarrow 2H_2SO_4 \quad (3)$$

$$S + 3Fe_2(SO_4)_3 + 4H_2O \rightarrow 6FeSO_4 + 4H_2SO_4 \quad (4)$$

$$FeS_2 + 14Fe^{3+} + 8H_2O \rightarrow 15Fe^{2+} + 2SO_4^{2-} + 16H^+ \quad (5)$$

Figure 1 The microbial oxidation of pyrite

MICRO-ORGANISMS INVOLVED

The principal micro-organisms involved in biological leaching form a group of highly specialised bacteria. Microbiologists refer to them as chemolithotrophes — that is, they are able to obtain energy from the oxidation of reduced inorganic substances, in this case ferrous iron (Fe^{2+}) and/or reduced sulphur, such as sulphide (S^{2-}). Many of them are also autotrophic — they obtain carbon by fixing carbon dioxide in much the same way as plants. Thus, they differ markedly from the majority of bacteria, which break down organic materials and use prefixed carbon units. Another important difference is their acidophilic nature, their pH range being around 1.5–4.0.

The best-known mineral leaching bacterium is *Thiobacillus ferrooxidans*. The name translates as 'iron-oxidising sulphur rod' and describes the ability of the microbe to oxidise both ferrous iron and reduced sulphur (reactions 2 and 3 in Figure 1). The oxidation reactions release small amounts of energy, which the bacterium is able to utilise. *T. ferrooxidans* was first described in 1951, and since then has been isolated from a range of highly acidic environments, such as acid mine drainage waters from disused mines and coal mine spoils.

Other bacteria involved in the dissolution of sulphidic minerals include:

(1) *Thiobacillus thiooxidans*: a bacterium similar in many ways to *T. ferrooxidans* but lacking the ability to oxidise ferrous iron. Mixed populations of *T. thiooxidans* and *T. ferrooxidans* occur naturally and are more effective at leaching some minerals than is pure *T. ferrooxidans*.

(2) *Leptospirillum ferrooxidans*: a relatively recently isolated bacterium (1972) which is able to oxidise ferrous iron, but not sulphur. Mixed cultures of *L. ferrooxidans* and sulphur-oxidisers are able to degrade sulphidic minerals, and it is probable that the micro-organism contributes to metal release in on-going leaching operations. Its greater tolerance to

acidity and higher temperatures compared with *T. ferrooxidans* may be of advantage in more efficient leaching schemes.

(3) Moderately thermophilic thiobacillus-like acidophiles: 'TH' strains-bacteria recently isolated from acidic hot springs and heating coal spoils which resemble *T. ferrooxidans* in many respects, but have a requirement for organic carbon compounds and temperature optima of around 50°. Leaching rates mediated by these iron-oxidisers can be very much greater than those of the more familiar thiobacilli.

(4) *Sulfolobus*-type thermophiles: these are the most thermophilic and acidophilic of the iron sulphur oxidisers yet isolated (temperature optima at around 70°). Although they have not been isolated from commercial leaching sites, the bacteria are probably active in the hotter zones in leach dumps and are postulated to have considerable potential in future operations, again due to their faster rate of leaching.

The rates of mineral degradation in leaching operations are undoubtedly affected by micro-organisms other than those so far mentioned. Eukaryotes, including protozoa, fungi and yeasts, and a number of heterotrophic bacteria, have been isolated from leaching sites and other extreme acid environments. The heterotrophic bacteria have a remarkable scavenging ability and can survive on the minuscule quantities of organic matter leaked by *T. ferrooxidans*. Work at Bangor and elsewhere has revealed that a considerable diversity of acidophilic heterotrophes exist, and their industrial potential goes beyond mineral leaching alone. However, by positively and negatively reacting with *T. ferrooxidans*, different isolates may be of use in stimulating leaching rates or inhibiting the chemolithotrophe where its activities are undesirable, such as in the production of acid mine drainage waters.

MECHANISMS OF MICROBIAL LEACHING

The objective of microbial leaching is the degradation of metal-containing minerals with the consequent release of soluble metal ions, the recovery of which is straightforward. Bacteria may achieve mineral dissolution by either of two mechanisms. These are described below, using the copper minerals chalcocite and chalcopyrite as illustrations.

(1) *Direct mechanism* Scanning electron microscopy has shown that thiobacilli attach specifically to the sulphide moieties of mineral rock surfaces and cause direct dissolution, as evidenced by the pitting of mineral surfaces. For chalcocite the oxidation reactions are:

$$2Cu_2S + O_2 + 2H_2SO_4 \rightarrow 2CuS + 2CuSO_4 + H_2O$$
$$CuS + 2O_2 \rightarrow CuSO_4$$

For chalcopyrite:

$$4CuFeS_2 + 17O_2 + 2H_2SO_4 \rightarrow 4CuSO_4 + 2Fe_2(SO_4)_3 + 2H_2O$$

(2) *Indirect mechanism* Sterile, acidic solutions of ferric sulphate are capable of degrading sulphidic minerals. Ferric ions oxidise the sulphide moiety to sulphate while themselves being reduced to ferrous (reactions 1 and 5, Figure 1). The microbial input is a ferrous to ferric regeneration, so allowing degradation to continue. Typical reactions are:

$$Cu_2S + 2Fe_2(SO_4)_3 \rightarrow 2CuSO_4 + 4FeSO_4 + S$$
$$CuFeS_2 + 2Fe_2(SO_4)_3 \rightarrow CuSO_4 + 5FeSO_4 + 2S$$

The sulphur generated in these reactions is oxidised by the thiobacilli to sulphuric acid, thus maintaining highly acidic solutions.

In practice, both mechanisms exist in leaching operations, although schemes exist in which exclusive use is made of indirect oxidation.

COMMERCIAL LEACHING OPERATIONS

To date, copper and uranium are the only metals which are commercially leached using bacteria. In the case of copper, the technique is used almost exclusively for the processing of low-grade ores (of the order of <1 per cent Cu). In 1954, when the bacterial involvement in leaching was first realised, only a few copper mining operations were making use of it in their overall schemes, but now most companies do so. It was estimated in 1982 that about 10 per cent of US copper production was mediated by bacteria.

There are a number of actual and proposed methods of microbial leaching. The two in most common use are dump and heap leaching. In commercial dump leaching operations, vast quantities of low-grade ores and waste rock are deposited on impermeable ground, frequently in valleys. Dumps vary in size, but can be up to 350 m high or more, holding millions of tons of rock; they are periodically irrigated with acidic solutions by spraying, flooding or injecting through vertical pipes. The liquor percolates through the dump, becoming increasingly laden with copper. Eventually the 'pregnant' liquor is collected from the base of the mound, processed to remove soluble copper and recycled back into the dump. The Kennecott Chino copper mine in New Mexico produces about 7.1×10^4 tons of mine waste each day. Leaching dumps constructed of this material are generally 25–30 m high at this particular mine; a schematic representation of such a mound is shown in Figure 2. A 'finger mound' design and sprinkling irrigation system both improve oxygen diffusion and wetting, thus stimulating biological oxidation. The removal of copper from preg-

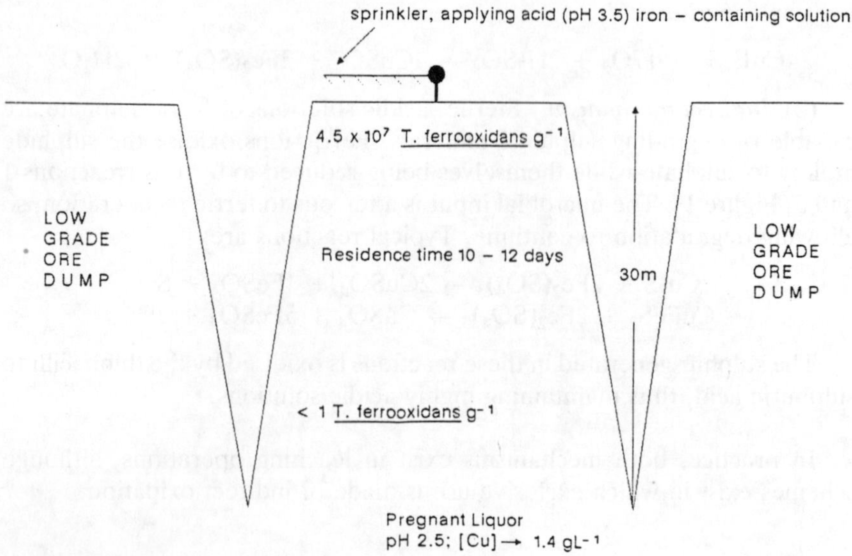

Figure 2 Schematic representations of a finger dump mound, showing typical figures

nant solutions is frequently achieved by addition of scrap iron (e.g. shredded cans), which causes surface deposition of copper by:

$$Fe^{\circ} + 2e^{-} = Fe^{2+}$$
$$Cu^{2+} + 2e^{-} = Cu^{\circ}$$

This 'cement copper' is sprayed off using high-pressure jets, and may be further purified and ingoted. The process produces 45–50 tons of cement copper each day at the Chino mine, from material which would otherwise be abandoned. Dump maintenance chiefly involves the periodic removal of surface ochreous materials with bulldozers. The operating life of a dump depends very much on its size — the ore in a very large dump may be processed for a number of years to recover as much copper as possible.

Heap leaching is a modification of dump leaching in which the use of higher-grade ores allows for greater capital input. Modifications involve the crushing of ores and the installation of aeration systems into mounds, both being designed to increase the rate of leaching, and the construction of impermeable pads to prevent seepage of pregnant liquor into the underlying soil.

In-place ore leaching is used to recover metals from abandoned underground operations, where up to 30 per cent of the metal values may still remain in the form of supporting pillars and walls. The leaching solution is applied directly to the roof and walls of the excavation, and the pregnant liquor is pumped out and processed. This method of leaching is also used to mine inaccessible low-grade copper orebodies and to recover uranium from low-grade deposits. Prior shattering of the orebody with a controlled underground explosion may be necessary.

Tank and vat leaching incur higher capital costs and thus need to utilise relatively high-grade ores. Although not currently operative with sulphidic minerals, economically viable schemes have been proposed.

URANIUM ORE LEACHING

Uranium does not occur as a sulphidic mineral, but principally as its oxide, UO_2, and frequently associated with pyritic minerals. Microbial leaching proceeds by the indirect mechanism, whereby acidic ferric sulphate oxidises UO_2 to soluble UO_2SO_4 (tetravalent to hexavalent uranium). Recently, however, it has been shown that *T. ferrooxidans* can itself oxidise uranium according to the equation

$$2UO_2 + O_2 + 2H_2SO_4 \rightarrow 2UO_2SO_4 + 2H_2O$$

The yield of uranium recovered from low-grade ores by microbial leaching is generally 30–80 per cent. However, by separating the ore leaching phase from the bioreactor, as in the scheme outlined in Figure 3, Manchee achieved a yield of over 90 per cent within 10 days.

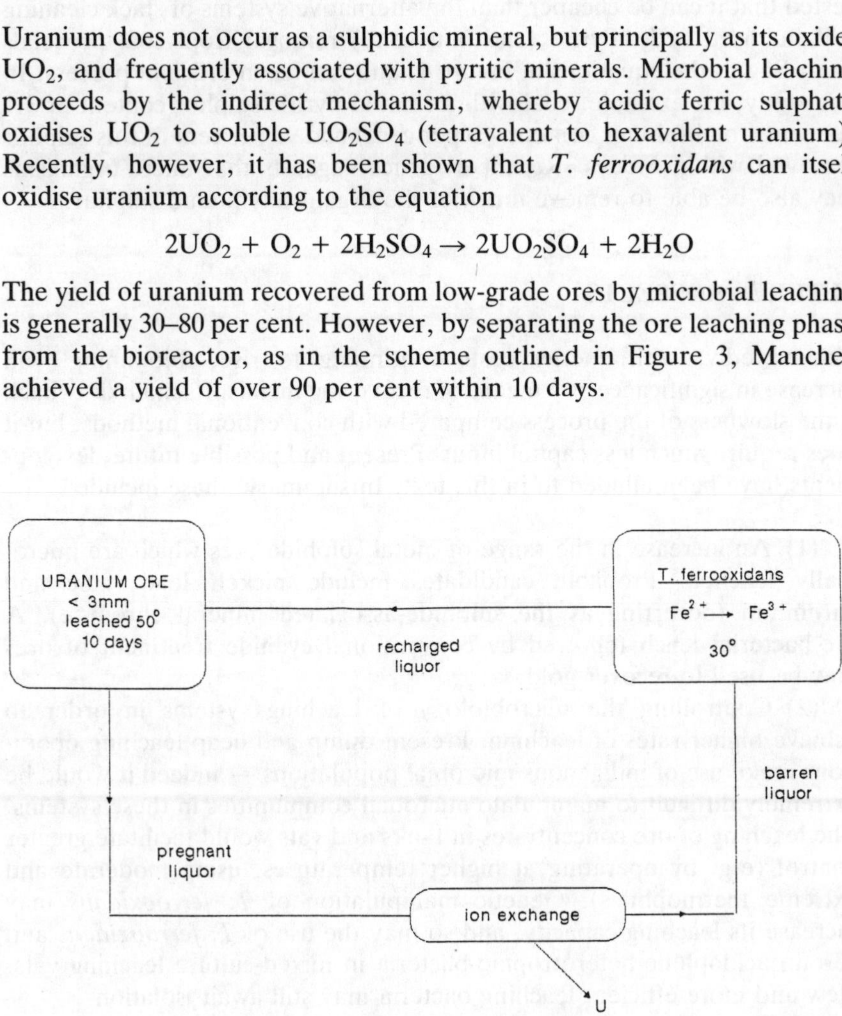

Figure 3 A scheme for uranium ore leaching (after Manchee, 1977)

MICROBIAL COAL DESULPHURISATION

A process closely akin to microbial leaching is that of using acidophilic iron-oxidising bacteria to remove pyritic sulphur from coal prior to combustion. The sulphur content of coal varies from less than 1 per cent to over 10 per cent — 2 per cent being common — in the form of pyritic and organic sulphur, and small quantities of sulphate. Combustion of coal releases the sulphur as SO_2 gas, and is frequently deemed a major contributor to acid rain pollution. Removal of sulphur from coal is an attractive means to reduce the hazard, and economic studies have suggested that it can be cheaper than the alternative systems of stack cleaning and fluidised beds. Pyrite (and marcasite) removal from coal proceeds according to the equations in Figure 1, and can be an efficient process. In one study, for example, coal with an initial pyritic sulphur content of 3.1 per cent (total sulphur content 4.6 per cent) had 97 per cent of this sulphur removed within 5 days. Recent evidence suggests that micro-organisms may also be able to remove much of the organically bound sulphur.

FUTURE PROSPECTS

Undoubtedly, with energy costs continuing to rise, 'biomining' will increase in significance as a means of recovering metals. Its main drawback is the slowness of the process compared with conventional methods, but it does require much less capital input. Present and possible future developments have been alluded to in this text. In summary, these include:

(1) An increase in the range of metal sulphide ores which are microbially leached. Probable candidates include nickel, lead, zinc and chromium (occurring as the sulphide-associated mineral chromite). A pre-bacterial leach followed by conventional cyanide treatment of ores may be used to recover gold.

(2) Controlling the microbiology of leaching systems in order to achieve higher rates of leaching. Present dump and heap leaching operations make use of indigenous microbial populations — indeed it would be extremely difficult to manipulate microbial communities in these systems. The leaching of ore concentrates in tanks and vats would facilitate greater control (e.g. by operating at higher temperatures, using moderate and extreme thermophiles). Genetic manipulation of *T. ferrooxidans* may increase its leaching capacity, and so may the use of *T. ferrooxidans* and certain acidophilic heterotrophic bacteria in mixed-culture leaching vats. New and more efficient leaching bacteria may still await isolation.

(3) The widespread use of microbial coal desulphurisation.

BIBLIOGRAPHY

1. Brierley, C. L. (1982). Microbiological mining. *Scientific American,* **247** (2), 42–51
2. Dugan, P. R. and Apel, W. A. (1978). Microbiological desulfurization of coal. In Murr, I. E., Torma, A. E. and Brierley, J. A. (Eds.), *Metallurgical Applications of Bacterial Leaching and Related Microbiological Phenomena*, pp. 223–250. Academic Press
3. Lundgren, D. G. and Silver, M. (1980). Ore leaching by bacteria. *Annual Review of Microbiology,* **34**, 263–283

10
THE USE OF MICRO-ORGANISMS FOR THE REMOVAL AND RECOVERY OF HEAVY METALS FROM AQUEOUS EFFLUENTS

I. S. Ross

INTRODUCTION

It has been well known for many years that some micro-organisms can take up large amounts of heavy metals such as copper, cadmium, lead and uranium, and this was discussed in a previous paper[18]. Attempts have been made to exploit this property to develop systems for the removal and recovery of metals from industrial effluents, both to reduce environmental damage from toxic metals and to recover those with a high commercial value. This chapter will review this work and examine the potential of microbial metal removal and recovery from effluents. For clarity, work on bacteria, algae and fungi will be discussed separately.

BACTERIA

The majority of studies of metal removal from solution using bacteria have involved the use of living cells, either immobilised or in continuous culture.

Silver and copper accumulation has been demonstrated in chemostat cultures of mixed populations of bacteria obtained using enrichment techniques[5, 9]. In the case of silver, a three-membered community consisting of *Pseudomonas maltophila*, an unidentified coryneform bacterium and *Staphylococcus aureus* removed 76 per cent of silver supplied as 1 mM silver nitrate (107.9 μg/ml silver) over a 2000 h period[5]. Copper was accumulated by a ten-membered community of bacteria isolated from activated sludge. This community was tolerant of up to 935 μg/ml of copper, and produced substantial growth on the walls of the chemostat vessel. Copper removal rose with increasing wall growth and reached a maximum of 92 per cent[9]. The stability of the wall growth depended on the nutrient limitation, and breakdown of wall growth resulted in a rapid drop in the efficiency of copper removal. No further information on development of these processes appears to have been published. A disadvantage of both is that they require a supply of nutrients to maintain culture growth, and these would presumably have to be added to any effluent to be treated.

Other workers have used immobilised bacterial cells with a view to the development of continuous or semi-continuous processes. Studies of uranium removal have been carried out using the actinomycete *Streptomyces viridochromogenes* immobilised in polyacrylamide gel and packed into columns with a diameter of 0.6 cm and a bed volume of 1.4 ml[17]. This system contained 97.6 mg dry weight of immobilised cells and removed 100 per cent of uranium from 200 ml of a 20 μg/ml solution at pH 6. The bound uranium could be released from the column with 10 ml 0.1 M sodium carbonate, and the uptake capacity of the column was retained through five cycles of loading and unloading of uranium. When the experiment was repeated with sea-water enriched with 10 μg/ml uranium, 80 per cent of the uranium was removed after four cycles. No further development of this work has appeared in the literature.

In another study, *Pseudomonas fluorescens* immobilised on soft polyvinyl choride granules and packed into columns at the rate of 588 g/l column volume was used primarily to remove nitrate, but was found to be capable of removing metals such as lead and zinc[21]. This system was stated to have been used as the basis of a water treatment plant for a Hungarian chemical factory producing an effluent containing zinc, chromium, barium, aluminium, nickel, lead and nitrate after the metal concentration had been reduced to about 1 μg/ml by precipitation with lime[21]. However, no information was given about the efficiency or reliability of this plant. In the same paper a system for removing plutonium from effluents using *Pseudomonas aeruginosa* immobilised on an oxygen plasma-treated polypropylene web was described[21]. Columns loaded with 1.5 g of web containing 0.3 g cells/g polypropylene were fed at the rate of 15 ml/h with plutonium at a concentration of 1.7 nCi/100 ml for 12 h. This gave removal efficiency of 75–80 per cent. If two columns were connected in series, efficiency increased to 98–99 per cent, with a useful column life of about 2 weeks. This system was not apparently tested on an industrial scale, and no attempt was made to recover the plutonium, it being suggested that plutonium-loaded biosorbent (bacteria plus polypropylene) could be packed into drums and dumped.

More recently, a system for the removal of cadmium and various other metals from aqueous effluents using a *Citrobacter* sp. immobilised in polyacrylamide gel has been described[14, 15]. Columns (10 cm × 3 cm) packed with immobilised cells were capable of removing cadmium from a solution containing 200 μg/ml Cd supplemented with 2 mM glycerol-2-phosphate with an efficiency of 65–66 per cent. The efficiency fell to 50 per cent after 10 litres of solution had been passed through the column, and remained at this level until the column became saturated after passage of 20–30 litres of solution. Column saturation capacities ranged from 2.5 to 3.5 g of cadmium. The use of three columns in series increased efficiency to 68 per cent and increased glycerol-2-phosphate concentration resulted in the removal of up to 90 per cent of the cadmium. The metal was readily released from the columns by citrate buffer (0.5 M, pH 5) and unloaded columns retained 84–85 per cent of initial activity on re-use.

Further studies with this system showed that removal of mixtures of cadmium, copper and lead was similar to that of single metals, reaching 87 per cent, 85 per cent and 83 per cent respectively, with an inflow of 1.12 μg/ml Cd, 0.63 μg/ml Cu and 2.1 μg/ml Pb[14]. If the total metal input was increased tenfold, the removal of cadmium and lead remained the same, but copper removal fell to 44 per cent. Copper removal was also depressed by addition of zinc sulphate, sodium arsenate, ammonium sulphate and potassium cyanide to synthetic effluents, whereas cadmium and lead removal was less affected. This system for metal removal and recovery would appear to have potential for commercial exploitation, especially as the mechanism involved (precipitation at the cell surface as metal phosphate) should, at least in principle, operate for any metal with an insoluble phosphate[15]. This system was recently shown to be capable of removing 95–96 per cent of uranium from a 200 μg/ml solution after passage of 8 litres of solution, and uranium removal continued with an efficiency of 91 per cent until 24 litres of solution had been treated, when the experiment was terminated[16]. At this point it was calculated that about 4.5 g of uranium had been accumulated by the column[16]. Uranium could be released from loaded columns by elution with 1 M citrate buffer at pH 4. About 3.3 g of uranium was released into 5 litres of eluant, and further elution was not regarded as worth while[16].

It is apparent that bacterial systems are capable of metal removal and recovery from effluents and in the case of the system developed by Macaskie and Dean[14–16] may have considerable potential, although, as far as I am aware, the latter system has not been tested on an industrial or pilot plant scale. Clearly, the use of polyacrylamide gel as a biomass support system would be uneconomic on a large scale, and a further disadvantage is the requirement for a glycerol-2-phosphate feed to the system. Also, as the process depends on the activity of a cell-bound phosphatase which remains active in resting cells, there is the possibility of enzyme inhibition by effluent components.

ALGAE

A number of investigations have shown that algal biomass is capable of binding considerable amounts of heavy metals, such as cadmium, copper and uranium[18], and the use of algal cells for metal removal and recovery has received some attention. Cells of *Chlorella regularis* immobilised in polyacrylamide gel and packed into columns have been tested for removal of uranium from sea-water containing 10 μg/ml, and were found to remove 100 per cent of the uranium from 100 ml of water in one passage through a column containing 771 mg dry weight of cells[17]. Bound uranium could be released by washing the columns with 0.1 M sodium carbonate. A German patent[11] describing the use of a green alga (not identified) for the removal of uranium from sea-water exists, but the author does not know whether a practical process was ever operated.

Recently, the possibility of selective recovery of gold and certain other metals using algal biomass has been studied[6]. Biomass of the alga *Chlorella vulgaris* was found to bind gold, silver and mercury at low pH (about pH 2–3) when binding of other metals was much reduced. Gold was bound effectively by *Chlorella vulgaris* biomass, and 98 500 μg/g dry weight of gold was taken up from a solution containing 19.7 μg/ml, a concentration factor of 5000[6]. When cells of *C. vulgaris* were immobilised in polyacrylamide gel, packed into a column and loaded with zinc, copper, mercury and gold from a mixed solution, zinc and copper were released by acetate buffer at pH 5 and pH 3.5, respectively, but mercury and gold were retained in the column. Further washing with mercaptoethanol at pH 5 released the gold, and mercury was released by the same reagent at pH 2[6]. This work suggests that algal biomass may provide a relatively simple and cheap means of separate recovery of certain metals, but more work is needed to confirm this.

It should be noted that algae have also been used for treatment of water, especially mine effluents, to reduce contamination with heavy metals in lagoons and meanders. These systems depend, not only on binding of metals by algal biomass, but also on sediment produced by decaying cells in which bacterial activity may be involved in metal immobilisation, and hydrogen sulphide generated as a result of bacterial activity may precipitate metals as sulphides. A brief review of the use of algal ponds in treatment of mine waste-water is available[1].

FUNGI

A considerable amount of work has been done on the ability of fungi to take up heavy metals, and this was discussed in a previous paper[18]. From the standpoint of metal removal and recovery from effluents, most interest has centred on the ability of non-growing mycelium of filamentous fungi to take up heavy metals. Studies using batch techniques have shown that a number of fungi have the ability to take up considerable amounts of uranium when non-growing mycelium is exposed to solutions of the metal[12], and some, notably *Rhizopus arrhizus* and *Penicillium chrysogenum*, may be more effective at binding uranium than activated charcoal and certain ion-exchange resins[27]. It is difficult to compare data from different workers, owing to great variation in uptake conditions such as pH, temperature, mycelial dry weight, etc., but an indication of possible binding capacities of various fungi is given in Table 1. A feature of uranium uptake by fungi is that the bound uranium is readily released by treatment of the metal-loaded mycelium with sodium carbonate or sodium bicarbonate[10, 24]. A detailed study of uranium release from loaded biomass of *R. arrhizus* showed that sodium bicarbonate was the most effective desorbent, since complete recovery of uranium was possible with high concentration factors. A further advantage was that this reagent did not damage *R. arrhizus* mycelium, which could be re-used in repeated adsorption–desorption cycles with 90 per cent of the original uranium uptake capacity[25]. This work suggests that fungal biomass might have considerable potential for removal and recovery of uranium from aqueous effluents, and a process for forming *Rhizopus* mycelium into pellets consisting of 5 per cent support material and 95 per cent biomass has recently been announced[25]. The biomass is apparently immobilised in a porous resin but, as the process is the subject of patent applications, no details are available.

Table 1 Uranium-binding capacities of some fungi

Fungus	*Uranium bound* (μg/g dry wt)
Aspergillus niger	46 530–80 730[a]
Aspergillus oryzae	57 640–63 420
Penicillium janthinellum	47 210–52 700
Penicillium chrysogenum	165 000[b]
Rhizopus arrhizus	> 180 000

[a] Data for several strains, from Ref. 12.
[b] Data from Ref. 27.

Uptake of metals other than uranium, including thorium[27], radium[26], chromium and lead[20] and copper, cadmium and zinc[22], by fungal biomass has also been demonstrated, but these studies were concerned largely with determination of metal uptake capacities rather than efficiency of metal removal. In a study with biomass of *Penicillium spinulosum* treated with a commercial detergent, it was found that over 90 per cent of copper could be removed from solutions containing 6.35 μg/ml of copper within 5 min in batch tests with 3 per cent fresh weight of biomass[19]. When the copper concentration was increased to 63.5 μg/ml, removal efficiency fell to 83 per cent. Metal removal was most effective at pH between 5 and 7.5. Fungal biomass not treated with detergent was much less effective. Cadmium was also removed but with lower efficiency[19].

Other workers have described possible processes for metal removal and recovery using fungal biomass. Two US patents exist[7, 8] which outline the use of growing and heat-killed fungal biomass for recovery of precious metals such as platinum and gold and some base metals such as copper and zinc. The examples given in both patents indicate that effective removal of precious metals from mixtures containing up to 11 metals can be achieved, with some apparent selectivity for precious metals (present at concentrations of 1–2 μg/ml) over base metals present at concentrations up to 13 μg/ml. It should be noted that the given examples of removal by growing mycelium[7] would, in the opinion of this author, reflect uptake by non-growing but living mycelium, as active growth would not be possible in the experimental conditions used. The biomass concentrations used in both cases were such that actual metal loading of the mycelium was very low, 0.01 per cent for precious metals and 0.01–0.02 per cent for base metals. These loadings suggest that metal recovery would not be economic, since it can be calculated that 17 000 g dry weight of fungus would yield 1 g of platinum and about 7800 g dry weight would yield 1 g of copper, assuming no loss during recovery from the fungus. As far as I am aware, no commercial processes were developed from these patents.

Removal of metals by fungus immobilised in columns has received little attention. In a study with the fungus *Trichoderma viride* immobilised in china clay[23] it was found that 68–75 per cent of the copper in a simulated leach effluent was removed in a system buffered at pH 5.5, and removal was independent of copper concentration over the range 5.94–127.5 μg/ml. However, in unbuffered systems, removal efficiency fell sharply with increasing copper concentration, and this was associated with a sharp drop in pH to 2.0 with 127.5 μg/ml copper. At copper concentrations of 14.4 μg/ml or less, pH changes from 5.5 to 3.6 did not affect removal efficiency. It is unlikely that removal efficiencies of this magnitude would be adequate in practice, and as biomass metal loadings were not determined, the potential for metal recovery cannot be estimated.

Removal of cadmium by biomass of *Aspergillus oryzae* immobilised on reticulated foam biomass support particles has been investigated[13] and removal efficiencies of 99 per cent have been obtained with a 5 min residence time in columns. A column containing 5.5 g dry weight of fungus removed cadmium from a 1 μg/ml solution, pH 5.5, with maximal efficiency until 50 ml of solution had been treated, and at 100 ml effluent volume, efficiency began to fall sharply. This was apparently due to saturation of biomass at the bottom (entry point) of the column with cadmium, thus progressively reducing the effective length[13].

CONCLUSION

It will be clear from the above discussion that the use of microbial biomass for metal removal and recovery requires further study, especially with respect to the development and testing of practical processes. Such processes will require low-cost biomass with the ability to remove at least 90 per cent of the metals present in the effluent to be treated and, if metal recovery is proposed, a high capacity for bound metal. The process will have to be capable of handling large volumes of effluent without the need to constantly change to biosorbent. The biosorbent itself will have to be convenient to handle and capable of being re-used after release of bound metals, and the released metals will have to be in a form suitable for chemical recovery and recycling. The experimental systems discussed above, except perhaps that utilising resin-immobilised *Rhizopus* biomass[25], are unlikely to meet these requirements.

Recently, however, a process has been described which meets a number of these criteria[2–4]. The biosorbent is a granular material produced from biomass, apparently of microbial origin, although, for commercial reasons, the exact nature of the biomass and details of its processing have not been revealed. This material consists of amorphous granules with a large area for metal–granule interaction and is capable of metal loadings of 32 mg/g silver, 214 mg/g cadmium, 140 mg/g copper, 661 mg/g lead and 131 mg/g zinc from solutions containing 10 μg/ml metal. Column tests showed that 6.5 g of the material removed 99 per cent of copper from solutions containing about 9 μg/ml copper and up to about 50 litres of solution could be treated before removal fell below 90 per cent. High efficiency of lead removal was also possible; 90 per cent of lead from a 100 μg/ml solution was removed after the passage of over 20 litres of solution[3]. Metal-loaded material could be treated to recover the bound metal and then re-used for metal removal. This material has been tested on a pilot plant scale using a fluidised-bed system to treat effluent from a secondary lead refining plant. The influent lead concentration varied from 0.01 to 4.3 μg/ml, and removal efficiencies of up to 99 per cent were obtained over a 39 day period[3]. The system has also been operated on a smaller scale, using canisters of the metal recovery agent. Precious metals, such as gold, could be removed from cyanide-containing liquors and recovered for recycling. This system for metal removal and recovery appears to have great potential and its developers clearly believe that the system will be commercially viable[4].

It will be apparent that although a considerable amount of laboratory work has been done on the use of microbial biomass for metal removal and recovery, few workers have developed practical processes suitable for scale-up to handle industrially useful volumes of effluent. Ultimately, the success of any process will be dependent on cost-effectiveness, and this is likely to be significantly affected by the ability to recover and recycle metals removed from treated effluent. This may be worth while for precious metals with a high commercial value, but seems likely to be uneconomic for many base metals at present. From an environmental standpoint, the ability of microbial systems to remove toxic metals from effluents containing relatively low concentrations in the region of 1–5 μg/ml may be important, but such systems will probably only become attractive if legislation on permissible levels of discharge is tightened.

REFERENCES

1. Brierley, J. A. and Brierley, C. L. (1983). Biological accumulation of some heavy metals — biotechnological applications. In *Biomineralisation and Biological Metal Accumulation* (P. Westbroek and E. W. de Jong, Eds.), Reidel, Dordrecht, pp. 499–509
2. Brierley, J. A., Brierley, C. L. and Goyak, G. M. (1985). AMT-Bioclaim^{TM} process for treatment of metalliferous wastewater from electroplating and other industries. Presented at *2nd Annual BIOMINET Meeting*, October 1985, Mississanga, Canada
3. Brierley, J. A., Brierley, C. L. and Goyak, G. M. (1986). AMT-Bioclaim^{TM}: A new wastewater treatment and metal recovery technology. In *Fundamental and Applied Biohydrometallurgy* (R. W. Lawrence, R. M. R. Branion and H. G. Ebner, Eds.), Elsevier, Amsterdam, pp. 291–304
4. Brierley, J. A., Goyak, G. M. and Brierley, C. L. (1986) Considerations for commercial use of natural products for metals recovery. Presented at a symposium *Immobilisation of Ions by Naturally Occurring Materials as Alternatives to Ion Exchange Resins*, Solvent Extraction and Ion Exchange Group, Society of Chemical Industry, April 1986, University of Lancaster. In *Immobilisation of Ions by Biosorption* (H. Eccles and S. Hunt, Eds.), Ellis Horwood, Chichester
5. Charley, R. C. and Bull, A. T. (1979). Bioaccumulation of silver by a multispecies community of bacteria. *Arch. Microbiol.*, **123**, 239–244
6. Darnall, D. W., Greene, B., Henzl, M. T., Hosea, J. M., McPherson, R. A., Sneddon, J. and Alexander, M. D. (1986). Selective recovery of gold and other

metal ions from an algal biomass. *Environ. Sci. Technol.,* **20**, 206–208
7. Drobot, W. (1981). Microbiological recovery of metals. US Patent 4,293,333
8. Drobot, W. and Lechavalier, H. A. (1981). Recovery of metals. US Patent 4,293,334
9. Dunn, G. M. and Bull, A. T. (1983). Bioaccumulation of copper by a defined community of activated sludge bacteria. *Eur. J. Appl. Microbiol. Biotechnol.,* **17**, 30–34
10. Galun, M., Keller, P., Feldstein, H., Galun, E., Siegel, S. and Siegel, B. (1983). Recovery of uranium (VI) from solution using fungi. II. Release from uranium-loaded *Penicillium* biomass. *Water, Air and Soil Pollut.,* **20**, 227–285
11. Heide, E. A., Paschke, M., Wagner, K. and Wald, M. (1975). Aus kultivierierbaren Mutanten von einzelligen Grunalgen bestehende Matrix sowie Verfahren zur Urangewinnung mittels dieser Matrix. German Federal Republic Patent 2,345,430
12. Horikoshi, T., Nakajima, A. and Sakaguchi, T. (1981). Studies on the accumulation of heavy metal elements in biological systems. XIX. Accumulation of uranium by microorganisms. *Eur. J. Appl. Microbiol. Biotechnol.,* **12**, 90–96
13. Kiff, R. J. and Little, D. R. (1986). Biosorption of heavy metals by immobilised fungal biomass. Presented at a symposium *Immobilisation of Ions by Naturally Occurring Materials as Alternatives to Ion Exchange Resins*, Solvent Extraction and Ion Exchange Group, Society of Chemical Industry, April 1986, University of Lancaster. In *Immobilisation of Ions by Biosorption* (H. Eccles and S. Hunt, Eds.), Ellis Horwood, Chichester
14. Macaskie, L. E. and Dean, A. C. R. (1984a). Heavy metal accumulation by immobilised cells of a *Citrobacter* sp. *Biotechnol. Lett.,* **6**, 71–76
15. Macaskie, L. E. and Dean, A. C. R. (1984b). Cadmium accumulation by immobilised cells of a *Citrobacter* sp. *Environ. Technol. Lett.,* **5**, 177–186
16. Macaskie, L. E. and Dean, A. C. R. (1985). Uranium accumulation by immobilised cells of a *Citrobacter* sp. *Biotechnol. Lett.*, **7**, 463–466
17. Nakajima, A., Horikoshi, T. and Sakaguchi, T. (1982). Recovery of uranium by immobilised microorganisms. *Eur. J. Appl. Microbiol. Biotechnol.,* **16**, 88–91
18. Ross, I. S. (1986). Uptake of heavy metals by microorganisms. *International Industrial Biotechnology,* **6** (6), 184–188
19. Ross, I. S. and Townsley, C. C. (1986). The uptake of heavy metals by filamentous fungi. Presented at a symposium *Immobilisation of Ions by Naturally Occurring Materials as Alternatives to Ion Exchange Resins*, Solvent Extraction and Ion Exchange Group, Society of Chemical Industry, April 1986, University of Lancaster. In *Immobilisation of Ions by Biosorption* (H. Eccles and S. Hunt, Eds.), Ellis Horwood, Chichester
20. Siegel, S., Keller, P., Galun, M., Lehr, H., Siegel, B. and Galun, E. (1986). Biosorption of lead and chromium by *Penicillium* preparations. *Water, Air and Soil Pollut.,* **27**, 69–75
21. Tengerdy, R. P., Johnson, J. E., Hollo, J. and Toth, J. (1981). Dentrification and removal of heavy metals from waste-water by immobilised microorganisms. *Appl. Biochem. Biotechnol.,* **6**, 3–13

22. Townsley, C. C. (1985). Heavy metal accumulation in filamentous fungi. PhD Thesis, University of Keele
23. Townsley, C. C., Ross, I. S. and Atkins, A. S. (1986). Copper removal from a simulated leach effluent using the filamentous fungus *Trichoderma viride*. Presented at a symposium *Immobilisation of Ions by Naturally Occurring Materials as Alternatives to Ion Exchange Resins*, Solvent Extraction and Ion Exchange Group, Society of Chemical Industry, April 1986, University of Lancaster. In *Immobilisation of Ions by Biosorption* (H. Eccles and S. Hunt, Eds.), Ellis Horwood, Chichester
24. Tsezos, M. (1984). Recovery of uranium from biological adsorbents — desorption equilibrium. *Biotechnol. Bioeng.*, **26**, 973–981
25. Tsezos, M. (1985). Biomass immobilised adsorptive applications in the mineral industry. Presented at *2nd Annual BIOMINET Meeting*, October 1985, Mississanga, Canada (cited in Ref. 4)
26. Tsezos, M. and Keller, P. (1983). Adsorption of radium-226 by biological absorbents. *Biotechnol. Bioeng.*, **25**, 201–215
27. Tsezos, M. and Volesky, B. (1981). Biosorption of uranium and thorium. *Biotechnol. Bioeng.*, **23**, 538–604

11
CELL ADHESION — AN INTRODUCTION

Stuart W. Shales

INTRODUCTION

The manner in which microbial cells can adhere to solid surfaces is of importance in many branches of biotechnology and there is much research in progress investigating the factors that may contribute to such adhesion. Some of the areas necessitating this work include the design of novel bioreactors, the development of microbial systems for enhanced oil recovery and ore leaching, the study of metal corrosion processes and the penetration of concrete structures by bacteria. Consideration of cell adhesion is also relevant to both health care and agriculture, in that the early stages of infection in human and plant diseases involve the interaction of microbes with a solid surface — for example, the lining of the gut or the surface of a leaf. There is much activity in the investigation of the processes that lead to tooth decay (dental caries is the most common human disease in the UK); this is yet another instance where the early stages may involve the adhesion of bacteria to a solid surface, namely the tooth. Researchers have accumulated considerable data on the physicochemical properties of microbial cell surfaces and are now in the position of relating this extensive information to adhesion processes. This chapter discusses some of the advances that have been made in the analysis of cell–surface features of micro-organisms and the possible role played in adhesion. The discussion will be extended to non-microbial systems where this is relevant to biotechnology.

BACTERIAL ADHESION TO SOLID SURFACES

By far the greatest amount of research into microbial adhesion to solid surfaces has involved bacterial systems[9]. The work has centred around many factors that have been thought to contribute to this, including the composition and nature of the bacterial cell surface, the physical forces involved in retention, the properties of the solid surfaces and the biosynthesis of certain polymeric substances (particularly polysaccharides) by bacteria which may aid cell retention. The results of this extensive research have led to a rather complicated picture which is still incomplete. It is apparent that many factors are involved in the adhesive processes, and these will be discussed in greater detail later.

The work of Zobell and colleagues in the late 1930s[10] may be considered to be a landmark in this research. Their research involved a study of the effect of storage in different vessels on micro-organisms in sea-water. They discovered that the growth and metabolic activity of these microbes was related to the surface area-to-volume ratio of the vessel; this indicated that there may be an interaction between the micro-organisms and the vessel walls. Their work has now been extended, and it is now appreciated that cell retention onto solid surfaces is a major factor in the growth and survival of micro-organisms in a diverse range of environments. The physicochemical properties of these surfaces vary considerably, even within the same ecosystem; this is demonstrated quite clearly in soil, which comprises such diverse components as clay and sand.

Not only do these particulates have different chemical features, but also they occur in widely differing size ranges (Table 1); hence, their surface area-to-volume ratios are dissimilar, and this may affect surface charge distribution and other properties. All of these factors could well contribute to variations in the retention of microbes at the surface of these soil

Table 1 Surface area of soil components

Particle type	*Diameter* (mm)	*Surface area* (cm^2/g)	*Particle surface area* (vol/cm)
Coarse sand	1.00–0.50	23	60
Medium sand	0.50–0.25	45	120
Fine sand	0.25–0.10	91	240
Silt	0.05–0.002	454	1200
Clay	<0.002	8 000 000	>30 000

Data from Alexander[2] with modification, after Foth and Turk[3].

particles and to differences in growth rate and metabolic activity. The physical and chemical nature of solid surfaces also varies between ecosystems, and this adds to the complexity in the interpretation of the requirements for microbial cell adhesion; the human gut lining, for example, is quite different from the surface of a tooth. Biotechnologists may be interested in immobilising micro-organisms on plastics supports and this presents yet another factor to consider: many plastics are hydrophobic.

The surface characteristics of bacteria also vary. In broad terms there are two classes of bacteria, namely Gram-positive and Gram-negative; these are typically represented by *Bacillus* species and *Escherichia coli*, respectively. This classification of bacteria is based on the composition and properties of the cell envelope; hence, the nature of the cell surface differs between the two groups. Gram-negative bacteria have a hydrophilic cell surface which has little or no charge; Gram-positive bacteria, on the other hand, exhibit hydrophobic surface characteristics with an overall negative charge. Differences in such features could well influence the retention of individual bacteria within a particular environment. It is becoming increasingly apparent that special features have also been evolved to enable certain bacteria to be retained preferentially at specific sites; this is especially true of pathogens, which in some instances have been shown to contain extrachromosomal genetic elements, called plasmids, which carry the genes that encode these features.

To summarise: it can be stated that there are characteristics, both of the bacterial cell surface and of the environment, that will contribute to the retention of the micro-organisms at solid surfaces. Some of the forces that have been considered as being important to cell adhesion include electrostatic, van der Waals and hydrophobic interactions. Bacteria are negatively or weakly negatively charged and would be expected to be repelled by a similarly charged surface such as glass, i.e. electrostatic repulsion. Such repulsive forces tend to be overcome as a result of the presence of ions in the aqueous environment of the cell which alter the charge characteristics of the cell surface; for example, it may be possible that the bivalent cations can form bridges between the cells and the solid surfaces. Similarly, the bacteria themselves may produce macromolecules which extend from the cells and which can interact with the solid surface, in effect acting as an anchor holding the bacteria in place. A full discussion of the physicochemical considerations of cell adhesion is complex and beyond the scope of an introductory chapter such as this.

THE PROCESS OF ADHESION

Generally the adhesion of microbes to solid surfaces occurs in two stages: the first is reversible and includes the initial approach of the cell to the surface; this is followed by a second phase in which the binding of the cell at the surface is irreversible[9].

The approach of the microbial cell to the solid surface has been considered in great depth, especially in relation to the forces involved. It has been elegantly modelled mathematically. There are two aspects to this: first, the chance of a collision between the cell and the surface, and second, the exact nature of the approach itself. There are some interesting conclusions to this work: as the cell approaches the solid surface, it encounters repulsive forces, but then the closer it gets, the situation suddenly reverses, so that the net forces favour attraction (Figure 1); the profile of the net attractive/repulsive forces in relation to spatial separation depends on the electrolytic concentration of the aqueous environment surrounding the cell. As implied in earlier statements, the retention of

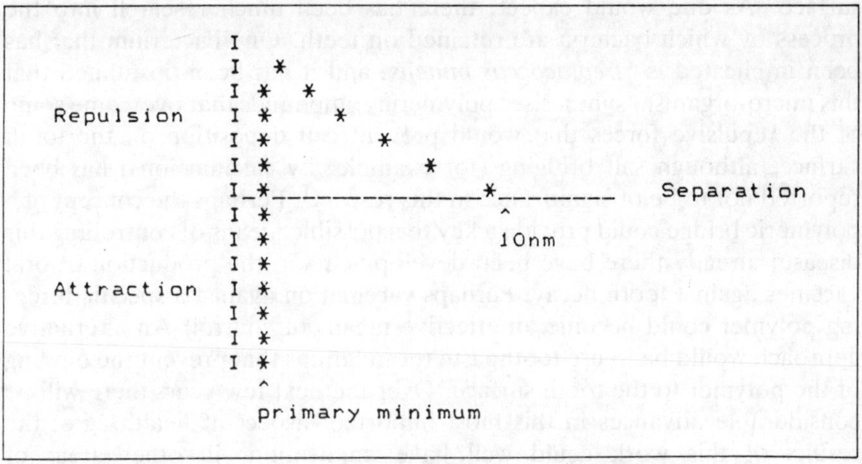

Figure 1 Representation of forces acting between a solid surface and a particle (e.g. a bacterial cell). The absolute values will depend on environmental conditions. The above is a much simplified example; there may be a secondary minimum at a distance of about 10 nm from the surface. Derived from Wardell *et al.*[9], after Marshall[5]

bacteria at solid surfaces is reversible in this early stage. Shortly after the bacteria have been retained at the surface, they start producing polymeric compounds (in particular, polysaccharides), which in effect act as adhesives and bind the cells in an irreversible manner; this represents the second stage of adhesion. Following this, the bacteria divide and form a monolayer film over the surface itself; such growth may continue, so that a thick layer of bacteria forms, from which cells may be released. This very simplistic explanation of the adhesion process does not justify the extensive research that has been undertaken by workers in this field. Some of the areas that have been investigated will be considered in greater detail later, as will the important aspect of the industrial applications of these developments. It is immediately apparent that much of this research has required interaction between scientists from many disciplines, including those of microbiology, biochemistry, chemistry and physics.

CELL ADHESION AND HEALTH CARE

Tooth decay costs the UK a considerable amount in man-days productivity lost through workers requiring treatment; therefore, there is much interest in establishing the exact mechanism of dental caries, so that new means of prevention can be established. Decay is generally attributed to bacteria, and the initial stages involve the adhesion of the microbes to the tooth surface. As one would expect, there has been much research into the process by which bacteria are retained on teeth. One bacterium that has been implicated is *Streptococcus mutans*, and it has been postulated that this micro-organism synthesises polymeric compounds that overcome some of the repulsive forces that would prevent cell deposition on the tooth surface, although salt bridging (for example, by calcium ions) has been reported not to be of significance in this respect[1]. Perhaps the concept of a polymeric bridge could provide a key to a possible means of controlling this disease; already there have been developments in the production of oral vaccines against tooth decay. Perhaps vaccination against a specific bridging polymer could become an effective means of control. An alternative approach would be to use toothpaste formulations that prevent the binding of the polymer to the tooth surface. Over the next few years there will be considerable advances in this most important aspect of health care; the results of this work could well have implications in other areas of biotechnology.

Cell adhesion to tissues is seen as one of the early stages in the process of pathogenicity of many bacterial infections. The mechanisms by which the microbial cells attach themselves vary considerably; in some instances specific appendages are produced on the cell surface. For example, fimbriae have been demonstrated to be required for the invasion of the gut by certain enteric pathogens and indeed the genes encoding fimbrial biosynthesis are often carried on plasmids in the pathogenic strains of bacteria. The use of appendages such as fimbriae represents a highly specialised means to aid cell attachment and has been shown to occur in other non-clinical environments. Since the appendages extend from the cells, they may overcome the effects of electrostatic repulsion between the cell and the solid surface. In Gram-negative clinical isolates the presence of lipopolysaccharide structures on the cell surface may also confer properties that enhance attachment processes.

CELL ADHESION AND AGRICULTURE

The problems of pathogenicity do, of course, extend to agriculture and animal husbandry; there is a considerable amount of information relating to the invasion processes of disease-causing organisms in the veterinary environment. Again many bacteria produce specific appendages that facilitate the attachment processes. As in the example of dental caries, these structures could provide a suitable target for preventive measures such as immunisation or the use of specific inhibitors.

Another area where cell adhesion is relevant to agriculture is the association of bacteria and plants; this may be considered not only in terms of pathogenic effects, but also with respect to beneficial aspects. The adhesion of *Agrobacterium* cells to host plant cell walls has been shown to be enhanced by the presence of lipopolysaccharide structures on the surface of the bacterial cells. *Agrobacterium* causes growth defects (for example, galls and other tumorous conditions) in many dicotyledonous plants. Bacteria from this genus do have a potential beneficial role in agriculture, in that they are being developed as a system for the genetic engineering of crop plants. *Rhizobium* bacteria form a highly specific symbiotic association with legumes; the early stages in the development of this relationship involve the invasion of the host plant roots by the bacterium. There appears to be a complex mechanism that facilitates the attachment of the bacteria to the plant roots; in part this involves the synthesis of polysaccharides by the microbial cells. The adhesive processes associated with plant pathogenicity are receiving much attention at present.

INDUSTRIAL CONSIDERATIONS

From the previous discussion it is apparent that cell adhesion to solid surfaces has implications in many situations where bacteria occur in the natural environment. The mechanisms of attachment are diverse and may in some instances involve the utilisation of specific structures on the bacterial cell surface. Adhesive polymers may also be synthesised by the attached micro-organisms as a second stage in the adhesion process. Biotechnologists are much concerned with the involvement of cell adhesion in industrial processes and activities. Some of these aspects include detrimental effects such as corrosion and fouling; others involve the exploitation of cell adhesion in the design of new processes.

Perhaps the best-known situation where cell adhesion has been exploited by Man is in the treatment of sewage and water purification. It is of interest to add that in terms of volumetric turnover these two activities still represent the greatest usage of biotechnological processes in the service of the human population. Early designs for trickle filters in sewage treatment and similar protocols for water purification involve the use of micro-organisms immobilised onto solid supports by natural adhesive processes. Coke was the most commonly used support phase in trickle filters in sewage plants, although different media have been tried, including polystyrene beads. A natural film of microbes builds up on the support and serves to reduce the organic matter content of the waste. Cell immobilisation is now being extensively applied in many biotechnological processes, and there is considerable interest in the manner in which bacteria and other micro-organisms can be attached to solid supports. Space precludes a full discussion of immobilised cell technology; however, it can be stated that there are several methods for immobilisation, including gel entrapment, covalent linking to solid support phases and adhesion. There has been much research into the adhesion of bacteria and other micro-organisms onto plastics, and this work could well be exploited for the development of novel industrial processes.

There is concern about the long-term behaviour of micro-organisms in continuous culture. One of the problems that may arise is genetic drift in the microbial population; this could cause changes in the ability of the microbes to adhere to solid surfaces within the reaction vessel and this may or may not be desirable, depending on the configuration of the system. If a process relies on the use of cells immobilised by adhesion onto solid supports, then the evolution of a population of non-adhering microbes would be deleterious to the operation. There is also much interest in the factors that lead to cell clumping, since such cell–cell adhesion can play an important role in industrial processes, in that this may allow of the easy removal of microbes from the fermentation broth. It is therefore apparent that consideration of adhesive processes is becoming increasingly important in the design of new-generation bioreactors.

Certain applications of microbial systems involve the introduction of micro-organisms into porous structures, and in these situations a knowledge of adhesive processes will be of importance. Examples of these applications include the use of bacteria in ore leaching and in the enhanced recovery of oil. In some of the methods proposed for these resource-winning processes, bacteria may physically be introduced into ore and reservoir formations. It is important that the microbes do not immediately adhere onto the solid surfaces within the porous structures; otherwise, undesirable clogging may occur. Oil reservoirs may also present extreme conditions of temperature and pressure as well as relatively high salt concentrations; therefore, any modelling of cell adhesion should take these factors into account.

Cell adhesion to solid surfaces can have important implications in other areas of industrial activity. Microbially mediated corrosion is of major concern in many situations, especially the offshore oil industry. It is possible that the early stages of such corrosion processes involve the adhesion of bacteria to metal structures. Eventually a film of micro-organisms forms and finally the metal pits, as a result of the action of products released from the microbial film. These processes could also be important in the penetration of concrete structures by bacteria. Again the early stages involve the adhesion of bacteria to the structure, which is then followed by film formation and pitting. The permeability of the concrete then increases, leading to further deterioration of the structure. This activity could cause problems in the dumping of toxic wastes at sea; concrete encapsulation is often used as a means of protecting containers of such wastes. Further research is required to investigate the potential hazard, particularly in view of the current desire to dump nuclear waste in the deep oceans.

BIOFOULING

The build-up of a film of bacteria, algae and higher organisms on the hulls of ships has a direct economic impact in that the extra drag imposed by the growth retards the passage of ships and as a result fuel consumption increases. The cost of biofouling is difficult to assess but figures indicating a doubling of fuel consumption have been quoted (Bio/Technology 3:38, 1985). In order to alleviate the biofouling problem, protective paints containing toxic metals (for example, organotin compounds) have been applied to the hulls of ships and other crafts. There has been much concern regarding the environmental impact of such coatings, since the toxic component may leach out. Research into new protective methods is a matter of priority. One possible alternative is to produce low-energy surfaces to which bacteria will not adhere (i.e. non-stick). It is questionable as to whether low-energy coatings will be totally effective against all species of bacteria. Reports of bacterial adhesion to several types of plastics are appearing in the scientific literature. If the increase in fuel consumption as a result of biofouling is as high as has been quoted, then the potential market for a novel anti-fouling agent will be tremendous: certainly there appears to be a financial stimulus for work in this field. An understanding of the physicochemical interactions between bacteria and surfaces, and of the genetics and physiology of bacterial cell envelopes, will be of much importance in developments in this field.

CONCLUSION AND FUTURE ASPECTS

A brief overview of cell adhesion such as this cannot possibly justify all the research that has been undertaken; it merely serves as an introduction to the potential applications of this work to industrial processes. In most instances where micro-organisms are utilised the mechanisms of cell adhesion have to be considered. In some cases such attachment processes are desirable, whereas in others the opposite is true. A complete understanding of the factors involved will be of benefit; however, owing to the diversity of micro-organisms and surfaces involved, no simple model is likely. In some situations specific appendages are produced by the microbes to enhance the attachment processes. Generally adhesion to solid surfaces occurs in two stages: the first involves the approach of the organisms to the surface and the second is often associated with the biosynthesis of adhesive polymers by the attached microbes. The physicochemical factors involved in adhesion are complex and may depend to a large extent on environmental conditions. Some of the areas of biotechnology in which cell adhesion may play an important role include health care, agriculture, resource winning, bioreactor design and corrosion prevention.

Owing to the importance of adhesive processes in many biotechnological activities, there is a considerable demand for continuing research into this field. This chapter has concentrated on the adhesion of bacteria to solid surfaces, primarily because most research has been undertaken on this subject. In industrial applications, however, higher organisms may be required, and at present there is much interest in the adhesive processes involving these. Fungi, including yeasts[8], are becoming important biocatalysts for many industrial processes and are therefore the subject of current investigations into adhesion, particularly with respect to immobilised cell technology. Similarly, plant and algal cells may be incorporated into new bioreactors, and information on their surface characteristics and properties will be of importance.

In relation to health care information, the attachment processes of pathogenic micro-organisms may lead to advances in preventive medicine — for example, through the development of novel vaccines. Oral vaccination against dental caries could save UK institutions a considerable amount of money in lost productivity due to workers requiring attention to dental problems.

POSTSCRIPT: ECONOMIC CONSIDERATIONS

The central theme of this chapter has been an introduction to the many areas in which cell adhesion plays an important role. Owing to the diverse nature of these, it is difficult, if not impossible, to attach hard financial facts and figures. A consideration of the 'pros and cons' will at least give an indication of the potential scope. There are many instances where the attachment of cells to surfaces leads to problems; these include the biofouling of ships' hulls, the corrosion of metal structures and, as just mentioned, dental caries. All of these aspects involve the expenditure of considerable amounts of money — for example, in the increased fuel consumption of ships, the replacement of corroded metal, and so on. Even open heart surgery suffers from problems associated with bacterial adhesion, in that in some cases artificial heart valves eventually become covered by a bacterial film and may have to be replaced. Such repeat operations not only are expensive, but also entail a considerable degree of personal suffering. In these instances the desired objective of the biotechnologist together with scientists from other disciplines is to find out how the microbes become stuck to surfaces and hopefully develop methods to prevent the early stages of adhesion. Other groups are expending their effort in ensuring that microbes *do* adhere to surfaces; immobilised cell bioreactors are an example of this approach. The potential financial reward for developments in this area may well be considerable, although there are other methods of immobilising cells than adhesion. To assess the overall value of research in cell adhesion, it is necessary to take into consideration all these individual facets. Certainly there will be much benefit to be gained from the activity that research groups world-wide are currently involved in.

REFERENCES

The literature relating to cell adhesion is extensive; the references listed below represent just a small proportion of such publications.

1. Abbott, A., Rutter, P. R. and Berkeley, R. C. W. (1983). The influence of ionic strength, pH and a protein layer on the interaction between *Streptococcus mutans* and glass surfaces. *Journal of General Microbiology*, **129**, 439–445
2. Alexander, M. (1977). *Introduction to Soil Microbiology*, 2nd edn. Wiley, New York. p.4
3. Foth, H. D. and Turk, L. M. (1972). *Fundamentals of Soil Science*. Wiley, New York
4. Jang, L. K., Chang, P. W., Findley, J. E. and Yen, T. F. (1983). Selection of bacteria with favourable transport properties through porous rock for the application of microbial enhanced oil recovery. *Applied and Environmental Microbiology*, **46**, 1066–1072
5. Marshall, K. C. (1976). *Interfaces in Microbial Ecology*. Harvard University Press, Cambridge, Mass.
6. New, P. B., Scott, J. J., Ireland, C. R., Farrand, S. K., Lippincott, B. B. and Lippincott, J. A. (1983). Plasmid pSa causes loss of LPS-mediated adherence in *Agrobacterium*. *Journal of General Microbiology,* **129**, 3657–3660
7. Pringle, J. H., Fletcher, M. and Ellwood, D. C. (1983). Selection of attachment mutants during the continuous culture of *Pseudomonas fluorescens* and the relationship between attachment ability and surface composition. *Journal of General Microbiology,* **129**, 2557–2569
8. Van Haecht, J. L., De Bremaeker, M. and Rouxhet, P. G. (1984). Immobiliza-

tion of yeast to a support without the use of a chemical agent. *Enzyme and Microbial Technology,* **6**, 221–227
9. Wardell, J. H., Brown, C. M. and Flannigan, B. (1983). Microbes and surfaces. In *Microbes in Their Natural Environments* (J. H. Slater, R. Whittenbury and J. W. T. Wimpenny, Eds.). Symposium 34, published for the Society for General Microbiology by Cambridge University Press, pp. 351–378
10. Zobell, C. E. and Anderson, D. Q. (1936). Observations on the multiplication of bacteria in different volumes of stored sea water and the influence of oxygen tension and solid surfaces. *Biological Bulletin,* **71**, 324–342

12
BACTERIAL PROTEIN EXPORT

Stuart W. Shales

INTRODUCTION

Consideration of protein export by bacteria is of major importance to many research institutions and industrial organisations currently involved in biotechnology. Bacteria (in particular, *Escherichia coli*), are used as hosts for the biosynthesis of proteins. Genes encoding these products may have been introduced into the cells by the elegant and powerful techniques of genetic engineering. It would be highly desirable if the bacterial host could export such proteins into the extracellular medium, namely the fermentation broth. There are two principal reasons for this: first, downstream processing (i.e. product separation, concentration and purification) would be simplified and, consequently, less expensive, and second, export would overcome the problem of product degradation by intracellular protease enzymes. Both of these factors could well influence the financial viability of a process and are, therefore, of significant concern to the biotechnologist. What is desired and what can be achieved are, however, entirely different and this is, in part, questioning the continuing use of existing bacterial strains in recombinant DNA technology. There are other factors relating to host suitability which need to be taken into account — for example, genetic instability, glycosylation, pyrogen and toxin production; these will be discussed later.

The bacterium *E. coli* continues to be the most desired host, primarily because our knowledge of its genetics is so far advanced compared with any other microbe. A wide range of plasmids, extrachromosomal genetic elements, are available to be used as vectors in recombinant DNA technology, and there is considerable experience in the cloning and expression of foreign genes in many strains of *E. coli*. One of the problems that limit the suitability of *E. coli* in commercial applications is that generally the bacterial cells do not export proteins into the extracellular growth medium. *E. coli* is a typical example of a Gram-negative bacterium and has a rather complex cell envelope structure. The cell contents are contained within a cytoplasmic membrane, a conventional phospholipid bilayer, which is surrounded by a thin layer of peptidoglycan and finally enclosed by a second membrane. This outer membrane contains many proteins, some of which are required for nutrient uptake; also the outer surface of this membrane, which is exposed to the external environment, includes lipopolysaccharide, the sugar chains of which extend from the cell and confer hydrophilic properties at the cell surface. There is a physical space between the cytoplasmic and outer membranes of Gram-negative bacteria and this is filled by an aqueous medium called the periplasm. Gram-positive bacteria (e.g. *Bacillus* spp.), in contrast, have a simpler cell envelope structure, consisting of a cytoplasmic membrane surrounded by a thick peptidoglycan layer; they have no outer membrane or periplasmic space. Generally Gram-negative bacteria export very few proteins extracellularly; instead they tend to accumulate in the periplasmic space (Figure 1). On the other hand, Gram-positive bacteria have been demonstrated to be capable of secreting many proteins (for example, proteases) into the growth medium. It would appear that the outer membrane of bacteria such as *E. coli* acts as a barrier to the total export of proteins from cells.

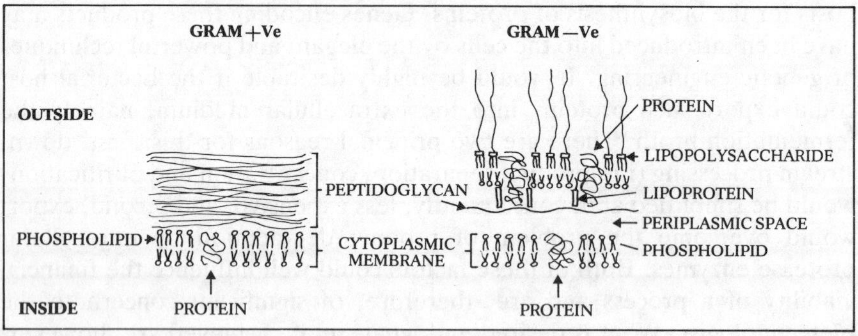

Figure 1 Cell envelope structure of Gram-positive and Gram-negative bacteria

MECHANISM OF PROTEIN EXPORT

The mechanism by which proteins are exported from cells or incorporated into envelope fractions has received much attention; of particular interest are the structural aspects of proteins which determine their ultimate location in or out of the cell. Early work by Palade[13] on protein secretion in eukaryotes showed that ribosomes occur in two states: either free in the cytoplasm or bound to endoplasmic reticulum. It was soon demonstrated that cytoplasmic proteins were synthesised by free ribosomes and that secreted proteins were produced by the bound ribosomes. This work suggested that proteins are exported as the peptide chains elongate (i.e. cotranslationally) rather than after protein synthesis and folding (i.e. post-translationally). No structural differences were found between bound and free ribosomes; thus, it was proposed that either the messenger RNA or the protein structure itself determined export. Further work suggested that special amino acid sequences at the start of secreted proteins (i.e. the N-terminus) have an affinity for membranes and initiate export. Indeed a large number of secreted and membrane proteins in eukaryotes and prokaryotes have been demonstrated to be synthesised as precursors with such leader (signal) sequences attached and are cleaved at some stage during the course of secretion. The absolute structure of signal sequences varies from protein to protein; however, in general terms they are about 20–25 amino acids long and add about 3000 to the molecular weight of the protein. The start of the signal sequence tends to be hydrophilic, whereas the bulk is hydrophobic (Figure 2). The mechanism of secretion appears to be similar for exported and membrane proteins; as the precursor protein is synthesised, the hydrophilic end of the signal peptide binds to the inner surface of the cytoplasmic membrane. Thus, the ribosomes will be attached to the membrane via the growing peptide chain. The hydrophobic part of the leader sequence then intercalates into the lipid domain of the membrane and in so doing initiates the translocation of the protein itself,

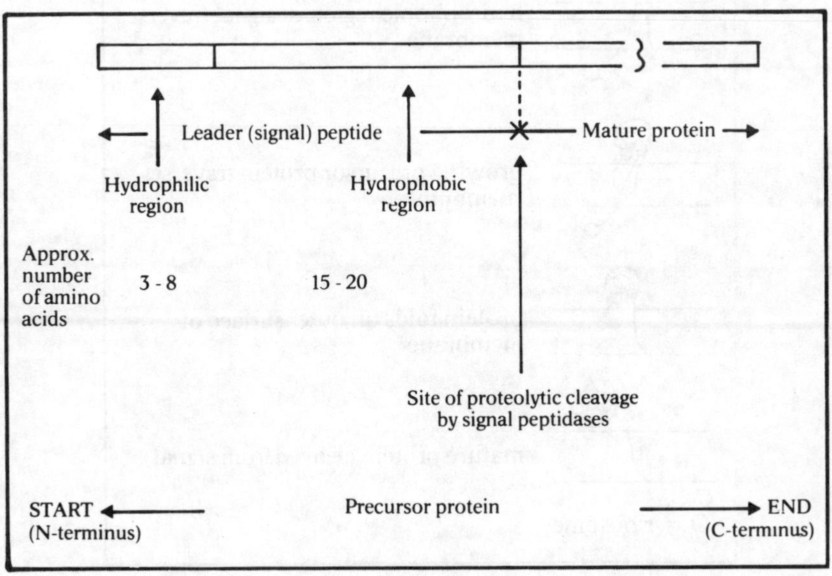

Figure 2 Structure of leader (signal) sequence

which then folds on the outer surface of the membrane (Figure 3). The mature protein is cleaved from the leader sequence by an endopeptidase (signal peptidase). Full details of the mechanism of secretion remain to be elucidated; of particular interest is the energetics of such a process. The secretion mechanism so described will result in protein export into the growth medium in the case of Gram-positive bacteria or into the periplasmic space of Gram-negative bacteria. There is much speculation as to how proteins become incorporated into the outer membrane of Gram-negative bacteria; it has been proposed that there are specific attachment sites between the cytoplasmic and outer membranes and that it is through these that such proteins translocate.

An alternative mechanism for protein secretion has been proposed, which involves post-translation export, in contrast to the cotranslational nature of the above system. The membrane trigger hypothesis proposes that the signal sequence allows the folding of the precursor protein into a form that is competent for translocation. When introduced into a hydrophobic environment, such as the cytoplasmic membrane, the folded precursor protein undergoes a conformational change that permits spontan-

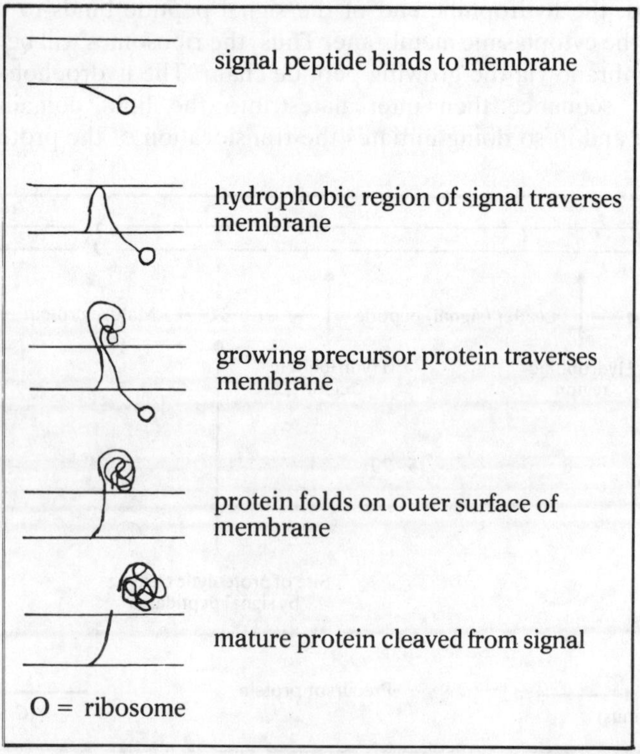

Figure 3 Possible mechanism for cotranslation protein secretion

eous incorporation. The process is driven by the final signal-processing stage. There is much evidence to support this model in both eukaryotes and prokaryotes.

Eventually, as our knowledge of the biophysical nature of membrane structure improves, a 'unified' theory for protein secretion will evolve which will incorporate aspects of both of the above models. It could be said that the exact mechanism of export is not of concern to the biotechnologist, but such an attitude is rather misguided, especially as there is considerable interest in the production of foreign proteins by bacteria. It will be essential to be aware of all the factors that contribute to protein secretion so that systems can be optimised for commercial operation.

GENETIC ANALYSIS OF PROTEIN TRANSLOCATION

So far in this chapter it has been stressed that a signal sequence is required for protein export; this forms the basis of both models discussed above. It is of interest to consider some of the genetic analysis that has been undertaken and how this work has been extended in the development of bacterial strains for the synthesis of foreign (in particular, eukaryotic) proteins. Much of this work has been undertaken using *E. coli* because its genetics are well defined and a considerable variety of mutants are available. Gene fusion is one of the powerful techniques that has been exploited; recombinant DNA methods are used to splice a gene that encodes a cytoplasmic protein into a gene for an exported protein. By this method it is possible to synthesise a cytoplasmic protein with a leader sequence; it is then of interest to observe the cellular location of the modified product. The β-galactosidase of *E. coli* is a suitable cytoplasmic protein for this work; it is encoded by the *lacZ* gene and can readily be assayed. Many *lacZ* gene fusions have been performed (for example, by Beckwith[10]) and these have demonstrated that with suitable leader sequences the protein can be exported into the cell envelope of *E. coli*, where it is not normally found. Interestingly, modified galactosidase does not generally accumulate in the periplasm, even if the leader sequence of known periplasmic protein is attached. Instead the protein remains in the inner (cytoplasmic) membrane fraction. This and other evidence suggests that specific sequences within the mature protein may be required for correct signal processing and protein localisation; this is in addition to the need for a leader sequence. In *E. coli* and other Gram-negative bacteria there are four possible locations for exported proteins: the cytoplasmic membrane, the periplasmic space, the outer membrane and, finally, the extracellular medium. The latter is rare. There has to be a mechanism to 'guide' particular proteins to specific locations; it may be that secondary structure and different signal peptidases (with affinity for a certain structure) may play an important role in such localisation.

An alternative approach to the genetic analysis of protein secretion is to isolate mutants[11] in which the leader sequence of a protein that is normally exported (for example, a membrane protein) is modified and to observe the fate of that protein during cell growth. This work has clearly shown that the central hydrophobic region of signal peptides is of major importance in export and that in many cases mutation in this area leads to accumulation of precursor protein in the cytoplasm. The role of signal peptidases in leader sequence cleavage has been investigated; the results are complicated and further investigations are required to elucidate the involvement of these enzymes and the secondary structure of the exported protein in the processes of secretion and localisation. It has been demonstrated that for certain periplasmic proteins in *E. coli* signal processing is necessary for release of the mature product into the periplasm, but it is not required for export of the protein from the cytoplasm; the non-processed protein accumulates in the cytoplasmic membrane as the precursor form.

In summary, genetic analysis has demonstrated that protein export occurs in various stages. First, a precursor form of the protein is translocated across the cytoplasmic membrane; the leader sequence plays an essential role in this process. The mature protein is then cleaved from the leader peptide by a signal processing protease[2,3]; the specificity of such peptidases for certain secondary structures may determine the final location of the mature protein within the cell envelope or in the extracellular medium. Full details of such genetic analysis are beyond the scope of this chapter.

BIOTECHNOLOGY AND EXPORT

Earlier sections have served to introduce the industrialist and other readers to some of the intricacies and complexities of bacterial protein export and to demonstrate some of the hurdles that have to be overcome when using bacteria as hosts for the production of foreign proteins. *E. coli* is still the most commonly used bacterial host in recombinant DNA technology; however, it suffers from the disadvantage that it exports few proteins into the extracellular environment. Many eukaryotic genes have been cloned into bacteria and it is of interest to survey some of these (Table 1) to assess the success of geneticists in overcoming barriers to protein export.

It is evident from Table 1 that when a gene encoding a foreign protein is introduced into *E. coli* in such a manner that a leader sequence is included, then that gene is transcribed and translated, and a precursor protein traverses the cytoplasmic membrane and may be successfully processed by the *E. coli* signal peptidases to release the mature product into the periplasmic space. Without a leader peptide, cloned gene products tend to

Table 1 The results of cloning of eukaryotic genes in bacteria

Gene system cloned	*Host bacterium*	*Host cell Gram reaction*	*Fate of cloned gene product*	*Ref.*
Preproinsulin	*Escherichia coli*	−ve	Some proinsulin exported into periplasm	5
Preproinsulin + β-lactamase signal sequence	*E. coli*	−ve	90% proinsulin exported into periplasm	5
Prepoinsulin	*Bacillus subtilis* (immobilised cells)	+ve	Proinsulin secreted into extracellular medium	12
Human growth factor (HGF)	*Pseudomonas aeruginosa*	−ve	HGF accumulated in cytoplasm	7
Precursor HGF	*P. aeruginosa*	−ve	Some HGF secreted into extracellular growth medium, but most HGF accumulated in periplasmic space	7

accumulate in the cytoplasm. To obtain the desired protein from the cytoplasm or periplasm, it is necessary to harvest the bacterial cells and to treat them with lytic agents or to mechanically disintegrate them. Such operations are very easy on a laboratory or pilot scale; however, when using large commercial fermenters (< 1000 litres), they could present logistic and expensive problems. It would be favourable if a host system were used that exported the cloned gene product directly into the extracellular growth medium. Gram-negative bacteria are rather limited in their ability to perform this function; there are, however, exceptions to this rule, including species of *Pseudomonas* and *Myxococcus*[1], which may be commercially exploitable in the future. Gram-positive bacteria are more ideally suited to export proteins extracellularly; *Bacillus subtilis* secretes several proteins and has been used in an immobilised form to synthesise proinsulin (Table 1).

FUTURE MICROBIAL HOSTS FOR BIOTECHNOLOGY

The biotechnology industry has to look forward to adopting new microbial hosts for the synthesis of cloned gene products. The question of protein export is just one consideration that is necessary. Gram-positive bacteria — for example, '*Bacillus species* — would appear to be ideal candidates. However, they do suffer from certain limitations: typically they produce extracellular proteases which would destroy the desired protein, so negating the benefit of their efficient export systems. It is feasible that non-protease mutants of *Bacillus* may be isolated, so overcoming this problem. Other limitations in exploiting new bacterial hosts include lack of knowledge relating to their genetics and problems of gene reorganisation and instability.

An alternative approach to this problem is to develop strains of *E. coli* which secrete proteins into the extracellular growth medium. Such 'leaky' mutants have been produced in the laboratory — for example, a strain of *E. coli* into which the genes for cellulose degradation were cloned produced the cellulase enzyme externally to the cell upon mutagenesis. The mutant also leaked several periplasmic proteins[6]. Leaky mutants of *E. coli* often result from gross alterations in the structure of the outer membrane and could well suffer from problems of cell fragility when used in fermenters.

In this chapter discussion has been limited to the question of bacterial export; there is in addition considerable interest in the use of yeast cells as hosts for gene cloning. Much research is in progress into the nature of protein secretion processes by yeasts — in particular, *Saccharomyces cerevisiae* (brewers' yeast). Yeasts offer several advantages: they do not produce toxins, they secrete proteins extracellularly, there is considerable experience in their use in fermentation processes, our knowledge of their genetics is quite advanced and, finally, they can glycosylate proteins. Glycosylation is a process by which sugar chains are added to proteins and may be essential for optimal activity; bacteria are unable to perform this function.

CONCLUSION

It is desirable in commercial processes to use host organisms that export cloned gene products into the extracellular growth medium, so reducing the complexity and cost of downstream processing. The mechanisms of protein secretion are not fully understood and require further elucidation. The bacterium *E. coli* is the host most commonly used in recombinant DNA technology; however, it is not efficient at exporting proteins extracellularly. New hosts are required but their utilisation may be restricted by other factors such as genetic instability, detrimental protease production and toxin biosynthesis.

Clearly there is much fundamental research to be undertaken and this is a typical instance where close liaison and collaboration between university departments and industry is of major importance.

REFERENCES AND BIBLIOGRAPHY

The literature relating to protein secretion is extensive; in order to conserve space, this list will be limited to reviews and a few representative supporting papers.

1. Breton, A. M., Micaud, J. M., Younes, G. and Guespin-Michel, J. F. (1984). *Myxococcus xantus*, a Gram-negative non-pathogenic bacterium, that secretes proteins into the extracellular growth medium, is a potential cloning host for excreted protein production. *Proceedings of the Third European Congress on Biotechnology, Munich,* pp.441–446
2. Date, T. (1983). Demonstration by a novel genetic technique that leader peptidase is an essential enzyme of *Escherichia coli. Journal of Bacteriology,* **154**, 76–83
3. Date, T. and Wickner, W. (1981). Isolation of the *Escherichia coli* leader peptidase gene and effects of leader peptidase overproduction *in vivo. Proceedings of the National Academy of Sciences, USA,* **78**, 6106–6110
4. Davies, B. D. and Tai, P. C. (1980). The mechanism of protein secretion across membranes. *Nature, Lond.,* **283**, 433–438
5. Emerick, A. W., Bertolani, B. L., Ben-Bassat, A., White, T. J. and Konrad, M. W. (1984). Expression of a β-lactamase preproinsulin fusion protein in *Escherichia coli. Biotechnology,* **2**, 165–168
6. Gilkes, N. R., Kilburn, D. G., Miller, R. C. Jr. and Warren, R. A. J. (1984). A mutant of *Escherichia coli* that leaks cellulase activity encoded by cloned

cellulase genes from *Cellulomonas fimi. Biotechnology,* **2**, 259–263
7. Gray, G. L., McKeown, J. A., Jones, A. J. S., Seeburg, P. H. and Heyneker, H. L. (1984). *Pseudomonas aeruginosa* secretes and correctly processes human growth hormone. *Biotechnology,* **2**, 161–165
8. Kingsman, S. M. and Kingsman, A. J. (1983). The production of interferon in bacteria and yeast. In *Interferons: From Molecular Biology to Clinical Applications* (D. C. Burke and A. G. Morris, Eds.). Symposium 35, published for the Society for General Microbiology by Cambridge University Press. pp.212–254
9. Michaelis, S. and Beckwith, J. (1982). Mechanisms of incorporation of cell envelope proteins in *Escherichia coli. Annual Reviews of Microbiology,* **36**, 435–465
10. Michaelis, S., Guarente, L. and Beckwith, J. (1983). *In vitro* construction and characterization of *phoA-lacZ* gene fusions in *Escherichia coli. Journal of Bacteriology,* **154**, 356–365
11. Michaelis, S., Inouye, H., Oliver, D. and Beckwith, J. (1983). Mutations that alter the signal sequence of alkaline phosphatase in *Escherichia coli. Journal of Bacteriology,* **154**, 366–374
12. Mosback, K., Birnbaum, S., Hardy, K., Davies, J. and Bulow, L. (1983). Formation of proinsulin by immobilized *Bacillus subtilis. Nature, Lond.,* **302**, 543–545
13. Palade, G. E. (1975). Intracellular aspects of the process of protein synthesis. *Science,* **189**, 347–358
14. Silhavy, T. J., Benson, S. A. and Emr, S. D. (1983). Mechanisms of protein localization. *Microbiological Reviews,* **47**, 313–344
15. Wickner, W. (1980). Assembly of proteins into membranes. *Science,* **210**, 861–868

13 THE BIOTECHNOLOGY OF MICROBIAL OILS AND FATS

Christopher A. Boulton

INTRODUCTION

The total world production of oils and fats currently stands at some 50 million tonnes per annum if dairy and related products are excluded. Of this output, some 40 million tonnes are annually consumed by humans in various forms of edible oils and fats. The remainder is used in a number of technical applications during the manufacture of products such as polishes, paints, detergents and soaps. Traditional sources of fats and oils are primarily plant seed-oils, which account for over 80 per cent of the total world market. The remaining 20 per cent is of animal and marine origin.

The search for alternative sources of oils and fats has been given impetus by two factors.

The first of these relates to present world trends in the production and consumption of these commodities. Currently the supply of oils and fats is more or less sufficient to meet the demand. However, it is not anticipated that this state of equilibrium will prevail indefinitely. Thus, it has been predicted that by the early 1990s, the demand for oils and fats will increase by some 20 million tonnes per annum. It is anticipated that only approximately half of this additional demand will be satisfied by an increase in the world production of oils and fats from traditional sources.

The second factor, which may provide a niche for a microbial product, relates to the increasing volatility of the oil and fat market. This is due in part to the vagaries of climate, which may at any time precipitate a crop failure and thereby cause short-term shortages with resultant price increases. Furthermore, political considerations are now playing an increasingly important role in determining commodity prices. Thus, those countries which are obliged to import oils and fats are of necessity at the mercy of the oilseed-producing nations. This is of particular relevance to Europe, which has no prospect of being self-sufficient in traditional sources of oils.

It would appear, therefore, that a strong case exists for research into the production of alternative sources of oils and fats. The following discussion will detail the potential of micro-organisms in these developments.

OLEAGINOUS MICRO-ORGANISMS

The particular group of micro-organisms which may provide commercial quantities of lipids have been termed 'oleaginous'. They may be defined as those which exhibit a propensity to accumulate greater than normal amounts of lipid when cultivated under the appropriate conditions. For purely practical reasons, micro-organisms are only classified in the group that may accumulate at least 25 per cent w/w lipid, although in some the value may be as high as 80 per cent w/w[1]. In accordance with the current terminology applied to protein derived from micro-organisms, the lipid extracted from oleaginous types is referred to as 'single-cell oil' (SCO).

The majority of oleaginous micro-organisms are eukaryotes, representative types being found among the yeasts, filamentous fungi and algae. Examples may also be found in a new prokaryotic algae species and there are a small number of oleaginous bacteria.

In common with plant and animal oils, the principal component of the microbial lipid is triacylglycerol (triglyceride), which may typically account for up to 95 per cent of the total. The remainder is usually phospholipid, with trace amounts of sterols, both free and esterified to fatty acids. The fatty acid composition of the triacylglycerol may be manipulated to some extent by varying the growth conditions, as will be described later. However, oleic acid (18:1) is usually predominant, with palmitic (16:0) and linoleic (18:2) acids following in order of abundance.

Some typical examples of oleaginous micro-organisms together with the fatty acid analysis of the extracted lipid are shown in Table 1. The data presented indicate that the fatty acid profiles, in many instances, are broadly similar to those of the commonly used seed-oils.

PATTERNS OF LIPID ACCUMULATION

In the majority of oleaginous micro-organisms so far examined, the prerequisite for lipid accumulation is to use a medium containing an excess of carbon over some other limiting nutrient. Media are usually constructed so that nitrogen is the limiting nutrient, but other nutrients, such as various metals, certain amino acids or phosphate, will also serve to facilitate lipid accumulation if they are present in growth-limiting amounts.

Batch cultivations of oleaginous micro-organisms usually exhibit the pattern of growth illustrated in Figure 1. Thus, there is an initial period of cell proliferation, where lipid levels may be modest, followed by a period of 'fattening', initiated by the exhaustion of the limiting nutrient. During this second phase, excess carbon continues to be transported into the cell, where it is utilised for lipid biosynthesis.

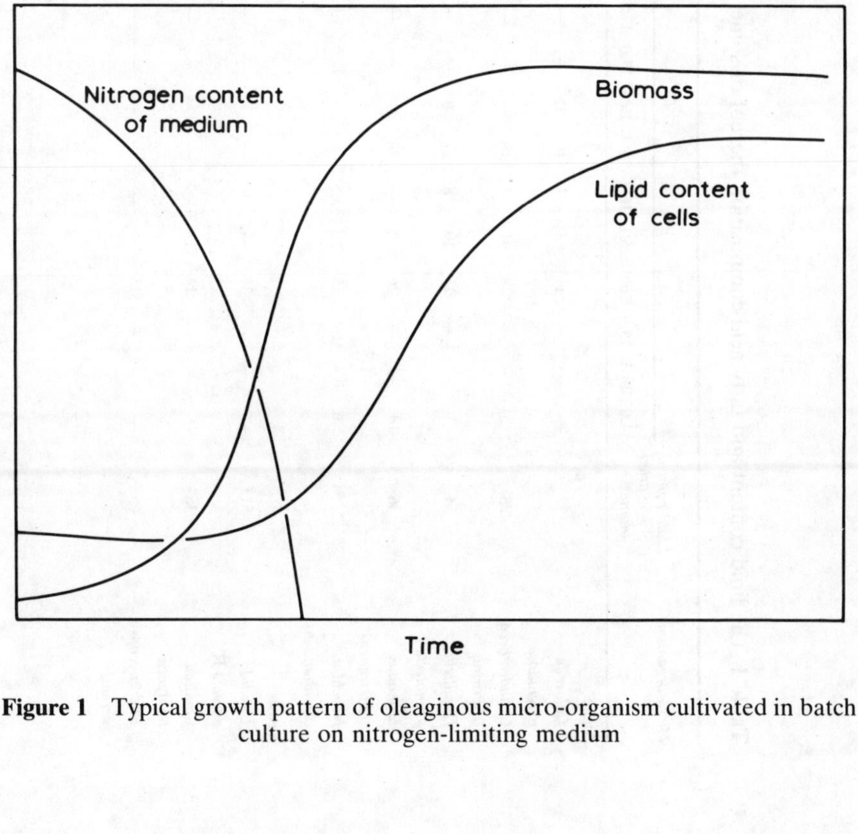

Figure 1 Typical growth pattern of oleaginous micro-organism cultivated in batch culture on nitrogen-limiting medium

Table 1 The lipid content and fatty acid composition thereof of some oleaginous micro-organisms

Micro-organisms	*Lipid content (% w/w dried biomass)*	*Fatty acid analysis (% total)* 14:0	14:1	15:0	16:0	16:1	16:2	16:3	16:4	17:0	17:1	18:0	18:1	18:2	18:3α	18:3γ	19:0	19:1	20:0	20:1	20:2	20:3	20:4	20:5	*Ref.*
Algae																									
Chlorella pyrenidosa	72	—	—	—	11	8	4	10	15	—	—	3	1	8	38	—	—	—	—	—	—	—	—	—	4
Cylindrotheca fusiformis	28	3	—	—	9	31	7	18	—	—	—	—	6	4	3		—	—	—	—	—	—	4	14	18
Dunaliella primolecta	54	5	—	—	11	10	8	7	6	—	—	—	6	6	10	2	—	—	—	1	7	1	2	10	18
Monodus subterraneus	36	10	—	—	26	2	—	—	—	—	—	1	19	9	—	6	—	—	—	6	1	—	10	8	18
Neochloris oleoabundans	35–54	1.6	0.4	1.4	15	2.5	3.5	—	—	11.7	1.0	11.0	36.0	7.4	—	—	0.8	0.1	2.1	2.5	—	—	—	—	16
Yeasts																									
Candida curvata R	51	—	—	—	31	—	—	—	—	—	—	12	51	6	—	—	—	—	—	—	—	—	—	—	7
Candida lipolytica	36	—	—	—	14	11	—	—	—	—	—	5	60	11	—	—	—	—	—	—	—	—	—	—	19
Endomycopsis vernalis	65	—	—	—	15	—	—	—	—	—	—	2	57	24	—	1	—	—	—	—	—	—	—	—	17

Lipomyces starkeyi	37	1	—	—	63	1	—	—	—	—	—	17	10	2		1		—	—	—	—	—	—	—	13
Rhodosporidium toruloides 1FO 0413	66	1	—	—	25	1	—	—	—	—	—	13	46	12		2		—	—	—	—	—	—	—	3
Rhodotorula gracilis	66	1	—	—	20	2	—	—	—	4	—	1	42	21		8		—	—	—	—	—	—	—	5
Trichosporan cutaneum sp 40	40	—	—	—	30	—	—	—	—	—	—	13	46	11	—	—	—	—	—	—	—	—	—	—	7
Filamentous fungi																									
Entomophthora coronata	43	34	—	—	12	—	—	—	—	—	—	1	30	11	—	2	—	—	—	—	—	—	—	—	8
Cunninghamella echinulata	45	1	—	—	17	1	—	—	—	—	—	7	55	11	6	—	—	—	—	—	—	—	—	—	6
Mucor mucedo	51	1	—	—	17	1	—	—	—	—	—	11	31	33	6	—	—	—	—	—	—	—	—	—	15
Pythium irregulare	30–42	6	—	—	26	16	—	—	—	—	—	5	26	5	6	—	—	—	—	—	—	—	—	—	11
Aspergillus flavus	35	—	—	—	13	—	—	—	—	—	—	18	37	32	—	—	—	—	—	—	—	—	—	—	14
Malbranchea pulchella	27	—	—	—	11	—	—	—	—	—	—	11	27	51	—	—	—	—	—	—	—	—	—	—	9

Some exceptions to the pattern have been reported, notably with the yeast *Cryptococcus terricolus* (now *C. albidus* var. *albidus*). This organism, which was originally described by Pedersen[10] and more recently re-examined in this laboratory, is capable of accumulating substantial amounts of intracellular lipid during the initial growth phase when there is no apparent limitation. The biochemical basis of this departure from the usual pattern is as yet unclear; however, this yeast would seem to afford some advantages in terms of shortening batch cultivation times for the production of SCO.

Oleaginous micro-organisms that may also accumulate lipid in continuous culture, provided that a medium with a high C:N ratio is used and the dilution rate is restricted to the range 0.02–02/h. Under these conditions, the growth rate of the organism is sufficient to exhaust the nitrogen in the growth vessel and the residence time is long enough for the excess carbon to be utilised for lipogenesis.

THE BIOCHEMISTRY OF OLEAGINICITY

Physiological studies with various oleaginous micro-organisms have allowed the development of a unifying hypothesis which appears to explain the mechanism by which lipid accumulation occurs. Briefly, this states that under conditions of nitrogen limitation, the biosynthesis of cell constituents dependent on this nutrient, such as nucleic acids and proteins, is restricted. However, the excess carbon may still be transported into the cell, where it is dissimilated via the usual glycolytic and pentose phosphate pathways. This results in an increase in the intracellular concentration of ATP with a concomitant decrease in AMP levels.

The intramitochondrial, NAD^+-dependent isocitrate dehydrogenases of all oleaginous micro-organisms so far examined have an absolute dependence on AMP for activity. Thus, it is postulated that the decrease in the concentration of this latter metabolite inactivates the NAD^+-dependent isocitrate dehydrogenase, resulting in the accumulation of intramitochondrial citrate via equilibration with aconitase. Citrate may then be transported into the cytosol in an exchange reaction with malate, a process catalysed by a specific translocase. In the cytosol, citrate is cleaved by ATP:citrate lyase to acetyl-CoA and oxaloacetate, thereby providing precursor acetyl units for lipogenesis. This enzyme has been shown to occur exclusively in oleaginous micro-organisms, and its presence may be used as a marker indicating a propensity for lipid accumulation.

The proposed reactions involved in lipid accumulation in oleaginous micro-organisms are summarised in Figure 2.

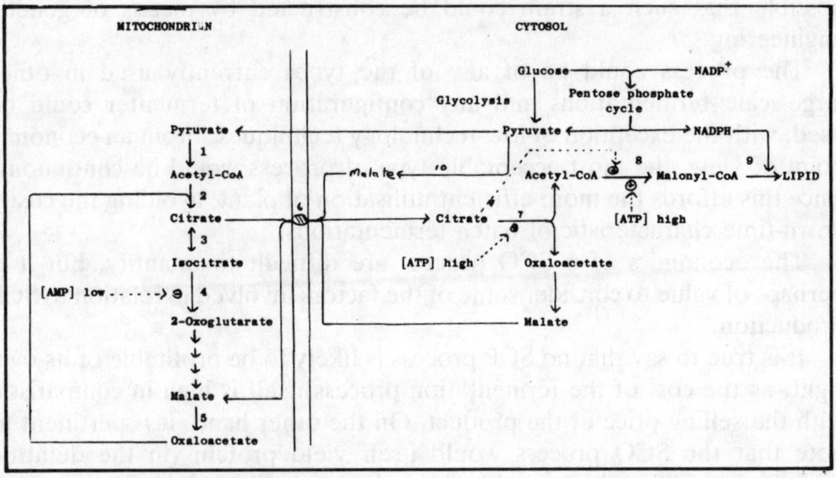

Figure 2 Glucose dissimilatory and lipogenic pathways in oleaginous micro-organisms. The proposed flow of carbon from glucose into lipid during lipogenesis in oleaginous micro-organisms. Enzymes: 1, pyruvate dehydrogenase; 2, citrate synthase; 3, aconitase; 4, isocitrate dehydrogenase; 5, malate dehydrogenase; 6, citrate-malate translocase; 7, ATP: citrate lyase; 8, acetyl-CoA carboxylase; 9, fatty acid synthetase; 10, malic enzyme

PROCESS FOR SCO PRODUCTION

The majority of traditional oils and fats are not particularly expensive. It follows, therefore, that if a microbial product is to be economically competitive, the process must be operated at maximum efficiency, using substrates of the lowest possible cost.

Potential substrates would include all of the commonly available bulk carbohydrates at present used in other large-scale fermentation. Presumably the major consideration would be cost and availability.

Apart from carbohydrates, some oleaginous strains may utilise hydrocarbons, notably n-alkanes, as a source of carbon. A full description of the hydrocarbon utilisation is beyond the scope of this chapter and the interested reader may obtain further information from a recent review[2]. However, it is noteworthy that n-alkanes and related substrates may be incorporated into microbial lipid with little modification. Thus, this offers the possibility of synthesising lipids with desired properties of potential high value.

Another abundant low-cost substrate is methanol, which indeed is currently being used for the production of 'single cell protein' (SCP) by ICI plc. Unfortunately, no oleaginous micro-organisms are known which can utilise this compound and still accumulate lipid. However, it is entirely possible that such a strain could be constructed by means of genetic engineering.

The process could be of any of the types currently used in other large-scale fermentations and any configuration of fermenter could be used, with the exception of low-technology techniques. From an economic point of view, the most acceptable type of process would be continuous, since this affords the more efficient utilisation of plant, avoiding the costly down-time characteristic of batch fermentations.

The economics of a SCO process are difficult to quantify, but it is perhaps of value to consider some of the factors involved in relation to SCP production.

It is true to say that no SCP process is likely to be profitable of its own right, as the cost of the fermentation process itself is high in comparison with the selling price of the product. On the other hand, it is pertinent to note that the SCO process would itself yield protein (in the defatted microbial residue) which would add to the total value of the fermentation product. This, however, must be countered to some extent by the additional processing costs involved in extracting the oil from the microbial cell, but the crux of the comparison resides in the relative values of SCP and SCO. Simple calculations show that 1 tonne of fermentable carbohydrate will yield approximately 0.55 tonne of biomass in both SCP and SCO processes. Of this biomass, the SCP fermentation will produce 0.3 tonne of protein, while the SCO process will produce some 0.22 tonne of oil and 0.13 tonne of protein. It follows, therefore, that the net relative productivity of each process is 0.22 tonne of oil in the case of SCO and 0.3 − 0.13 (i.e. 0.17) tonne of protein in the case of SCP. Since the cost of even the lowest-value conventional plant seed-oil is greater than that of an equivalent amount of a typical animal feed protein, it can be seen that the SCO process will always be more economically viable than the SCP process. This statement, of course, has even greater veracity if the microbial oil cells are in the middle to higher price bracket.

CONCLUSIONS

The foregoing discussion has indicated that the large-scale cultivation of micro-organisms for the production of oils and fats is theoretically and practically feasible. Whether or not SCO can ever capture a significant portion of the oils and fats market, is a moot point. Clearly, the current practice, in Europe, of applying agricultural subsidies for the production of seed-oils leads to artificial prices for these commodities, which would not allow the microbial product to compete fairly. Perhaps, at this time, it is better to retain SCO as a form of insurance should the political climate alter in such a way that conventional oil and fat supplies become prohibitively expensive. Conversely, it may be more appropriate to concentrate on the development of more specialised microbial oils whose high value, and possibly unique properties, would repay the costly investment in sophisticated biotechnology plant.

REFERENCES

1. Boulton, C. A. and Ratledge, C. (1984a). *Cryptococcus terricolus,* an oleaginous yeast re-appraised. *European Journal of Applied Microbiology and Biotechnology,* **20**, 72
2. Boulton, C. A. and Ratledge, C. (1984b). The physiology of hydrocarbon-utilizing micro-organisms. *Topics in Enzyme and Fermentation Biotechnology,* **9**, 11
3. Fuji Oil Co. Ltd (1977). A process for the production of microbial oil. US Patent 4,032,405
4. Iwata, I. (1964). *Agricultural Biological Chemistry,* **28**, 610
5. Kessel, R. H. J. (1968). The fatty acids of *Rhodotorula gracilis:* fat production in submerged culture and the particular effect of pH value. *Journal of Applied Bacteriology,* **31**, 220
6. Kiseleva, S. I., Gerasimova, N. M. and Bokhtereva, M. N. (1972). Lipid composition of some (+) and (−) strains of *Cunninghamella elegans. Microbiologiya,* 269
7. Moon, N. J. and Hammond, E. G. (1978). Oil production by fermentation of lactose. *Journal of the American Oil Chemists Society,* **55**, 683
8. Mumma, R. O. and Bruszewski, T. E. (1970). The fatty acids of *Entomophthora coronata. Lipids,* **5**, 915
9. Mumma, R. O., Fergus, C. L. and Sekura, R. D. (1970). The lipids of thermophilic fungi. *Lipids,* **5**, 100
10. Pedersen, T. A. (1961). Lipid formation in *Cryptococcus terricolus.* I. Nitrogen nutrition and lipid formation. *Acta Chemica Scandinavica,* **15**, 651
11. Raheja, R. K., Gupta, B. K. and Bhatia, I. S. (1977). *Indian Journal of Experimental Biology,* **15**, 1094
12. Ratledge, C. (1982). Microbial oils and fats: an assessment of their commercial potential. *Progress in Industrial Microbiology,* **18**, 119
13. Roy, M. K., Vadalkar, K., Baruah, B., Misra, U., Bhagat, S. D. and Baruah, J. N. (1978). Production of intracellular fat by the yeast *Lipomyces starkeyi. Indian Journal of Experimental Biology,* **16**, 511
14. Shaw, R. (1965). The occurrence of linolenic acid in fungi. *Biochimica Biophysica Acta,* **98**, 230
15. Summer, A. J. and Morgan, E. D. (1969). The fatty acid composition of sporangiospores and vegetative mycelium of temperature-adapted fungi in the Order Mucorales. *Journal of General Microbiology,* **59**, 215
16. Tornabene, T. G. and Holzer, G. (1983). Lipid composition of the nitrogen starved green alga *Neochloris oleobundans. Enzyme and Microbial Technology,* **5**, 435

17. Wither, B., Debuch, H. and Steiner, M. (1974). Der lipide von *Endomycopsis verralis* bei verschidener stickstoff-ernahrung. *Archives for Microbiology,* **101**, 321
18. Wood, B. J. B. (1974). Fatty acids and saponifiable lipids. In *Algal Physiology and Biochemistry* (Ed. W. D. P. Stewart). Blackwell Scientific, Oxford, p.236
19. Zuyagintseva, I. S. and Pitryuk, I. A. (1976). Lipid synthesis in *Candida lipolytica. Mikrobiologiya,* **45**, 470

14
SCP* — THE BP PROTEIN PROCESS

Gilbert Jenkins

INTRODUCTION

The entry of BP into the nutrition business represents an interesting case study for all those associated with biotechnology. Following a technical breakthrough in the late 1950s, successful R & D over the next two decades was stymied by a failure to get the product licensed in the country of production, Italy. Nevertheless, the associated market developments provided BP with a substantial business in the nutrition industry; in 1985 a turnover of nearly £1000 million was achieved, with a profit of £37 million.

This chapter outlines the technical development programme, discusses the related planning environment and summarises the main features of the present BP Nutrition business.

TECHNICAL DEVELOPMENT

In the post-World War II period, international organisations, such as the United Nations, have taken an increasing interest in the world's protein shortage. It is estimated that two-thirds of the world population has a protein-deficient diet. With a world population forecast to increase to six billion by the year 2000, only a worsening protein deficit situation can be foreseen. Therefore, a technical development which could make additional protein available seemed justified.

* *Editor's note*: SCP, the abbreviation for single-cell protein, was first coined by Prof Carrol Wilson at the Massachusetts Institute of Technology, in May 1986, to avoid the unpleasant connotation of 'bacterial' or 'microbial' terms in food usage. It is now taken to mean all forms of micro-organisms and micro-life used as protein food.

A potential new source of protein may be provided by micro-organisms, in a man-controlled environment, to produce high protein content materials for animal feeding.

In 1959, at Lavera in Southern France, a French BP research team under Alfred Champagnat observed that the yeast micro-organisms they were studying grew well on a substrate of the normal paraffins in gas oil, metabolising n-paraffins in the C_{10} to C_{30} range. A process was developed at Lavera using gas oil distillate and fed to a continuous fermentation stage where the yeast metabolised a proportion of the feedstock. The remaining hydrocarbons were separated from the biomass in a subsequent harvesting stage and then returned to the oil refinery.

As work progressed on this process, another research project was set up by BP at Grangemouth, UK, to explore a complementary route using a pure normal paraffin substrate. In this process the n-paraffins were first separated from an appropriate petroleum distillate.

By 1965, the pilot plants at Lavera and Grangemouth were producing yeasts. Further development units, with design capacities of 4000 tonnes per annum and 20 000 tonnes per annum were built at Grangemouth and Lavera. The first 100 000 tonnes per annum plant, using the pure n-paraffin process, was completed in 1976 at Sarroch, Sardinia, for operation by Italproteine SpA, a joint BP/ANIC company.

The n-paraffins and gas oil protein processes differ in three major respects. In the n-paraffin process, the fermentation is carried out under aseptic conditions, so that all organisms are excluded except the specific one inoculated into the system. In the gas oil process, the selected organism is permitted to dominate the fermentation by correct choice of the operating temperature and pH.

The relatively expensive pure n-paraffin feedstock is almost totally assimilated in the fermentation. This simplifies the separation of the cells from the fermenter broth and avoids the need for subsequent solvent purification of the biomass product. The gas oil feedstock, on the other hand, is relatively cheap, but as it is not all assimilated, it must be separated from the cells and they, in turn, must be solvent-leached to produce a satisfactory final protein product.

n-PARAFFINS FEEDSTOCK PROCESS DEVELOPMENT

A 4000 tonne per annum plant was built to demonstrate the feasibility of large-scale continuous hydrocarbon fermentation and to produce sufficient material for widespread customer evaluation. Production was from a single fermenter and product separation and recovery was achieved in a single stream.

Aseptic conditions are necessary only as far as the fermenter. Thereafter normal precautions, as exemplified by general food industry practice, were observed to maintain a high standard of cleanliness.

Fermentation

Mineral nutrients and growth factors, essential for cell propagation, were fed continously at a controlled rate to the fermenter, as an aqueous solution. A metered stream of n-paraffins joins this aqueous feed prior to passing to the medium sterilisation system. The combined medium was heated and held at a high temperature sufficiently long to ensure complete 'kill' of all micro-organisms. The cooled sterilised feed passed directly to the fermenter.

The oxygen necessary for cell growth was provided from a continuous stream of sterilised air blown into the fermenter. Excess air and the carbon dioxide formed in the fermentation passed to the atmosphere. Nitrogen, essential for cell growth, was provided by ammonia gas fed continuously into the air supply. The ammonia served also to maintain the desired pH of the fermenter contents, the rate of addition being automatically controlled according to the pH.

Transfer of hydrocarbon, essential minerals and oxygen to the cells was achieved in the fermenter by intimate mixing of the four phases air, water, oil and yeast, using mechanical agitation.

The high heat release during the fermentation, together with the relatively low fermentation temperature of about 30°C, necessitated external cooling. Fermenter broth was, therefore, de-aerated and pumped through an external exchanger, the fermentation temperature being controlled by regulating the flow of cooling water to the exchanger. A product stream was withdrawn continuously from this cooling circuit to maintain a constant hold-up in the fermenter.

Harvesting

The product stream of fermenter broth was processed in a yeast/water centrifugal separator to yield a separate aqueous phase and concentrated yeast cream. Fermentation processes require a considerable volume of water, so care was taken to re-use the water.

The yeast cream, at about 15 per cent weight solids, then passed to the final spray-drying stage. The spray-dried powder, containing 5–6 per cent weight moisture, was discharged from the drier chamber and cyclone separators into a pneumatic conveying system. Here it was cooled as it was conveyed to bulk storage hoppers. The final BP protein concentrate was bagged, or loaded in bulk directly from these hoppers for dispatch.

GAS OIL FEEDSTOCK PROCESS DEVELOPMENT

This is similar to the n-paraffin process except that there is no sterilisation section, but there is an additional solvent-leaching, product-purification section. The process as described was scaled up from a continuously operated pilot unit, producing 130 tonnes per annum.

Fermentation

The fermentation was carried out in large vessels through which air was sprayed at a high rate to provide both the oxygen essential to sustain metabolic synthesis of biomass and the necessary energy for vigorous agitation.

The temperature was controlled at about 30 °C by heat-exchanging, circulating fermenter broth against cooling water, and the pH was controlled automatically by the injection of ammonia with the air.

Harvesting

A product stream was withdrawn continuously under automatic control from the fermenter and transferred to a decanting vessel, where a large part of the water phase was separated for re-use. The remaining supernatant broth, containing cells, gas oil and water, was then de-aerated.

The resulting cream was next pumped to a mechanical separating stage where centrifuges were used to recover separate streams of product gas oil, a water phase and a slurry of cells concentrated to about 15 per cent weight dry solids, with some unmetabolised oil. The slurry then passed to a spray-drying stage.

Purification

This intermediate product biomass was fed to a continuous countercurrent solvent-leaching section where residual gas oil was removed to an extremely low level, together with the greater part of the lipids.

INTERNATIONAL APPROVAL

All the toxicological and nutritional information was made available to appropriate government departments, with the result that the product obtained from both the processes was permitted for use in animal feed in most EEC countries and in Switzerland, Venezuela and South Africa. Italy, however, did not give approval.

BP's yeast was marketed under the brand name Toprina for the animal feedstuffs industry. Its nutritional characteristics are similar to those of soya.

Toprina was marketed for five years for use in pig and poultry feeds and as a skim milk replacer for calves and lambs.

PRODUCT QUALITY ASSESSMENT

BP protein concentrate made from n-paraffins has a crude protein content (dry basis) of 60 per cent weight, while that from gas oil contains 68–70 per cent weight. More detailed analyses are given in Table 1.

The amino acid composition is similar to that of soyabean and fish meals, except that it has a slightly lower methionine content than fish meal.

Table 1 Typical analyses of BP protein concentrates produced from hydrocarbons

Analysis	*n-Paraffin process*	*Gas oil process*
Moisture % wt	<7	<7
Crude protein (N × 6.25) % wt dry basis	60	68–70
Lipids % wt dry basis (after acid hydrolysis)	10	1.5–2.1
Ash % wt dry basis	6	7.9
	g/16 g N	
Amino acids	*n-Paraffin process*	*Gas oil process*
Isoleucine	5.1	5.3
Leucine	7.6	7.8
Lysine	7.2	7.8
Phenylalanine	4.9	4.8
Tyrosine	3.5	4.0
Cystine	1.0	0.9
Methionine	1.8	1.6
Threonine	5.1	5.4
Tryptophane	1.3	1.3
Valine	6.5	5.8
Arginine	5.1	5.0
Histidine	2.0	2.1
Total S.acids	2.8	2.5

TOXICITY TESTING

Yeasts grown on hydrocarbons are in a situation which, while not unique, is different from that of most other food or feed materials.

BP took the view that these products should be shown to be safe by the most stringent testing procedures that could reasonably be applied. When testing was started in 1963, there were no internationally recognised testing schemes for materials of this nature, so it was necessary to devise them.

A toxicity testing programme was devised in collaboration with the CIVO (Central Institute of Nutrition and Food Research) at Zeist in Holland. CIVO is part of the Dutch TNO organisation; it commands State support and is completely independent of BP. It has a high reputation internationally and has worked for United Nations agencies as well as for the Dutch and other governments. Testing of BP products started in 1964.

The basis of the testing scheme was to show that the yeasts grown on hydrocarbons by the BP processes had no ill effects when fed at high levels in the diet of laboratory animals for short, medium or long periods of time. In addition to these tests, because of the nature of the substrates, special long-term studies were carried out to prove that the yeasts had no tendency to produce cancer. To complete the programme, a separate study was made to ensure that the yeasts had no bad effects upon reproduction. It was also shown that the yeasts had no mutagenic effect. Nutritional evaluation was also carried out and proved completely satisfactory.

PLANNING AND UNCERTAINTY

Many biotechnology projects will require development over some years before a product is sold, to provide a return on the investment. Since the early 1960s, when the BP Protein process was first planned, planning methodology has developed considerably. The world environment for energy and food also changed considerably from the early 1960s to the late 1970s, and this change in the environment had a major impact on BP's food business.

By the early 1960s, international organisations such as OECD and UN had made substantial progress in providing good databases on economic activity and energy consumption. Multinational companies, BP included, were also active in this area, and the BP Statistical Review of the World Oil Industry provided a detailed world-wide database of oil resources, production, supply and demand.

With the availability of good databases, forecasting methodology and the interest in forecasting developed rapidly.

Major special forecasting exercises were sponsored by OEEC in 1955 (Hartley) and in 1960 (Robinson), and, by the early 1960s, the major oil companies were making annual forecasts of world oil supply and demand for up to 6 years ahead. Even though there were shocks such as the Suez Crisis of 1967, these long-term projections of oil supply and demand enjoyed a high credibility through the early 1970s.

World oil supplies grew rapidly through the 1960s, and the main criticism levied against the oil demand and supply forecasts was that oil demand forecasts were always too cautious. During the 1960s, world oil supplies were abundant and there was pressure on all the major oil companies by the producing countries to produce as much oil as possible. In fact, the pressure on the major oil companies throughout the 1960s was similar to that on OPEC in the 1985/1986 period: too much oil supply for the current levels of demand.

At the early stages of planning to manufacture food from oil, little consideration was given to the desirability of manufacturing a renewable source from a non-renewable source. Uncertainty was appreciated concerning the cost of producing single-cell protein from oil, and the ability of certain potential markets to pay an economic price, but these were considered to be justifiable commercial risks.

When the BP Protein development work was substantially complete in the early 1970s, there was a change in the environment. Long-term forecasting methodology had become discredited and there was an apparent shortage of oil in the world. The apparent shortage of oil was corrected by price increases. Throughout the 1960s the cost of oil to BP was $1 per barrel and it was sold on the open market at $2 per barrel. The price increased to $3 per barrel in 1973, $12 per barrel in 1974 and $40 per barrel in 1979 before falling back in the 1980s. This substantial price increase of oil in the 1970s corrected the apparent shortage and also changed the planning environment.

The current state of the art with long-term forecasting is an awareness that we cannot precisely forecast the world oil supply and demand and that any forecast of the price of oil has practically no credibility. Similarly, we are aware that we cannot forecast the open market price of soyabeans. We could, therefore, have little confidence in a forecast which would indicate that single-cell protein produced from mineral oil could always compete with vegetable protein in world markets. Furthermore, the current emphasis is on producing energy from renewable resources, and not vice versa.

THE BP NUTRITION BUSINESS

The BP Nutrition group of companies provides services to the farming industry. Principally the companies supply feedstuffs and breeding stock, but increasingly they are engaging in the processing and marketing of poultry and other meats.

Breeding stock production comprises day-old chicks, turkeys and young pigs, which are sold to the farming industry. The feedstuffs business covers the milling and compounding of bulk and specialist feeds.

There are operating companies in most European countries, with more than 40 units producing over two million tonnes of animal feeds. Outside Europe there are operations in Canada, Singapore and the USA.

BP Nutrition is a leader in world markets for breeding stock. More than 15 per cent of the world's broiler chickens and 10 per cent of the world's egg-laying hens come from the breeding operations. The extensive feed-milling in Europe covers virtually every type of feed, with a total of more than 100 formulations. Formula feeds are supplied for dairy and beef cattle, pigs and chickens. Vitamin/mineral pre-mixes are supplied to other companies in the feed industry and to farmers who do their own mixing. Special computer programs assist the production of formulations at minimum cost.

Through its range of technical, advisory and management services, BP Nutrition works closely with the whole farming sector. Apart from the breeding operations, it is not, as a matter of policy, involved significantly in livestock production itself.

Also marketed is a range of speciality feeds for dogs, cats and zoo animals, and milk replacers for young calves, to replace cows' milk. Feeds for farmed fish, such as trout and salmon, are supplied in Europe.

The main operating companies within BP Nutrition are the Hendrix International Group and, in Britain, BP Nutrition (UK). Hendrix is strongly based in animal feed blending throughout Europe and runs the international breeding operation. Hendrix's business was strengthened by the purchase of the NANTA Company in Spain, and in 1984 by acquiring the NORIA/UFAC group in France. Both companies are leaders in animal feed markets in their respective countries. One of its subsidiaries, Trouw, concentrates its activities on lower-volume specialist vitamin and mineral supplements and pet foods. Cooked meat products are processed, packed and supplied to shops by other Hendrix subsidiaries. Oven-ready birds, portions, fillets, and delicatessen products such as sausages and patés are produced.

Although the in-house production of single-cell protein is not now part of BP's business, nutrition interests continue to increase in importance to the BP Group.

CONCLUSIONS

Novel biotechnology research needs careful planning, not only for the research, but also for the business of which it is to be part. Biotechnological research remains a high-risk area and, as such, needs the encouragement it is getting with substantial government help.

Business is likely to retain a strong interest in biotechnology, but the major pay-off for business will remain in developing technology which fits well with its existing operations or with a sector in which it intends to become established.

BIBLIOGRAPHY

1. Champagnat, A. and Llewelyn, D. A. B. (1962). Protein from petroleum. *New Scientist,* **16**, 612–613
2. Champagnat, A. (1971). Future of the petroleum industry as a source of foods. *8th World Petroleum Congress, Moscow*
3. Llewelyn, D. A. B., interviewed by D. Fishlock, *Financial Times*, 20 April 1983

15
INDUSTRIAL ENZYME APPLICATIONS

G. M. Frost

HISTORICAL REVIEW

Probably the first artificial use of an enzyme was the addition of calf stomach extract to milk for coagulation and subsequent separation of cheese. This process was used for thousands of years before it was understood that the coagulation was caused by the proteolytic enzyme rennin, which occurs in calf stomachs. Similarly, many processes of fermentation of foods and drinks have been used for preservation or enhancement of palatability over a very long period. It has only very recently been established that these fermentation processes are catalysed by complex systems of enzymes operating under the genetic and regulatory control of the organisms involved. In more recent years certain enzymes have been separated from the producing organisms for use in controlled and limited reactions. The production of enzymes for separate application has become an important part of the activities of the fermentation industry.

During the early part of this century two pioneers, Rohm and Takamine, of German and Japanese origin, respectively, developed the modern enzyme industry in the West. Industrial enzymes production was greatest for some years in the USA, where Miles, Rohm and Haas, and Wallerstein were major producers; but in more recent years the focus of activity has tended to shift back to Europe, and Novo Industri of Denmark and Gist-Brocades of The Netherlands have become the world's largest enzyme producers. There has also been a thriving enzyme production industry in Japan, based initially on the traditional semisolid fermentation processes for miso and shoyu (soy sauce).

An approximate chronology for establishment of some of the more important industrial enzyme uses is given in Table 1. The list, while not

Table 1 Approximate dates for establishment of modern enzyme applications

Decade	*Enzyme*	*Use*
–1920	Fungal α-amylase	Bread improvement
1920–30	Fungal α-amylase	Fruit juice clarification
	Pectinase	Fruit juices and wine: extraction and clarification
	Bacterial α-amlyase	Viscosity reduction, chocolate syrup
1930–40	Bacterial α-amylase	Production of brewers' wort
	Bacterial proteases Fungal α-amylase Fungal α-amylase	Syrups from starch
1940–50	Amyloglucosidase	Production of brewers' wort
	Fungal α-amylase	Production of distillers' mash
	Proteases	Tenderising meat
	Lipases	Cheese flavour improvement
1950–60	Bacterial α-amylase	Sugar production from starch
	Proteases	Manufacture of biscuits
	Glucose oxidase	Stabilisation of powdered egg Oxygen removal from drinks
1960–70	Fungal proteases	Manufacture of cheese (calf rennet substitute)
	Bacterial proteases	Use in detergents
	Amyloglucosidase	Production of glucose from starch
1970–	Glucose isomerase	High-fructose syrups from starch
	Pullulanase	Improvement of yield of sugar from starch
	β-Glucanase	Improvement of brewers' wort
	Raffinase	Improvement of yields from sugar-beet
	Lactase	Low-lactose milk products

comprehensive, illustrates the steady growth both in numbers of enzymes used and in variety of applications for each enzyme. The world market for enzymes is now around £200 million and can be expected to grow at a rate of at least 10–15 per cent for some years to come[22].

DEFINITION AND NATURE OF ENZYMES

Enzymes are most simply defined as biological catalysts. The rates of the hundreds of reactions which make up the metabolism of living cells and organisms are determined by the activities of enzymes, each catalysing a specific reaction. The substance whose reaction is catalysed by a particular enzyme is termed its substrate. The synthesis and catalytic capacity of the individual enzymes are in turn controlled by the various regulatory mechanisms of the cell or organism, so that growth and other activities of the organism proceed in an orderly and balanced way.

Enzymes are a large and diverse class of proteins. Proteins are complex polymeric molecules composed of alpha amino acid units linked through their alpha carbons in a polypeptide chain (see Figure 1). Twenty different amino acids are used in protein structures. The precise sequence of the amino acids is specific to the particular protein and is controlled genetically. The sequence of amino acids in a protein is called the primary structure. Enzymes vary in molecular weight, but all are relatively large molecules. Most fall within the range 10 000–200 000 daltons, which means that there are from about 100 to 2000 amino acid units in the polypeptide chain. The basic structure is always a protein, but there may be other components bound into the enzyme molecule. Of particular importance are the smaller organic molecules called coenzymes, which are located at the region of the molecule known as the active site, where the catalytic reaction occurs. Coenzymes take part in the reaction. Many enzyme reactions do not require coenzymes; of those which do, some utilise readily dissociable coenzymes in equilibrium with the solution; others have the coenzymes covalently bound into the enzyme molecule. Carbohydrate is also bound covalently into some enzyme structure and often renders the enzyme more stable chemically.

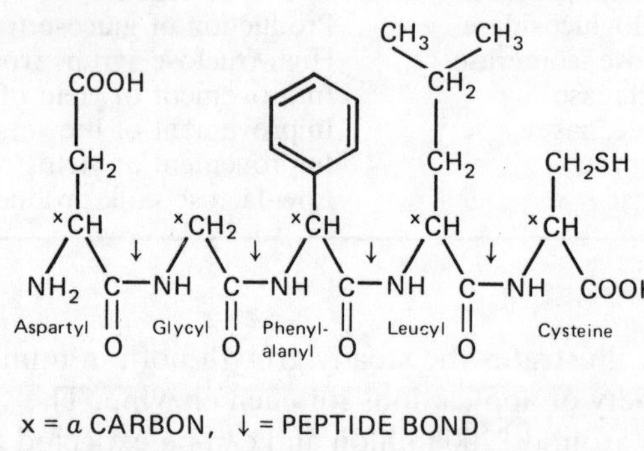

Figure 1 A protein polypeptide chain containing five amino acids

The polypeptide chain is usually mainly coiled into a helical structure stabilised by hydrogen bonds between amino acids fairly close together in the primary structure. This, coiling, again, is specific to the particular protein, and is called the secondary structure. The coiled chain is further folded in a manner characteristic of the particular protein, termed the tertiary structure. This is determined by interactions between amino acids in more remote parts of the chain. Most enzyme molecules are globular in shape. Usually there is a central hydrophobic region containing many non-polar amino acid side-chains, and an outer shell containing most of the charged or polar groups protruding into the aqueous phase and heavily hydrated.

A functional enzyme often contains a definite arrangement of subunits, which may all be the same or may be different in structure. Each contains separate polypeptide chains. The subunits are usually capable of separating into single units in response to specific stimuli or relatively minor changes in conditions. Subunit associations are frequently involved in control of enzyme action. For example, a four-subunit aggregate may be active catalytically but a controlled dissociation into the four inactive subunits may occur.

The arrangement of subunits in a protein is called the quaternary structure.

CLASSIFICATION OF ENZYMES

Enzymes are classified and named according to the reactions which they catalyse. The International Union of Biochemistry (IUB) periodically publishes a complete list of well-characterised enzymes subdivided according to the classification scheme[28]. The six major groups of enzymes are listed in Table 2.

Table 2 Classification of enzymes into major categories

1. Oxidoreductases	Oxidise a hydrogen donor while reducing a hydrogen acceptor
2. Transferases	Transfer a group from one molecule to another
3. Hydrolases	Cleave a bond by adding the elements of water across it
4. Lyases	Cleave a bond by eliminating a group or add to a double bond
5. Isomerases	Catalyse change of internal structure of a molecule
6. Ligases (synthetases)	Join molecules together, usually coupling with hydrolysis of a pyrophosphate bond

Each major category is subdivided into two further levels of subdivision, and each enzyme has a classification number divided into four elements, of which the last is specific for a particular reaction. Thus, the enzyme lactase (β-galactosidase) catalyses the splitting of the sugar present in milk (lactose) into its two component monosaccharides, glucose and galactose, and has the classification number 3.2.1.23. This is explained in Table 3.

Table 3 Classification for lactase

Number	*Denotes*
3	Major classification—hydrolase
2	Secondary classification—hydrolase acting on glycosyl compounds (one of 11 subclasses of hydrolase)
1	Tertiary classification—glycosyl hydrolase acting on an *o*-glycosyl compound (one of 3 subclasses of glycosylases)
23	Specific) *o*-glycosyl hydrolase reaction catalysed by this enzyme (89 enzymes are listed in this category)

The IUB system, in fact, classifies reactions catalysed by enzymes. For each individual entry in the classification there will be a variety of proteins differing to a greater or lesser extent in structure and detailed properties. Thus, the major metabolic enzymes will occur in nearly all species of organisms but will differ in detailed structure between any two organisms, depending on the degree of evolutionary divergence. The diversity of enzymes existing in nature is truly staggering. The existence of different enzymes catalysing the same reaction can be of value in broadening the practical range of conditions for use and, therefore, the number of practical applications. For example, yeast lactase from *Kluyveromyces fragilis* operates best at the natural pH of around 6.5 found in milk, while fungal lactases from *Aspergillus* moulds have optimum activity at the pH of cheese whey (around 4.5).

MANUFACTURE OF ENZYMES

Industrial methods for the preparation of enzymes have recently been reviewed by Lamber and Meers[33].

Enzymes used in industry are obtained from a wide variety of sources. Bacteria, yeasts, moulds, and plant and animal organs are all used as sources for industrially important enzymes. Animal enzymes are by-products from animals reared for meat, obtained from unwanted or low-value organs such as the stomach, pancreas or liver. Plants are sometimes grown on a relatively small scale specifically for production of enzymes — for example, horseradish, from whose root the enzyme peroxidase, used in analysis, is purified; and papaya, which contains a high level of the proteolytic enzyme complex papain in the fruit[3]. The higher organisms tended to be used earliest for enzyme production because of the relative ease of extraction, but their relatively slow growth rates and the long time-scale for strain improvement has caused the enzyme industry to turn to micro-organisms as the major group for enzyme production. Techniques for recovery and purification, methods for formulation and general properties of enzymes for all types of organisms are essentially similar.

Two basic types of fermentation process are used for enzyme production. In semisolid fermentations a solid substrate, usually aerated, irrigated and gently mixed, is inoculated with mould spores. The mould grows over a period of several days and the enzymes are secreted into the surface moisture of the solid mass. They are extracted by washing with an aqueous solution at the end of the process. The semisolid method has the disadvantages that growth is relatively slow; the process is difficult to scale up and automate; and maintenance of a strict monoculture of the production strain is very difficult to achieve.

In the submerged fermentation process the medium is a sterilised homogeneous aqueous solution (perhaps containing some suspended solids) contained in a sealed, previously sterilised, fermenter. The production organism is added as a liquid culture or spore suspension and is maintained in an aerated stirred condition for a period from about 12 h to 1 week, according to the process involved. Submerged fermentations tend to proceed faster and are easier to protect from infection. Most of the enzymes currently produced are made by submerged processes. Semisolid fermentation tends to produce a broad spectrum of hydrolytic activities and is of value in producing enzymes for digestion or liquefaction of complex mixtures (e.g. pectinases for fruit juice production).

Most enzymes produced by submerged fermentation are extracellular[55] — that is, they are secreted by the organisms into the medium. Extracellular enzymes can easily be recovered from a culture in a relatively pure state, and their downstream processing to a saleable form is much simpler than that of cell-bound (intracellular) enzymes, which need to be released from the cells and are mixed with a wide variety of other intracellular proteins, from which they may need to be separated. However, intracellular enzymes have the advantage that the relatively easy process of separating the organism from the broth gives a high concentration factor for the enzymes, so that the volume to be treated in further steps is much less than for extracellular enzymes.

Enzymes are sold as solids or solutions of widely varying strengths and purity. For applications where it is essential to use a low-cost product, a stabilised crude extract or filtered fermentation broth may be sufficiently pure. At the other extreme, for analytical, pharmaceutical and some food uses, essentially pure enzyme products may be required. Most commercial enzymes contain significant quantities of enzymically inactive protein from the source organism.

Figure 2 shows the sequence of operations used in typical enzyme recovery processes. The enzymes made from higher organisms are usually secreted into the extracellular fluids, so that cell disruption is not required, or they are contained in cells which are very easily disrupted by stirring or autolysis. Cell separation from deep fermentation of unicellular organisms (bacteria and yeasts) is usually achieved using flow-through centrifuges, generally the stacked-disc self-desludging type. Filamentous moulds are more coveniently separated using a filter. Filter-presses and vacuum filters are both employed in different processes.

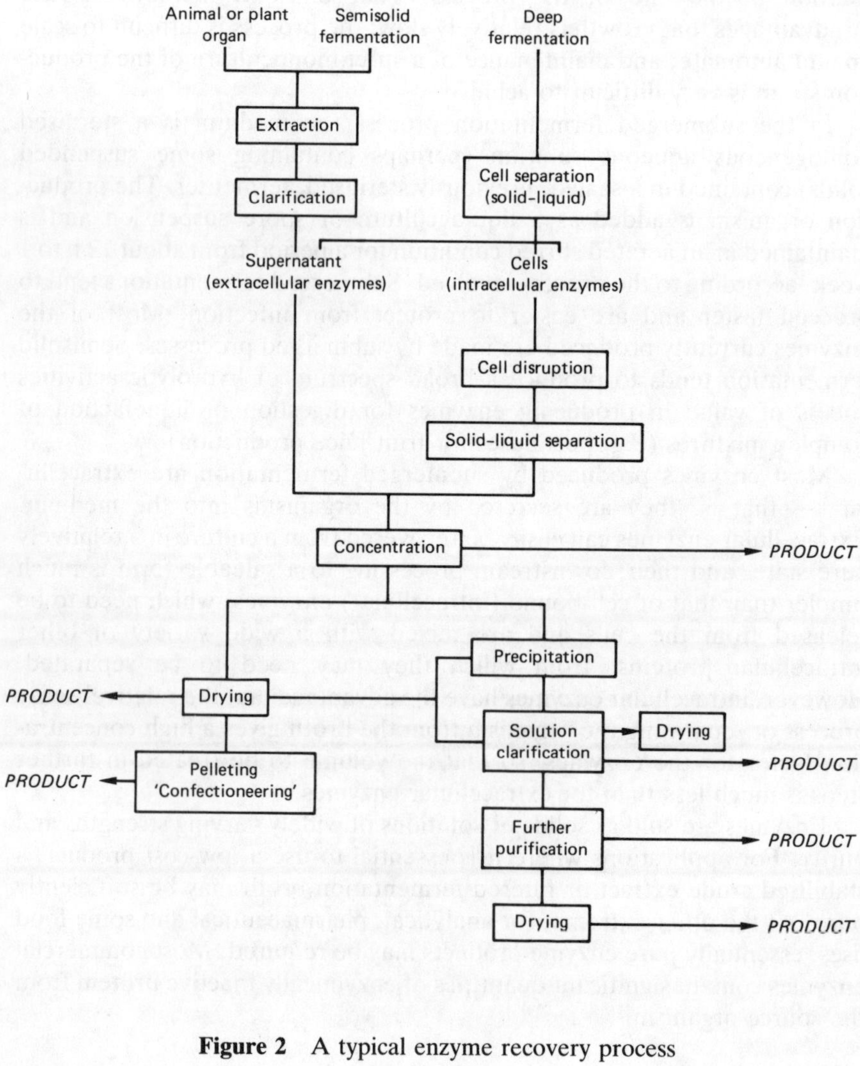

Figure 2 A typical enzyme recovery process

Cell disruption is often a difficult stage in large-scale enzyme processing. In general, each enzyme is released by a different process. Lysis by chemical[19] or enzyme treatment or mechanical disruption using high-pressure homogenisers[20,67] or various types of mill[36,60] are all used. Following cell disruption, the released intracellular contents have to be separated from the solid cell fragments and organelles. This is more difficult than separation of whole cells from broth, because of the smaller particle sizes involved; a similar range of equipment is used but flow rates tend to be slower than for broth separations.

Culture filtrates containing extracellular enzymes must nearly always be concentrated for sale or further processing. Cell extracts may need to be concentrated but can frequently be used directly. The principal method for concentration of enzymes used to be vacuum evaporation but this has largely been supplanted by ultrafiltration, which has the advantages of being less temperature-dependent and less energy-consuming.

For preparation of solid products spray-drying[44] or drying in a vacuum oven at moderate temperatures or, for higher-value low-volume products, freeze-drying[63] is employed. Spray-drying is the most common route; enzymes are surprisingly temperature-stable during this process, principally because the solids content of the feed solution is relatively high and the exposure time in the liquid state at the drying temperature is very short. Freeze-drying and spray-drying produce a finely powdered solid which generates enzyme dust. For proteases, in particular, which constitute a health hazard to workers who inhale significant quantities of dust, more complex processes of treatment to yield smooth non-abrading particles on drying are used.

To obtain more purified grades of enzymes, a variety of precipitation techniques can be used[17]. The most popular is alcohol precipitation, usually with isopropyl alcohol. This is generally preferred because the solvent is cheap, is relatively non-toxic and can readily be recovered for re-use by distillation. Some enzymes are unstable to alcohols; for these, salt precipitation, usually with ammonium sulphate, has been used in industrial production. Salt precipitation has several disadvantages. The increase in density of the solution containing the necessary high levels of salts makes recovery by centrifuges more difficult. It is not economic to repurify the salt, which then presents a problem in effluent disposal. Precipitates are frequently separated using tubular-bowl flow-through centrifuges (Sharples design). These machines are limited in possible scale and newer developments in stacked-disc centrifuges make it seem probable that they will ultimately replace the tubular-bowl type for large-scale enzyme recovery.

A variety of chromatographic separation techniques are available for enzymes. These are only applied to the small sector of the enzyme production industry which supplies highly purified enzymes for diagnostic and pharmaceutical uses.

IMMOBILISED ENZYMES

Enzymes are theoretically capable of continuing to catalyse their reactions indefinitely. In practice, they are normally used in homogeneous aqueous solutions or suspensions from which it is not economically possible to recover them for re-use. After the substrate present in the process stream has been converted by enzyme catalysis, the enzyme remains in the stream and passes into the product as a catalytically active molecule or, if a heat treatment has been used, as an inert protein.

To allow more productive use, to avoid contamination of the product by presence of added enzymes, and sometimes also to control the extent of reaction, so-called immobilised enzymes can be used. Immobilised enzymes are either bound to solids in an insoluble form or mechanically confined so that the process stream can flow over or through a reactor containing the enzyme.

Very extensive work has been done on manufacture and properties of immobilised enzymes[8,11], but despite much early enthusiasm only a few instances of practical application have become established. Of these, the most important example is use of glucose isomerase to make sweet fructose-containing syrups from hydrolysed starch, where the immobilised enzyme is used in packed-bed reactors[10]. One reason why immobilised enzymes have failed to replace soluble enzymes is that they are much more complicated to apply, requiring special plant and control systems, whereas soluble enzymes merely require to be poured into the batch or metered into the stream. Furthermore, unless the enzyme concerned is unusually stable, there may be insufficient benefit from repeated re-use to pay for the additional manufacturing costs and process losses involved in converting the enzyme into an immobilised form.

A further difficulty is that most enzymes are only stable over long periods of use at fairly moderate temperatures and pH values, so that in most instances microbiological contamination in the process is a more severe problem than it is for single-use soluble enzymes. Immobilised enzymes are best suited to use with fairly pure process streams containing low molecular weight dissolved substrates. If the substrate is a large molecule or an aggregate, there is usually insufficient accessibility of the immobilised enzyme to allow the reaction to proceed fast or far enough.

TYPES OF ENZYMES USED IN INDUSTRY

Enzymes with world markets of more than about £1 million are listed in Table 4 with the major areas of their use. Hydrolytic enzymes account for over 90 per cent of the world market, and, of these, proteases account for over half and glycosidases for nearly one-third[22]. The other five major categories of enzymes account for less than 10 per cent of the world market and include only two with relatively large markets.

ENZYME USAGES CLASSIFIED BY INDUSTRY

Detergents

Rohm took out a patent for use of enzymes in detergents in 1913. Presoak detergents containing animal pancreas extracts, where the active constituents were the proteases trypsin and chymotrypsin, were used for many years, but markets were relatively small until the early 1960s, when the more stable bacterial enzymes were produced in quantity by fermentation. These are suitable for use under the normal conditions for laundering with detergents and were rapidly adopted for major household detergent brand production in Europe and the USA. The market suffered a severe setback around 1969, when it became clear that there was a health risk to process workers manufacturing the enzymes[70]. The proteolytic action of these enzymes appears to be able to potentiate their allergenic effects, and respiratory symptoms, generally reversible, were observed. A study commissioned by the USFDA[1] concluded that there was no significant risk to the health of consumers. Meanwhile the producer companies had introduced methods for producing smooth-coated enzyme particles of uniform size. Products treated in this way contain virtually no dust and any hazard to health caused by handling them is minimal.

While there has been doubt about the effectiveness of enzymes in improving the general performance of household detergents, it is clear that biological stains, especially those of blood, are more effectively removed by enzyme-containing products. Thus, there is a definite benefit in using enzymes. Markets for detergent enzymes have recovered from the difficulties caused by concern about health risks. With the new interest in washing at cooler temperatures[47] promoted by a desire to save energy, it is probable that the importance of enzymes as ingredients in washing formulations will increase.

Starch Processing

The starch processing industry is the largest user of enzymes on both a tonnage and a cost basis. An immense tonnage of starch-producing plant crops is grown each year, and a proportion of this crop is used for purification and subsequent processing of starch for use in a variety of industrial applications.

Manufacture of soluble hydrolysis products from starch was introduced in the 1930s using an acid hydrolysis process in pressure vessels at 120°[58].

Table 4 Enzymes with significant industrial markets

Enzyme	*Reaction catalysed*	*Source species*	*IUB-number*	*Principal uses*
Proteases	Hydrolysis of peptide bonds in protein to yield shorter peptides and α-amino acids	*Bacillus subtillis* (alkaline)	3.4.21.14	Detergents
microbial		*B. subtillis* (neutral)	3.4.24.4	Brewing, baking, leather
		Aspergillus spp.	3.4.23.6	Baking
		Mucor spp.	3.4.23.6	Milk coagulation in cheese manfacture
plant		Papaya fruit (papain)	3.4.22.2	Beer haze prevention; meat tenderising
animal		Pig and ox pancreas (trypsin)	3.4.21.4	Food; digestive aids; medical uses
		(chymotrypsin)	3.4.21.1	Food; digestive aids; medical uses
		Calf stomach (rennet)	3.4.23.4	Milk coagulation in cheese manufacture
		Pig and ox stomach (pepsin)	3.4.23.1	Milk coagulation in cheese manufacture; digestive aids
α-Amylase	Hydrolysis of α,1,4 glucosidic bonds in starch. Attacks randomly along chain to produce short oligoglucosides	*B. subtilis*	3.2.1.1	Starch processing
		Bacillus licheniformis		
		Aspergillus oryzae	3.2.11	Bread improvement
Amyloglucosidase	Hydrolysis of α,1,4 glucosidic bonds in starch. Splits glucose sequentially from non-reducing chain ends. Acts slowly on α,1,6 glucosidic bonds	*Aspergillus niger*	3.2.1.3	Production of dextrose from thinned starch
				Beer wort production; low-sugar beers
β-Amylase	Hydrolysis of α,1,4 glucosic bonds in starch. Splits maltone sequentially from non-reducing chain ends. No action on α,1,6 bonds	Barley	3.2.1.2	Production of high-maltose syrups from thinned starch

Glucose isomerase[a]	Glucose $\rightleftharpoons$ fructose (xylose $\rightleftharpoons$ xylulose)	Various micro-organisms	5.3.1.5	Production of sweet high-fructose syrups from starch hydrolysate
Pectinase[b]	Digests pectin by hydrolysis of methyl substituents from the polygalacturonate units, followed by hydrolysis of the glycoside linked chain	*A. niger*	3.1.1.11 3.2.1.15	Improvement of juice and purée yields from plant crops Clarification of fruit and vegetable juices
Glucose oxidase	Glucose + O_2 → glucomate + H_2O_2	*A. niger*	1.1.3.4	Stabilisation of foods and beverages; determination of glucose
Lipase[d]	Triglycerides (lipids) lower glycerides + fatty acids	*Rhizopus arrhizus* *A. niger*	3.1.1.3	Food flavouring; modification of fats
Cellulase	Hydrolysis of α,1,4 glucosidic bonds in cellulose	*A. niger* *Trichoderma spp.* *Penicillium* spp.	3.2.1.4 3.2.1.21	Extracts from plants; food modification; oil (petroleum) extraction
Invertase	Sucrose → glucose + fructose	*Saccharomyces* spp.	3.2.1.26	Manufacture of soft-centre confectionery

[a] IUB classify this enzyme as xylose isomerase, because some of the better-characterised sources produce enzymes which act more effectively on xylose than on glucose. However, some commercial sources act at least as well on glucose as on xylose.[9]

[b] Some preparations also contain pectate transeliminase (4.2.2.2) and pectin lyase (4.2.2.10), which break the polymer chain by an elimination reaction.

[c] Preparations for food use also contain catalase (1.1.1.6), which decomposes the hydrogen peroxide produced in the glycose oxidase reaction.

[d] Most lipases also hydrolyse the lower glycerides ultimately to produce glycerol and fatty acids.

This process had shortcomings. Neutralisation produced high levels of salt in the products and the conditions used promoted a reverse reaction yielding reversion products, so that yields of glucose never exceeded about 80 per cent.

Enzymes were first introduced to increase the sugar level following an initial acid-thinning and partial hydrolysis. For example, an acid syrup of DE 40* was treated with fungal α-amylase to produce a DE 63 syrup containing a high level of maltose[13]. Subsequent introduction of amyloglucosidase and reduction of extent of the initial acid hydrolysis allowed higher yields of dextrose than were obtainable by acid hydrolysis. These processes, where acid treatment and neutralisation is followed by enzyme action, are called acid–enzyme processes. Acid–enzyme processes have now been replaced almost entirely by enzyme–enzyme processes, where a treatment with bacterial α-amylase is used instead of the initial acid treatment and this is followed by digestion with a sugar-producing exoglucanase enzyme (β-amylase or amyloglucosidase).

The action of the various digestive enzymes on starch is represented diagrammatically in Figure 3. Two types of glucose polymer are present in starches. Amylose is an unbranched α,1,4-linked glucose polymer. Amylo-

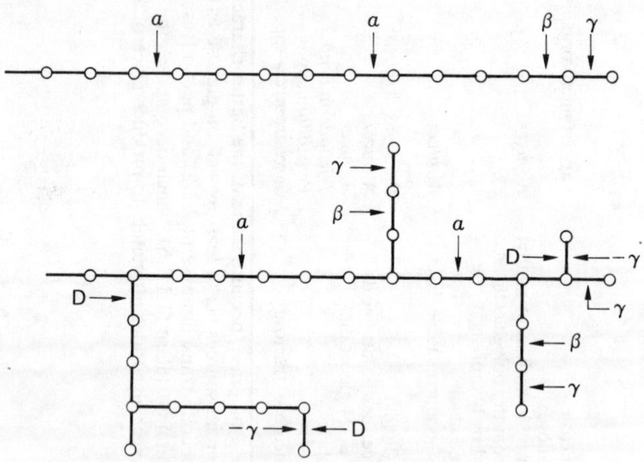

Figure 3 Diagrammatic representation of various digestive enzymes on starch. α = α-amylase (endo-amylases); β = β-amylase (exo-amylases); γ = γ-amylase (amyloglucosidase); D = debranching enzymes (pullulanase)

*DE, dextrose equivalent, is the reducing group value as a percentage of total solids calculated as glucose equivalent.

pectin has a highly complex branched structure where occasional glucose residues are also attached with a 1,6 bond to form a branch point. The α-amylases (endoamylases) split α,1,4 bands anywhere in the polymer chains. Exhaustive digestion of starch by bacterial α-amylases produces mainly fragments containing 3–6 glucose residues[48]. The proportions of the fragments varies between the two commonly used bacterial α-amylases from *Bacillus subtilis* and *B.licheniformis*. Fungal α-amylase is also capable of attacking the chain internally, but the main product from exhaustive digestion is the disaccharide maltose. β-Amylases split successive maltose units from the non-reducing chain ends to produce effectively 100 per cent maltose from amylose, but with amylopectin action stops when a branch point is reached. Thus, a so-called β-amylase limit dextrin is produced which is the core of an amylopectin molecule with a 1,6 link at each non-reducing chain end. If α-amylase is present with the β-amylase, the amylopectin is also converted largely to maltose. Amyloglucosidase or γ-amylase attacks the non-reducing chain ends, splitting off glucose and maltose residues; it also hydrolyses the α,1,6 branching links but only at a very slow rate, so that ultimately glucose is produced in nearly 100 per cent yield from amylose and amylopectin. Enzymes exist which specifically attack α,1,6 bonds (starch-debranching enzymes). Pullulanase from the bacterium *Klebsiella aerogenes* degrades the mixed α,1,4–α,1,6 polymer pullulan and also acts efficiently on amylopectin branch points. This is used in conjunction with β-amylases or fungal α-amylase to produce high maltose syrups and can also accelerate amyloglucosidase action.

Figure 4 gives an outline of the enzyme-utilising processes by which the various starch products are made. Several reviews of enzyme and acid–enzyme processes from starch are available[39,48,56,60].

To render starch susceptible to extensive enzyme attack, it must be heated to destroy the crystalline structure of the natural starch granule (gelatinisation). On cooling gelatinised starch slurries, very high viscosities are reached, so an acid or enzyme thinning before the starch is allowed to cool is essential. Total disruption of the granule structure of most starches is achieved at 105°. The original bacterial α-amylase used for thinning (*Bacillus subtilis*) is stable only up to about 90° in starch slurries when stabilised by calcium. Using this enzyme, it is usual to make two additions: the first before high-temperature cooking, to effect partial thinning; the second after cooling, to carry the reaction to the desired stage. The discovery[42] of a more thermostable α-amylase from *Bacillus licheniformis*, which is capable of surviving short periods of cooking at 105°, allowed development of processes where a single enzyme addition could be made. A variety of products with DE values of from 3 to about 20 are made by allowing the α-amylase action to reach the desired point and then using a short heat treatment at 140–150° to inactivate the enzyme. The products may be sold as slurries or syrups, or they may be dried to powders for sale.

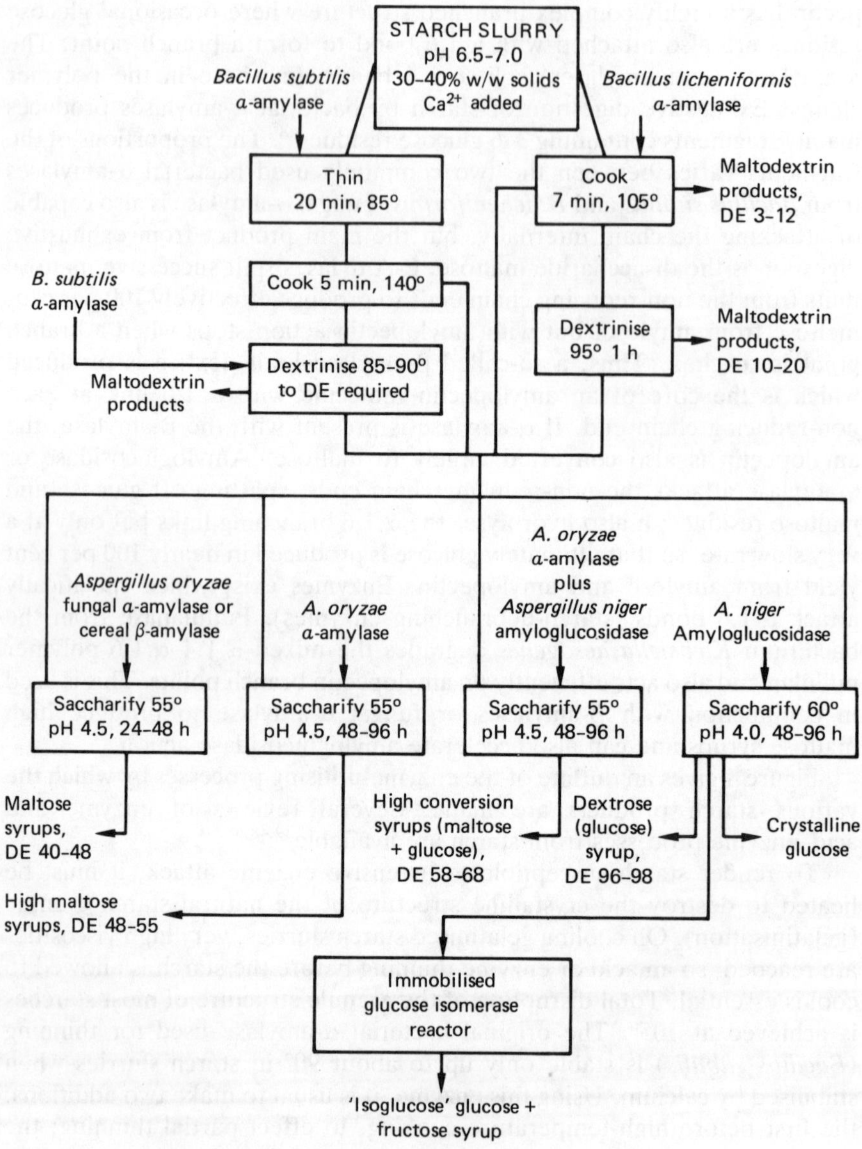

Figure 4 Enzymic processes in starch utilisation

They are used as glues and gums, or as fillers, thickeners and stabilisers, in a wide variety of applications. For use in further processing to sugar syrups, the starch is subjected to the minimum degree of α-amylase action required to give a stable liquid of moderate viscosity at 55–60°. This is done because by-products of α-amylase can limit the maximum conversion obtainable with the saccharifying enzymes.

Different enzymes or mixtures of enzymes are used for further hydrolysis (saccharification) according to the required DE and composition of the intended product. Fungal α-amylases and cereal β-amylases yield syrups containing predominantly maltose. The reaction is controlled at pH 4.5 and 55° and takes 24–96 h. The digests are heated to inactivate the enzymes and evaporated to about 70 per cent solids. The products are fairly sweet syrups and powders made from them with DE values of 40–55. They are used in a variety of ways in the food industry and also as supplementary sources of sugar in brewing. A mixture of amyloglucosidase and fungal α-amylase can be used to produce mixed glucose–maltose syrups of somewhat higher DE. The content of glucose is limited, to avoid crystallisation, and the maximum DE obtainable in a stable syrup is about 68.

The *Aspergillus niger* amyloglucosidase is slightly more thermostable than the fungal α-amylase. It is used to produce essentially pure glucose from dextrinised starch at 60° and pH 4.0. Maximum glucose yield of 95–97 per cent is reached in 48–96 h. Evaporated glucose syrups are sold as food ingredients and to the fermentation industry as raw materials. They suffer the shortcoming that they must be transported and stored in heated tanks at 50° to avoid crystallisation. High-DE glucose syrups are also converted into sweeter, non-crystallising, isomerised syrups containing mixtures of glucose and fructose, by using the enzyme glucose isomerase. When it was first developed as an industrial enzyme, glucose isomerase was produced in relatively poor yields. Systems were therefore developed for immobilising it in column reactors to yield a high re-use factor. Several forms of immobilised glucose isomerase, each from a different bacterial or fungal source, have been commercialised[8] and there is a vast accumulation of patents on production by fermentation, immobilisation and use of glucose isomerases. Fructose syrups have been received enthusiastically by the market. The use of glucose isomerase began in 1970 and by 1980 world production of isomerised syrups was estimated at three million tonnes. This growth has also led to increased demand for the bacterial α-amylase and amyloglucosidase used upstream in the process.

Glucose isomerase catalyses achievement of an equilibrium between glucose and fructose. In practice it is more economic to use the enzyme to reach 42 per cent fructose, and most of the product sold contains this level. The 42 per cent fructose mixture is slightly less sweet than invert sucrose.

Chromatographic separation processes have been developed which yield a high glucose fraction for recycling and a 90 per cent fructose fraction. This can be used to mix with 42 per cent material to increase fructose to 55 per cent, where the mixture is equal in sweetness to invert sucrose, or sold as an ultra-high-fructose product.

Production of sugars by any of the saccharifying enzymes is limited to a greater or lesser degree by slow reaction of the α,1,6 glucosidic linkages at the branch points of amylopectin. Debranching enzymes, which specifically attack these linkages, have been available commercially for several years but have not found widespread application, despite the fact that improved yields or faster reactions can be demonstrated when they are used in conjunction with saccharifying enzymes. Pullulanase from *Klebsiella aerogenes*[5] has been used to a limited extent to produce high-glucose syrups with amyloglucosidase. Isoamylase, from *Pseudomonas amylodermosa*, which has slightly different properties from pullulanase in terms of the low molecular weight sugars which it hydrolyses[26], has been used for enhancing maltose yields in conjunction with fungal α-amylase. It has been claimed[49] that the *K. aerogenes* enzyme is unsuitable for efficient use with amyloglucosidase, because its thermal and acid stability are insufficient to allow use of the optimum conditions for the amyloglucosidase.

Fuel Alcohol

Increasing petroleum prices and concern over the ultimate depletion of fossil fuels has generated interest in alternative sources of liquid fuels, especially for road vehicles. Alcohol is an excellent fuel for internal combustion engines. An attractive route to its production is the use of renewable carbohydrate from photosynthesis as a fermentation substrate to produce ethanol which can be distilled to yield a usable fuel[32].

In free market economic terms, alcohol for fuel is unlikely to compete with petroleum for many years, but certain governments have offered subsidies or tax incentives to encourage development of the technology and to establish some production.

The major photosynthetic carbohydrate sources are sugar (sucrose), starch and cellulose. The best-known, and probably the most useful, organisms for ethanol production are yeasts. These ferment sugar readily without prior treatment, but the more complex polysaccharides need to be degraded to simple sugars. In the case of starch the enzyme processes outlined above, especially the α-amylase–amyloglucosidase route to high-glucose syrup, are used to produce fermentable substrates for yeast. Significant fuel alcohol production from starch is at present confined to the USA, where the source crop used is corn. In Brazil sugar-cane juice is used, but an analysis has suggested[4] that indigenous starch sources such as cassava or babassu nut could advantageously be used.

Cellulose is not at present used for ethanol production but has been the subject of much research work. It has certain theoretical advantages over starch[32]. It is not yet certain whether enzyme or acid hydrolysis will be preferable as a route to sugars for fermentation or, indeed, whether direct fermentation of cellulose to alcohol might become viable. Cellulose has evolved as a structural carbohydrate and is therefore much more resistant to enzyme attack than is starch. Enzyme costs per tonne of glucose produced from cellulose are therefore very much higher than those for glucose from starch. Offset against this is the fact that there are many unused cellulosic materials which are theoretically available for the cost of their transport from the farm to the fermentation plant.

Fuel alcohol production from starch in the USA increased by 50 per cent between 1983 and 1984 to a production level of 2000 million tonnes, with further subsequent increases. Longer-term prospects for fuel alcohol, both in the USA and elsewhere, are the subject of divergent views. Other Western governments, notably that of France, are spending money on evaluation and research.

Brewing

The traditional process for beer manufacture involves germination of a mass of barley grains (malting), followed by extraction with warm water (mashing), boiling of the aqueous extract (wort) with hops, and fermentation with yeast.

During malting the content of hydrolytic enzymes in the barley grains[54] increases, so that the protein and carbohydrate reserves in the grains are largely solubilised at the mashing stage of nourishment of the yeast during fermentation. Yields and qualities of worts are greatly affected by the quality of the barley and of the malt produced from it. Furthermore, the malting process is expensive and time-consuming. Unmalted barley cannot be used for wort production, but if the appropriate microbial enzymes are added to finely ground barley, the mashing process will yield a wort which ferments satisfactorily[40]. In the 1960s and early 1970s much work was done on the use of barley with added enzymes in place of malt, and it was expected that this method would be adopted for commercial production. However, the taste and properties of the product are affected by such a change, and in more recent years public opinion has dictated a conservative approach. Most beer is still brewed largely from malt and this situation now appears unlikely to change. However, the use of a small proportion of unmalted barley in the mixture is common, and this addition of minor amounts of 'adjuncts', auxiliary sources of fermentation materials such as rice or wheat flours, is also a common practice in some areas, particularly the USA. The use of these supplementary ingredients and the variability in quality of barley available for malting have led to the need for the addition of microbial enzymes to supplement the malt enzymes.

Table 5 lists the enzymes used in mashing and their effects. They are either sold to the brewing industry as individual enzymes for use by the

Table 5 Uses of enzymes in mashing

Enzyme	*Effects*
Fungal α-amylase } Amyloglucosidase }	Increase yield of fermentable sugar (e.g. where low-diastase malt or unmalted barley is used)
Bacterial α-amylase	Improves yield of wort (e.g. where starchy adjuncts used). Allows use of materials containing high-temperature gelling starch
Bacterial β-glucanase	Reduces viscosity of wort and improves filterability (e.g. where barley or high-glucan malts are used)
Bacterial neutral protease	Increases yield of assimilable nitrogen in wort (e.g. where some unmalted barley is used)

knowledgeable brewer to his own specification or offered as blends for particular purposes by specialised marketing companies.

The enzymes may be included in the initial mix for brewing — for example, when high-quality malt is not available or where adjuncts or unmalted barley are included in the ingredients — or they may be used as 'first aid' expedients when a particular batch shows evidence of poor yield, slow filtration or unsatisfactory composition.

If breakdown of the starch is inadequate at the mashing stage, the wort may contain an excessive amount of unfermentable dextrins and the fermentation may not proceed to the desired extent. In this case addition of fungal α-amylase or amyloglucosidase can be used to produce maltose and glucose from the dextrins, which allows the yeast to continue fermenting.

A growing use for amyloglucosidase is the production of low-calorie 'diet' beers or the related diabetic beers. Normal beer contains a proportion of unfermented dextrin which is metabolised by the consumer as glucose. Amyloglucosidase can break down these materials to glucose, which ferments to yield a beer which is essentially free of carbohydrate. As in industrial starch degradation, the limiting factor in breaking down the dextrins is breaking of the α,1,6 branch linkages derived from amylopectin. Pullulanase is used alone or together with amyloglucosidase to attack the α,1,6 bonds and ensure efficient conversion of the dextrins to glucose. In a different and less widespread use amyloglucosidase is added to certain dark beers after pasteurisation, to produce glucose to sweeten the beer. This practice is confined largely to the UK, where sugar added for sweetening is subject to excise (alcohol) tax, which is levied on the total dissolved solids added throughout the process. Thus, the unfermentable dextrins have been paid for in duty assessment and converting them to simpler, sweeter, sugars avoids additional payment for added sweeteners.

A long-standing problem with packaged lager beers is the formation of chill haze during low-temperature storage. Oxidative changes promote the formation of a protein–tannin complex which precipitates, to damage the appearance of the beer. This can be avoided in either of two ways: by partially digesting the protein with a proteolytic enzyme or by totally removing the small amount of oxygen present with glucose oxidase. The proteolytic method, if digestion proceeds too far, causes damage to foam formation and stability in the product. The most-used enzyme for this application is papain, which digests the beer proteins only partially, to yield fragments which retain their foam-promoting properties but are too small to yield insoluble complexes with tannin. The related protease bromelain, obtained from pineapples, is also used to some extent.

The use of glucose oxidase–catalase preparations to prevent haze by removing oxygen has been proved effective technically[52,61] but has not yet found favour commercially. However, oxygen in beer also has deleterious effects on flavour, the lager types brewed in the USA being particularly sensitive even to levels of oxygen as low as 0.1 ppm. The use of glucose oxidase preparations in the beer at the final filtration stage and in the final product is a promising method for ensuring and maintaining a low oxygen content. Beer normally contains sufficient glucose to react with the levels of oxygen present, but, if necessary, glucose can be added or amyloglucosidase can be used, to produce glucose from dextrins. Most of the work to date with glucose oxidase in beer has involved addition of the soluble enzyme to the batch. Modern brewing practice is well suited to the application of immobilised enzyme reactors. An immobilised enzyme preparation and reactor designed specially for brewing use have been successfully tested on several brands of beer[27]. This reactor contains glucose oxidase and catalase, to reduce the dissolved oxygen content of the beer to negligible levels. Amyloglucosidase can be incorporated into the system to supply additional glucose for the glucose oxidase reaction or the system can be adapted to convert more glucose for sweetening purposes by using amyloglucosidase on its own.

Dairy Products

The formation of cheese curd is initiated by a proteolytic enzyme reaction. Proteolytic action of added rennets splits the κ-casein fraction of the milk protein in a highly specific reaction, to yield a modified protein with a high capacity for aggregation under the mediation of calcium ions. The protein coagulates, also trapping the fat globules. When the curd so formed is cut, the liquid whey, containing salts, lactose and a small proportion of the soluble proteins, drains. The curd is worked into shape and allowed to mature into a finished cheese. Calf stomach is the traditional source of rennet and still accounts for a large proportion of the world market. With modern dairy practice the supply of calves for slaughter has, however, become limited and alternative sources of acid proteases have been exploited to make rennets. Extracts from stomachs of adult mammals contain the enzyme pepsin, which is also effective in coagulating milk, and in the last 10–15 years increasing use has been made of bovine and pig pepsins, either alone or in mixtures with calf rennet. The shortage of calf rennet also stimulated a vigorous search for microbial sources which could be used to produce rennets by fermentation. Enzymes from the fungi *Mucor pusillus* and *Mucor meihei* have become established as substitutes for the mammalian enzymes and now dominate the American markets.

While the clotting action of rennet is proteolytic, too much proteolysis leads to flavour defects in the cheese, which may become excessively bitter, or there may be loss of consistency and liquefaction. However, the rennets are also believed to contribute to the development of the characteristic desirable flavours of cheese by continuing to catalyse limited protein breakdown during maturing[66]. Most proteases are unsuitable for cheese-making because of excessive general proteolytic activity, but some may be insufficiently active to give optimum flavour in long-maturing cheeses.

During separation some of the rennet is lost in the whey, while a proportion remains in the cheese. With increasing use of whey or treated whey as a food ingredient, a disadvantage of the *Mucor meihei* enzyme has been discovered. This enzyme is more thermostable than calf rennin and is not inactivated by pasteurisation. As a result it may remain active in foods containing whey or whey protein concentrate. These include baby food formulations based on whole milk, and instances have occurred where these coagulate on mixing. This problem has been circumvented by marketing of enzymes which have been chemically modified to reduce their stability to pasteurisation[45,46,50,51]

The fungal rennets have slightly different properties from those of the calf enzyme, and some sectors of the market have proved resistant to changing from the traditional source. A new approach to the shortage of calf stomachs has become possible with the development of gene manipulation techniques — that is, to clone the calf rennin gene in micro-organisms to produce the enzyme by fermentation. Several research groups have worked on this and at least one patent application has been published[21].

As mentioned above, enzymes are involved in the development of cheese flavour. In traditional manufacture of Italian cheeses the method of preparing the rennet ensured that certain lipases originating in pregastric glands of the animals were present in the rennets and these gave characteristic flavours to the cheeses by selectively releasing certain fatty acids from the milk lipids. These lipases are now manufactured and sold separately for use with other rennets to give the characteristic Italian cheese flavours.

For long-maturing cheeses such as Cheddar, proteinases and peptidases from the starter culture bacteria, the rennet, and probably from the milk are involved in flavour development[34]. Lipolytic activities may also be involved. The maturing process for these cheeses is costly and reduces flexibility of supply. This has stimulated work on the acceleration of flavour development by using additional enzymes. Promising results have been obtained[35] but it is not yet clear whether this application will become important commercially.

Many cheeses, such as Gouda, Emmental and Edam, are ripened by a relatively short high-temperature storage which involves growth of propionic acid bacteria. If the milk is contaminated with butyric acid bacteria, these also proliferate during maturation and cause off-flavours and excessive gas production. On occasions these bacteria have caused cheeses to explode. The glycosidase lysozyme is a natural bacteriocidal enzyme which attacks cell walls of some bacteria, including the butyric acid type[68]. Commercial lysozyme, prepared from egg-white, does not attack the desirable species of propionic acid bacteria. Lysozyme is finding an increasing market for addition to cheese milk to prevent subsequent bacterial spoilage.

A lysozyme of similar specificity to that of the egg-white enzyme is present in human milk. This functions to restrict the range of bacteria in the intestines of young infants. If baby food formulations based on cows' milk are used, the intestinal flora is affected, because cows' milk contains very little lysozyme and the infants may be more susceptible to diarrhoea and poor nutrition. To avoid this, egg lysozyme is added to some baby milk powder formulations.

An important development for the dairy industry is the availability of the enzyme lactase, which hydrolyses the sugar lactose, present in milk, to its two component monosaccharides, glucose and galactose. The monosaccharides are sweeter, more soluble and more readily fermentable than lactose. Two types of lactases are commercially available. The yeast enzymes (*Kluyveromyces* spp.) are suited to use in milk or skim milk without pH or other adjustments, but are relatively unstable, having a practical temperature range for use of 0–40 °C. The mould enzymes (*Aspergillus* spp.) are stable up to 50–60 °C and have more acid pH optima. They can be used in cheese whey without pH adjustment. Both types of enzymes have also been produced in immobilised form for use in reactors through which milk or whey can be fed for conversion[14,43].

The use of lactase in milk and milk products has three principal advantages. Because of the relatively low solubility of lactose, concentrates of milk or whey can only reach around 30 per cent solids before lactose begins to crystallise. At this solids level, the concentrates are still microbiologically unstable. Use of lactase allows the materials to be concentrated further. The hydrolysis also causes significant sweetening, giving a saving on added sweeteners if the milk is to be used in sweetened products. Lactose is not always readily metabolised by mammals. Genetic deficiencies may occur in lactose hydrolysis or assimilation, and these can be overcome by ensuring that the lactose is hydrolysed before feeding. This is useful in both human and animal nutrition. Many individuals in Third World populations are capable of hydrolysing lactose as infants but lose their intestinal lactase in childhood following weaning. Lactase is of value in treating Western milk surpluses intended for supply to the Third World in food aid programmes.

By prevention of lactose crystallisation, lactase has been shown to make it possible to obtain high-quality ice-cream at lower cost by increasing the whey powder content of the mix[24]. Use in yoghurt can lead to a product of improved quality[25].

Fruit Juice and Wine

Yields of juice from fruits and vegetables can be increased by using pectolytic enzymes (pectinases)[53]. This is an old-established use of enzymes and a large number of products are available on the market. Pectin is a polymer of galacturonic acid with methyl substituents on some of the hydroxyl groups of the galacturonic acid residues. It functions in plants to bind the cells together into a tissue. The main mechanism for degradation of pectin involves splitting off the methyl groups by pectin methylesterase, followed by hydrolysis of glycosidic links by polygalacturonases. As with degradation of starch by amylases, there are two types of glycosidases attacking pectin: some attack randomly along the chain (endopolygalacturonases); others split single residues from chain ends (exopolygalacturonases). In a separate mechanism pectin lyases (pectin transeliminases) split the pectin chain by elimination, leaving a double bond. Commercial pectinase enzymes usually contain all four of the above types of enzymes[29].

Pectinases are almost universally used as an aid to extraction of fruit juice. Following pulping of the fruit, the pectinase is mixed in and the mixture is held at a warm temperature for up to a few hours. The viscosity of the pulp at first increases, owing to solubilisation of insoluble pectin, then starts to decrease; by the time the original viscosity of the pulp has been restored, the pulp can be pressed and will give a higher juice yield than untreated pulp. Pectinases are also used to reduce viscosity of fruit and vegetable purées and pulps which are to be used directly. The pressed juices may be turbid and very viscous, owing to the presence of partly degraded soluble pectins; in such cases pectinases can be added to the juice to yield a clear liquid of moderate viscosity.

In practice vegetable material contains other structural polysaccharides and proteins which contribute to viscosity of the pulps and turbidity in the juices. Pectinase preparations are usually manufactured by semi-solid fermentation processes with moulds on a complex medium containing residues from crop plants. Under these conditions a wide variety of other hydrolytic enzymes is produced, and these persist in the product. In many cases these side-activities contribute to the effectiveness of the pectinase treatment. Because of the complexity of both the enzyme and the material being treated, it is extremely difficult to analyse all the mechanisms involved, but manufacturers frequently top up the side-activities by adding other enzymes to the pectinase. The most common additives are amylases, proteases and cellulases. Amylases are needed when certain varieties of apples are used, especially if they are immature, since their juices contain starch which precipitates in the juice. Cellulases, alone or with pectinase, are particularly effective for liquefying whole fruits to be sold as 'nectars' or purées, and can help to achieve high juice yields when sophisticated mechanical equipment is not available[62].

Grapes to be used for wine production are often infected with *Botrytis*, a mould which produces a β,1,3 glucan polymer[16] which causes problems in clarification and filtration. β,1,3 Glucanases have been shown to be effective in eliminating these problems[18,69].

The colour and flavour of wines and soft drinks containing natural colours or oils can be affected by oxidative deterioration. Oxygen levels in bottles and other containers are usually very low but can be reduced by using inert gas purges during filling. However, the small amount of dissolved and headspace oxygen which remains can be sufficient to cause colour and flavour changes. A glucose oxidase–catalase enzyme preparation is effective in removing oxygen.

Baking

Action of amylases and proteases is needed in baking to ensure the desired qualities in the product. Amylases produce sugars from starch in the flour, and it is the fermentation of these sugars which causes the bread to rise. The sugars also give the characteristic colour to baked products by participating in a Maillard browning reaction with amino acids released by proteases. The proteins of flours (glutens) determine the physical properties of the doughs and batters and products made from them. A controlled degree of breakdown of the gluten by proteases is needed to give products of even quality.

The natural levels of amylases and proteases present in flours is variable, according to the variety of wheat, the climate in which it is grown and the conditions of storage and transport. Both these enzymes may be supplemented by adding preparations from microbial sources[57]. An optimum level of amylase exists; excess amylase may cause poor loaf volume and sticky texture. Hard wheats imported from North America are low in α-amylase and it is normal for the millers to supplement this by adding fungal α-amylase. Further additions may be made at the baking stage, depending on the type of bread process being used. It is essential that amylase action be stopped at the last stages of baking, since, otherwise, continued action after baking causes spoiling of the bread. Any added amylase must, therefore, be susceptible to thermal inactivation at baking temperatures. Fungal α-amylases from *Aspergillus* are similar in thermal stability to the cereal α-amylases and are used in preference to bacterial α-amylases, which are too stable.

In the manufacture of biscuits, a 'soft' flour is needed to allow working or pumping of the dough or batter. If the gluten content of the flour is high, giving a 'hard' highly cross-linked structure to the flour, proteases are added to break up the structure and soften the dough. This is rarely necessary in the UK, where the indigenous flour is soft, but is widely practised in Europe. In North America the flours are very hard, and protease is added to bread flour as well as biscuit flours. The proteases used must, like the amylases, be susceptible to inactivation during baking. *Aspergillus* proteases and *Bacillus* neutral protease have the required properties.

Other Food Uses

Invertase has a long-established use in manufacture of soft-centre chocolates. Mixtures of filling materials containing invertase and sucrose are made with the correct moisture content to allow working and shaping. The prepared centres are enrobed and the breakdown of the sucrose to yield more soluble invert sugar causes the centre to liquefy.

The enzyme raffinase (α-galactosidase) is used in recovery of sugar from beet[37]. Raffinose is a trisaccharide comprising a sucrose molecule substituted in the 4 position of the glucose moiety with galactose. α-Galactosidases hydrolyse it to yield sucrose plus galactose. In the sugar-refining process raffinose interferes with precipitation to reduce sucrose yields. Use of raffinase at an appropriate stage of the process yields additional sucrose from raffinose and also alleviates precipitation losses. Enzyme preparations for this use must, of course, be essentially free of invertase. Since yields of the enzyme are relatively low, crude immobilised preparations have been developed to allow a higher enzyme productivity in use. The content of raffinose in beet is variable and the use of raffinase is confined to beet-growing areas where the soil and climate favour raffinose formation; the most important is Colorado, USA.

Powdered whole or separated egg is subject to colour and flavour deterioration caused by the Maillard reaction between the 0.1–0.5 per cent glucose present and the egg proteins. The same glucose oxidase–catalase preparations used for removing oxygen from packages can be employed to remove the glucose from egg. This has been practised for about 30 years. In egg albumen (dried separated egg-white) the Maillard reaction could adversely affect the essential functional properties of the product (foam volume and stability), and nearly all albumen is treated. The enzyme is added to the bulk liquid egg and hydrogen peroxide is added continuously or intermittently; catalase from the enzyme preparation generates oxygen from H_2O_2 and this is used by the glucose oxidase in a reaction with glucose to yield gluconic acid. Similar processes can be used for reducing glucose levels in other materials for storage, such as potato powders or coconut, but these have not been commercialised.

Development of tenderness in meat is due to the action of endogenous proteolytic enzymes. This action can be augmented by adding protease[15]. The plant sulphhydryl proteases are most used for this purpose, since they have properties which suit the conditions during cooking. Papain, from papaya fruit, is the most popular tenderising enzyme. Bromelains, from pineapple fruits or stems, and ficin, from figs, can also be used. Methods for ante-mortem or post-mortem addition of tenderisers have been developed. They are also offered for retail sale and can be sprinkled on meat just prior to cooking. Although existing methods are effective, the ideal enzyme for meat tenderising has yet to be discovered. Such an enzyme would attack the heavily cross-linked regions of collagen molecules but leave the helical regions substantially intact, since it is the cross-linked regions which cause toughness in meat; however, if the helical regions are significantly attacked, the texture of the meat is adversely affected.

Paper, Textiles and Leather

Starch is extensively used in the paper industry to improve the strength and stiffness of the products. Native starch is, however, too viscous to be applied to paper surfaces, and some degradation is necessary. α-Amylases are used to modify the starch for this purpose.

Starch is the most commonly used 'size' in the textile process, where it protects the warp against breakage during weaving. The size must subsequently be removed to allow efficient dyeing, bleaching or texture modification; this is achieved by passing the cloth through a bath containing bacterial α-amylase. This is one of the oldest-established industrial uses of enzymes. Originally malt extracts containing cereal amylase were used; these were replaced by animal pancreatic enzymes and then by bacterial enzymes, which are cheaper and more thermally stable.

Modification of animal skins during the process of leather manufacture uses enzymes in several ways[23]. Water uptake in the soaking stage is promoted by the use of bacterial alkaline proteases and these enzymes are also used to aid dehairing of hides. At the bating stage, during which the hide is softened by partial degradation of the skin fibres, the traditional dog faeces have been replaced by mixtures containing enzymes. Enzymes from different sources or blends are used, depending on the type of skin or type of leather being produced. Soft leathers use 'strong' bates of relatively high enzyme activity. Rigid leathers use 'weak' bates with low enzyme activity. The main activity involved is that of protease. Pancreatic preparations containing trypsin and chymotrypsin are most commonly used; lipase activity may also contribute to the effect of the bate. Bacterial neutral and alkaline proteases and fungal proteases can replace pancreatic enzymes.

Analytical Uses

Over the past thirty or so years enzymes have been used in an increasingly large number of analytical methods[6]. The principal attraction in the use of enzymes for analytical purposes is their very high specificity of action; thus, an enzyme can be used to determine concentration of a specific isomer in a mixture containing several molecules of very closely related structure. Most of the enzymes used in analysis, unlike the industrial enzymes used for bulk conversions, are highly purified, to eliminate other enzymes present in their biological source which might interfere in the assays in which they are used. The vast majority of enzymatic assays have only relatively small-scale uses. The enzymes required may be produced in the necessary quantities within the laboratory where they are used, or they may be made on the laboratory scale by chemical suppliers and sold at a relatively high price. However, the enzymes used in some clinical screening tests are required in relatively large quantities and are manufactured on an industrial scale.

Most of this market is in test kits, sometimes sold alongside a particular instrument for which they are tailored. The four tests which account for most of the market are listed in Table 6. The market in diagnostic test kits is of the order of £200 million but only about 10 per cent of this represents enzyme costs. Various alternative procedures exist for each of these tests: for example, glucose can be assayed using a glucose oxidase–peroxidase coupled reaction, by measuring oxygen uptake by glucose oxidase, by a coupled reaction with hexokinase and glucose-6-phosphate dehydrogenase or with glucose dehydrogenase. Glucose oxidase methods are the most popular, and this is the largest-volume enzyme used in the diagnostic field.

The market for enzyme diagnostic kits grew very rapidly throughout the 1970s, in line with a general increase in spending on health care in the

Table 6 Enzymatic blood analyses

Test	*Enzymes*
Glucose (diabetes)	Glucose oxidase Glucose dehydrogenase Glucose-6-phosphate dehydrogenase Hexokinase Peroxidase Catalase
Urea (BUN)	Urease
Cholestrol	Cholesterol oxidase Peroxidase Cholesterol esterase
Triglycerides (blood lipid)	Glycerol dehydrogenase Glycerokinase Glycerol phosphate dehydrogenase Lipase Pyruvate kinase Lactate dehydrogenase

developed world. This growth has been checked more recently, owing to economic pressures and the introduction of discrete sample analysers which use much less reagent than the first-generation flow-through automatic analysers. It seems probable that the market may decline in future.

Medical and Drug Uses

Enzyme preparations have been offered for sale to the public as digestive aids for many years. It is, however, doubtful that these have any value except in rare cases of specific enzyme deficiency. In the USA the FDA now permits such preparations to be sold but does not permit them to be advertised, because no satisfactory proof of their effectiveness has been forthcoming. Originally digestive aids were powdered animal pancreatic extracts containing mainly amylase, lipase and protease activities, but these have been supplemented or replaced by microbial enzymes, and mould cellulases have been added to some formulations.

Cleaning of wounds and sores using proteolytic enzymes has been practised for some time. It has even been reported[64] that enzyme administered orally has a generalised anti-inflammatory effect. Enzymes are also used for treatment of blood clots by local or generalised application. These are special proteases which have high activity on the blood-clotting protein fibrin. Various microbiological sources are known but the most-used enzyme is urokinase, purified from human urine.

Intravenously injected enzymes as anticancer agents have had limited success but at present are not used. Cancer cells require some nutrients, such as asparagine and glutamine, which are synthesised by healthy mammalian cells but not by tumour cells. If these substances can be removed from the bloodstream by injection of appropriate enzymes, the cancer cells are starved. The use of injected asparaginase or glutaminase does ameliorate leukaemia and cancers of the lymph glands. However, these treatments never effect a complete cure, because the cancer cells readapt to synthesise these metabolites, while the enzymes provoke an increasingly strong immunological response from the body and are more and more rapidly destroyed in the blood as treatment continues.

In the field of drug manufacture penicillin acylase is used to remove the side chain from penicillin G to produce 6-amino penicillanic acid. This is used as an intermediate for manufacture of semisynthetic penicillins by introducing new substituents on the amino group, e.g. by condensation with phenyl glycine to yield ampicillin (α-aminobenzylpenicillin). The enzyme is usually made in-house by the antibiotic manufacturers and is used in the form of a crude immobilised whole-cell preparation[7].

OTHER APPLICATIONS

The hydrolytic enzyme lipase is unusual in that it normally acts at the surfaces of oil droplets in aqueous emulsion, remaining in solution in the aqueous phase and binding molecules present in the lipid phase. The normal mode of action is hydrolytic, so that the fat molecules, glycerol esters, are converted to fatty acids and glycerol, usually by a sequential hydrolysis of the three ester bonds in a triglyceride. In conditions of restricted availability of water the lipase increasingly catalyses a transfer reaction to hydroxyl groups in the lipid phase, and this can be utilised to exchange fatty acids in natural oils to improve their properties and, hence, their commercial value[41]. The enzyme is absorbed on a solid inert support containing a controlled amount of moisture, and this can be packed into a column through which the substrate stream containing a mixture of fatty acid and natural glyceride oil can flow.

An example where the unique capacity of enzymes to act solely on a particular stereoisomer is exploited commercially is the production of amino acids for pharmaceutical or animal feed uses[12]. Mammalian metabolism only utilises the α-L-amino acids but chemical synthetic processes normally produce a racemic mixture with 50 per cent each of α-D and α-L. The acetyl ester of the DL-amino acid racemate is prepared chemically and passed through a column containing an immobilised amino-acylase which hydrolyses the L-amino acid ester but not the D-amino acid ester. The free L-amino acid can be crystallised from the mixture and the acetyl D-amino acid fed back into the start of the process. Alanine, methionine, phenylalanine, tryptophan and valine have all been prepared successfully on an industrial scale using this method.

THE FUTURE OF ENZYMES AS INDUSTRIAL PRODUCTS

Most of the existing markets for enzymes seem reasonably secure for the forseeable future and many promise a healthy degree of expansion. The history of the enzyme industry in the last few years is one of strong technical competition, with the result that costs of enzymes in real terms have declined. This process seems likely to continue, and the reduced costs, as in the past, should make it possible for existing enzyme products to become economical for use in a stream of new uses which will compensate for possible reduced money turnover in existing applications as the enzymes become cheaper.

Of the huge variety of enzymes which exist in nature, only relatively few have been exploited commercially. It is not surprising that the majority of these are the hydrolytic digestive type of enzymes. Since these are adapted to operate under conditions outside the controlled environment of the cell, they are commonly much more stable and robust than are other enzymes. A different class of enzymes which already have some applications and whose use seems likely to expand is the oxidoreductases. For example, a wide variety of uses have been postulated for hydrogenases[30]. The most immediately interesting of these is perhaps the separation of hydrogen isotopes, including purification of waste waters containing tritium from nuclear plants[31]. A further possible application which ultimately could become very important is the possible use of chloroplasts together with hydrogenase for the direct production of hydrogen from water by photolysis with solar energy. This system has been shown to function, but so far at impractically low efficiencies. However, steady progress in improving it is being made[30]. When petroleum reserves ultimately become depleted, their role as fuel might be taken over by hydrogen produced in solar reactors using chloroplasts and hydrogenase.

It was widely predicted that enzyme technology would be harnessed to synthesise complex chemicals using the reactions which occur *in vivo*. Thus, the goal of separating anabolic enzymes and immobilising them together or sequentially in reactors to catalyse multistep reactions was adopted. In practice this goal has proved elusive because of several difficulties. Biosynthesis usually is linked to the energy-producing mechanisms of the cell, which are often highly organised spatially and cannot yet be reassembled artificially once the cell is disrupted. Synthetic enzymes are frequently quite unstable, and only limited capability exists to modify enzymes artificially to improve stability. The synthetic reactions nearly always involve small reactive molecules, the coenzymes, which are very often even less stable than the enzymes. Coenzymes function as catalytic intermediates and economics dictate that they must be re-used in the system. Re-use requires that they also must be immobilised, and this has proved difficult, although practical laboratory methods have been described[38]. Finally, the kinetics of complex reactions involving several enzymes and one or more coenzymes tend to dictate extremely slow rates in immobilised reactors, where diffusion of substrates and intermediates severely limits overall rates. It now seems doubtful that the immobilised enzyme reactor will be used in more than a few cases for synthesis.

Attention has now switched to the use of immobilised whole cells, which can be viable and have their energy-producing systems intact. It is arguable whether these systems represent a development of enzyme technology or of fermentation technology.

The development of recombinant DNA technology offers the possibility of increasing yields in enzyme fermentation processes by multiplying gene copies and of transferring genes from difficult to more tractable organisms. The cloning of the calf rennin gene in *E. coli* has already been mentioned[21]. As recently as 1979 Aunstrup[2] gave the authoritative opinion that it would be a long time before practical application of 'genetic engineering' to enzyme technology would occur. It is clear that there has been much progress since then. Long-established enzymes, whose manufacture processes have been very heavily optimised, are less likely to be subjects for successful application of these techniques, but with continuing improvement in methods it seems probable that gene multiplication and transfer will become normal procedures for application to new enzymes.

An exciting development in the same field is the advent of 'enzyme engineering'. Using DNA manipulation and transfer techniques, it has become possible to induce an organism to synthesise a protein whose structure is altered from the natural form in a predetermined way. It is anticipated that the results of modifying enzyme primary structure and determining the effect on higher orders of structure and on properties will lead rapidly to a greatly increased understanding of the rules governing structure–function relationships for enzymes[57]. This could offer the possibility ultimately to design molecules to catalyse any desired reaction or to have specific desired properties, to synthesise the appropriate DNA, and to clone the resulting gene in a stock enzyme-producing organism. The ultimate realisation of this rather awesome objective may be a long way off, but minor modifications to improve properties of existing enzymes for industrial use should soon be within reach.

Objectives in enzyme engineering which would be commercially useful have broad scope, but the science will have to develop further before realistic probabilities of success can be assessed. It may prove possible to stabilise many naturally unstable intracellular enzymes so that they are stable enough to be isolated and used. If at the same time they could be modified so that coenzymes were bound in the structure tightly or to use alternative cheaper or more stable coenzymes, the use of enzyme reactors for complex industrial chemical synthesis might become possible. In other cases modification to cause inactivation at specific useful temperatures could clearly be valuable. Reduction in sensitivity to troublesome inhibitors or denaturing agents in the intended process streams could also be valuable.

With the advent of the new techniques of genetic and enzyme engineering, as well as the more mundane improvements to current technology which are continuing, it seems certain that more and more applications for enzymes in industry will be found. With the already healthy state of markets in existing applications, it seems certain that enzymes will assume increasing commercial importance in the future.

REFERENCES

1. Anon. (1971). The Ad Hoc Committee on Enzyme Detergents: *Report PB 204 118*, Division of Medical Sciences, National Academy of Sciences – National Research Council, Enzyme-containing laundering compounds and consumer health
2. Aunstrup, K. (1980). *Proc. Microbial Ind. Alim. Conf. Int. (1979)* 77–96, Micro-organisms as producers of enzymes
3. Balls, A. K. (1941). U.S. Dept. Agriculture, Circular No. 631, Protein-digesting enzymes of papaya and pineapple, Washington DC
4. Barretts De Menezes, T. J. (1982). *Process Biochem.*, **17**(3), 32–35, Starchy materials for alcohol fuel production
5. Bender, H. and Wallenfels, K. (1961). *Biochem. Z.*, **334**, 79–95, Pullulan II specific decomposition by a bacterial enzyme
6. Bergmeyer, H. U. (Ed.) (1983). *Methods of Enzymatic Analysis,* 3rd edn, 10 vols, Verlag Chemie, Weinheim
7. Brodelius, P. (1978). In *Advances in Biochemical Engineering* (Eds Ghose, T. K., Fiechter, A. and Blakeborough, N.), Springer Verlag, **10**, 76–129, Industrial applications of immobilised biocatalysts
8. Buchholz, K. (Ed.) (1979). *Characterisation of Immobilised Biocatalysts,* Dechema Monographs, Vol.84, Verlag Chemie
9. Bucke, C. (1981). In *Enzymes and Food Processing* (Eds Birch, G. G., Blakeborough, N. and Parker, K. J.), Applied Science, London, 51–72, Enzymes in fructose manufacture
10. Bucke, C. (1983). In *Microbial Enzymes and Biotechnology* (Ed. Fogarty, W.M.), Applied Science, London, 93–129, Glucose-transforming enzymes
11. Chibata, I. (Ed.) (1978). *Immobilised Enzymes, Research and Development*, Halsted Press, NY
12. Chibata, I. (1981). *Immobilised Enzymes. Research and Development,* Wiley, Section 4.2, 163–206, Chemical processes using immobilised enzymes

13. Dale, J. K. and Langlois, D. P. (1940). U.S. Patent 2 201 609, Syrup and method of making same
14. Dohan, L. A., Baret, J. L., Pain, S. and De Delala, P. (1980) *Enzyme Engng*, **5**, Proceedings of the International Enzyme Engineering Conference of New Hampshire, 1979, Plenum Press, 279–294, Lactose hydrolysis by immobilised lactase. Semi-industrial experience
15. Dransfield, E. and Etherington, D. (1981). In *Enzymes and Food Processing* (Eds Birch, G. G., Blakeborough, N. and Parker, K. J.), Applied Science, London, 177–194, Enzymes in the tenderisation of meat
16. Dubourdieu, D., Ribereau-Gayon, P. and Fournet, B. (1981). *Carbohydrate Res.*, **93**, 294–299, Structure of the extracellular β-D-glucan from *Botrytis cinerea*
17. Dunnill, P. (1983). *Process Biochemistry,* **18**, 9–13, Trends in downstream processing of proteins and enzymes
18. Felix, R. and Villettaz, J. C. (1983). In *Industrial Enzymology* (Eds Godfrey, T. and Reichelt, J.), Macmillan, 410–421, Wine
19. Fenton, D. M. (1982). *Enzyme Microb. Technol.,* **4**, 229–232, Solvent treatment for β-D-galactosidase release from yeast cells
20. Follows, M., Hetherington, P. J., Dunnill, P. and Lilly, M. D. (1971). *Biotechnol. Bioengng.,* **13**, 549–560, Release of enzymes from bakers' yeast by disruption in an industrial homogeniser
21. U.K. Patent Application 2 123 005 A 13.04.83, Genex Corporation, Bovine calf chymosin
22. Godfrey, T. and Reichelt, J. R. (Eds.) (1983a). *Industrial Enzymology*, Macmillan, p.1
23. Godfrey, T. and Reichelt, J. R. (1983b). In *Industrial Enzymology* (Eds Godfrey, T. and Reichelt, J. R.), Macmillan, 321–330, Leather
24. Gregory, K. W. (1982). *Util. Enzymes Technol. Aliment.*, Symp. Int., 249–252 (Ed. Dupuy, P.), Tech. Doc. Lavoisier, Paris, The application of yeast lactase to the production of ice-cream
25. Guenther, E. and Buerger, E. (1982). *Util. Enzymes Technol. Aliment.*, Symp. Int., 243–248 (Ed. Dupuy, P.), Tech. Doc. Lavoisier, Paris, A new method of manufacturing lactose-hydrolysed yoghurt by means of β-galactosidase
26. Harada, T., Yokobayashi, K. and Misaki, A. (1968). *Appl. Microbiol.,* **16**, 1439–1444, Formation of isoamylase by Pseudomonas
27. Hartmeier, W. and Willox, I. C. (1981). *Technical Quarterly,* Master Brewers' Association of America, **18**, 145–149, Immobilised glucose oxidase and its use for oxygen removal from beer
28. I.U.B. (1979). *Enzyme Nomenclature 1978*, Academic Press
29. Kilara, A. (1982). *Process Biochem.,* **17** (4), 35–41, Enzymes and their uses in the processed apple industry. A review
30. Klibanov, A. M. (1983). *Process Biochemistry,* **18** (8), 13–16, Biotechnological potential of the enzyme hydrogenase
31. Klibanov, A. M. and Huber, J. (1981). *Biotechnol. Bioengng,* **23**, 1537–1551, Applications of immobilised hydrogenases for the detritiation of water
32. Kosaric N., Ng, D. C. M., Russell, I. and Stewart, G. S. (1980). In *Advances in Applied Microbiology,* **26**, 147–227, Ethanol production by fermentation: an alternative liquid fuel

33. Lambert, P. W. and Meers, J. L. (1983). *Phil. Trans. Roy. Soc. Lond. B,* **300**, 262–282, The production of industrial enzymes
34. Law, B. A. (1981). *Dairy Science Abstracts*, **43**, 143–153, The formation of aroma and flavour compounds in fermented dairy products
35. Law, B. A. and Wigmore, A. S. (1982). *J. Soc. Dairy Technol.,* **35**, 75–77, Microbial proteinases as agents for accelerated cheese ripening
36. Limon-Lason, J., Hoare, M., Orsborn, C. B., Doyle, D. J. and Dunnill, P. (1979). *Biotechnol. Bioengng.,* **21**, 745–774, Reactor properties of a high-speed bead mill for microbiological cell rupture
37. Linden, J. C. (1982). *Enzyme and Microb. Technol.,* **4**, 130–136, Immobilised α-galactosidase in the sugar beet industry
38. Lowe, C. R. (1981). *Topics in Enzyme and Fermentation Biotechnology* (Ed. Wiseman, A.), Wiley, **5**, 13–137, Immobilised co-enzymes
39. MacAllister, R. V., Wardrip, E. K. and Schnyder, B. J. (1975). In Reed, G. (Ed.) *Enzymes in Food Processing,* 2nd edn, Academic Press, 331–359, Modified starches, corn syrups containing glucose and fructose, and crystalline dextrose
40. Macey, A., Stowell, K. C. and White, H. B. (1967). *Proc. Eur. Brewing Conv.,* 283–294, Experimental brewing procedures using unmalted cereals and enzymes
41. MacRae, A. R. (1983). *J. Am. Oil Chem. Soc.,* **60**, 291–294, Lipase-catalysed interesterification of oils and fats
42. Madsen, G. B., Norman, B. E. and Slott, S. (1973). *Staerke,* **25**, 304–308, A new, heat stable bacterial amylase and its use in high temperature liquefaction
43. Marconi, W., Bartoli, F., Morisi, F. and Marani, A. (1980). *Enzyme Engineering,* **5**, Proceedings of the International Enzyme Engineering Conference of New Hampshire, 1979, Plenum Press, 269–278, Improved whey treatment by immobilised lactase
44. Masters, K. (1979). *Spray Drying Handbook,* George Goodwin, London
45. U.K. Patent Application 2 024 828 A 16.01.80, Miles Laboratories, Process for decreasing the thermal stability of microbial rennet
46. U.K. Patent Application 2 058 082 A 08.04.81, Miles Laboratories, Process for decreasing the thermal stability of microbial rennet
47. Nielsen, M. H., Jepsen, S. J. and Outtrup, H. (1981). *J. Am. Oil Chem. Soc.,* **58**, 644–649, Enzymes for low temperature washing
48. Norman, B. E. (1981a). In *Enzymes and Food Processing* (Eds Birch, G. G., Blakeborough, N. and Parker, K. J.), Applied Science, London, 15–50, New developments in starch syrup technology
49. Norman, B. E. (1981b). U.K. Patent Application 2 074 167, Saccharification of starch hydrolysates
50. U.K. Patent Application 2 038 339 A 23.07.80, Novo Industri A/S, Thermal destabilisation of microbial rennet by acylation thereof
51. U.K. Patent Application 2 045 773 A 05.11.80, Novo Industri A/S, A method of modifying the thermal destabilisation of microbial rennet, rennet so modified and a method of cheese making using rennet so modified
52. Ohlemeyer, D. W. (1957). *Food Technol.*, **11**, 503–507, Use of glucose oxidase to stabilise beer
53. Pilnik, W. (1982). *Util. Enzymes Technol. Aliment.,* Symp. Int.,425–450 (ed.

Dupuy, P.), Tech. Doc. Lavoisier, Paris, Enzymes in the beverage industry (fruit juices, nectar, wine, spirits and beer)
54. Prentice, N. (1976). *Technical Quarterly,* Master Brewers' Association of America, **13**, 91–101, Development of enzymes during malting and their function in mashing
55. Priest, F. G. (1983). In *Microbial Enzymes and Biotechnology* (Ed. Fogarty, W. M.), Applied Science Publishers, 319–366, Enzyme synthesis: regulation and process of secretion by micro-organisms
56. Radley, J. A. (1976). *Starch Production Technology*, Applied Science, 311–397, The manufacture of dextrose sweetening agents from starch and derived products
57. Rastetter, W. H. (1983). *Appl. Biochem. Biotech.*, **8**, 423–436, Review: enzyme engineering
58. Reed, G. (1966). *Enzymes in Food Processing,* 1st Ed, Academic Press, Chapter 12, 255–272, Starch, starch syrups and dextrose
59. Rehacek, J. and Schaefer, J. (1977). *Biotechnol. Bioengng.,* **19**, 1523–1534, Disintegration of micro-organisms in an industrial mill of novel design
60. Reichelt, J. R. (1983). In *Industrial Enzymology* (Eds Godfrey, J. and Reichelt, J. R.), Macmillan, 375–396, Starch
61. Reinke, H. G., Hoag, L. E. and Kincaid, C. M. (1963). *Am. Soc. Brew. Chem. Proc.,* 175, Effect of antioxidants and oxygen scavengers on the shelf-life of canned beer
62. Rombouts, F. M. and Pilnik, W. (1978). *Process Biochem.,* **13**(8), 9–13, Enzymes in fruit and vegetable juice technology
63. Rowe, T. W. G. (1960). *Annals of the New York Academy of Science,* **85**, 641–681, The theory and practice of freeze-drying
64. Sizer, I. W. (1972). In *Advances in Applied Microbiology* (Ed. Perlman, D.), **15**, 1–11, Medical applications of microbial enzymes
65. Ter Haseborg, E. (1981). *Process Biochem.,* **16**(5), 16–19, Enzymes in flour and baking applications especially waffle batters
66. Visser, S. (1981). *Neth. Milk Dairy J.,* **35**, 65–88, Proteolytic enzymes and their action on milk proteins. A review
67. Wang, D. I. C., Cooney, C. L., Demain, A. L., Dunnill, P., Humphrey, A. E. and Lilly, M. D. (1979). *Fermentation and Enzyme Technology* (Ed. Heden, C. G.), 238–310, Wiley, Enzyme isolation
68. Wasserfall, F. and Teuber, M. (1979). *Applied and Environmental Microbiology,* **38**, 197–199, Action of egg white lysozyme on *Clostridium tyrobutyricum*
69. Wucherpfennig, K. and Dietrich, H. (1982). *Die Weinwirtschaft,* **23**, 598–603, Verbesserung der Filtrierfahigkeit von Weinen durch enzymatischen Abbau von Kohlenhydrathaltigen Kolloiden
70. Wuthrich, B. and Ott, F. (1969). *Schweiz Med. Wochenschrift,* **99**, 1584–1586, Berufsasthma durch Proteasen in der Waschmittelindustrie

16
THE SECOND GENERATION OF COMMERCIAL ENZYMES

Jane Llewelyn

INTRODUCTION

This chapter briefly reviews the enzyme industry and its growth to a multi-million dollar business based, in just over 50 years, on the sales of five major enzymes. Then it examines the potential scope for new enzymes, and provides illustrations of the current scope available for improving existing enzyme systems and developing new opportunities in waste treatment, health care and fine chemicals.

In the next 20 years enzymes produced by biotechnology could revolutionise industrial practice by enabling ambient temperatures and low-pressure conversions in the manufacture of raw chemicals, drugs and foodstuffs. Even our energy requirements and waste treatment systems could be managed by the judicious design of 'enzyme-like' molecules. Industrial landscapes could ideally be restored to non-polluted greenery-planted parks. Green surroundings are known to be relaxing and therefore help to reduce stress. Since we are entering the era of stress-related diseases, this would benefit local inhabitants and workers alike. In the meantime, there are plenty of opportunities for improving the current range of enzymes and the conversion processes concerned.

GROWTH OF THE CURRENT INDUSTRY

Of the 2000 or so enzymes categorised, only 18 or so are commercially available on a large scale. Many of these date back to the 1930s and are mostly hydrolases. In 1926, Sumner had isolated urease, identifying it as proteinaceous, not microbial or just a chemical. By 1940 Takamine was developing microbial enzymes, notably diastases. Commercially, *Bacillus* proteases used in detergents have been one of the most successful microbial enzymes. In under 50 years the commercialisation of these biological catalysts has led to a world-wide business worth a conservative $350 million.

There are only five major commercial enzyme manufacturers: Novo (Danish); Gist Brocades (Belgian); Rohm (German); Miles (American); and Hansens (Dutch). Novo dominate the market with a 41 per cent share, followed by Gist with just over 20 per cent.

Industrial enzymes are currently produced in bulk quantities and can be considered to be commodities. The commercial enzymes business is still dependent on five enzymes (see Table 1).

Table 1 Tonnages and end use of the five major commercial enzymes

Amlyases	10 000 t/a	Starch industry
Proteases (bacterial)	500 t/a	Detergents
Papain	100 t/a	Dehazing beer, tenderising meat
Glucose isomerase	70 t/a	Corn syrup production
Rennin	25 t/a	Dairy industry

MAJOR DEVELOPMENTS SINCE THE 1940s

In 40 years the industry has witnessed two major developments: bacterial proteases in detergents and immobilised glucose isomerase for corn syrup production. The use of enzymes in detergents has been beset with allergy problems; this led to bio-detergents being removed from the market and reformulated in the 1960s. Recently, renewed interest in the problem of detergent-related allergies has been voiced in the scientific press.

Immobilised Enzymes

Until recently, enzymes had limited applications as catalysts in industrial reactors, because of their sensitivity to high temperature and organic solvents. The advent of immobilised enzymes has now permitted their use in non-aqueous as well as aqueous phases. By attachment to inert supports enzymes can be recycled and kept separate from the end-product. The first examples of this were carboxypeptidase, amylase and ribonuclease. Grubhofer and Scleith first bound them to polyamide resin. Although immobilised enzymes can operate for long periods with organic solvents, some damage to the supports and the enzymes still occurs. So far, immobilised glucose isomerase is the only immobilised enzyme to become established on an industrial scale. The economics of corn syrup production are such that it can effectively compete with invert sugar from cane or beet. Lactase and cellulase are two other strong candidates for immobilisation. Dow and the Milk Marketing Board are said to be developing an immobilised lactase.

Immobilised Cells

A logical development is the immobilisation of whole cells in polymer lattices. This overcomes the problems of multi-enzyme systems. The combination of one or more enzyme systems — for example, malate dehydrogenase–citrate synthase (to mimic mitochondria) and amyloglucosidase–glucose isomerase for corn syrup production to convert starch to fructose in one stage — is a predictable future development. In diagnostic kits the coupled enzyme system glucose oxidase–peroxidase is used in the reagent for the determination of blood glucose.

Improved Detergents

Novo have shown no signs of expanding the enzyme product range. Development work at Novo is centred on improving the existing range. For example, published patents suggest that Novo are developing lipases for detergent use. Lipases are useful for dissolving grease spots on shirts of the cotton–polyester type. They are still not available in sufficient quantities at economic prices. Strain selection for yield enhancement, temperature and pH optimisation dominates the industry's efforts. Nevertheless, enzymes continue to show healthy growth. Novo are expanding plant capacity in the USA and Japan. Miles, too, are expanding production capacity in the USA, where the biggest increased use in enzymes is predicted over the next few years.

Genetically Engineered Enzymes

Obviously there is little incentive for the large enzyme companies to look at producing new enzymes, or indeed to look at new production methods such as genetic engineering. The scarcity of calf chymosin, however, coupled with the inferior quality claimed for the microbial rennets, led to a flurry of activity among biotechnology companies and others to clone chymosin. Dow Corning and Genentech, collaborating as Genencor, Celltech and Unilever were all favourites in the race. Gradually various problems (patent rights, clearance from the regulatory authorities, technical feasibility and market demand) have impeded the success of this pioneering work.

In the food and drink industry there are many options for genetically engineered enzymes. Early cloning work in *E. coli* will need to be transferred to more acceptable host models such as food yeasts before the industry at large will find the genetically engineered enzyme attractive. The regulatory authorities are most concerned about stability of strain and toxic residues. The present clearance system is based largely on the industries' own submissions containing confidential details about production methods, quality controls and strain purity. Much of the safety data for traditional enzymes is anecdotal, based on their history of safe use. Therefore, the most likely source of new developments derives from the enzyme users who can best identify their current and future requirements.

GROWTH OF THE INDUSTRIAL ENZYME INDUSTRY

There is plenty of scope to develop the present range of industrial enzymes. Enzymes are required in a purer form to reduce bulk and at a cost comparable to that of current sources. Many are required to be more or less thermotolerant and to have the pH specificity shifted. Enzymes are required which are not inhibited by metal ions. The food industry would prefer enzymes expressed in yeasts instead of bacteria.

The brewing industry, for instance, would love to have brewing yeasts with their own complement of saccharifying enzyme activity. Guinness is sponsoring research at Trinity College, Dublin, to develop a yeast capable of hydrolysing β-glucans. This would reduce the filtration problems and the tendency for beer to become hazy on storage. The Brewing Research Foundation has isolated the gene coding for dextranase from a non-brewing yeast into a brewing yeast and increased its activity sixfold.

There is scope for shortening the time of cheese ripening, by modifying the neutrases responsible. Currently, mature Cheddar development takes up to a year. Miles is attempting to modify rennin so that it loses its heat resistance.

Other examples where enzymes could be made commercially viable for flavour development are:

Esterases, important in Italian cheese flavour.

Nariginase, which removes the bitter peel flavours, especially of grapefruit.

Phenolases, important in flavour production in cider, dates, tea and coffee; also responsible for browning in bananas and deepening the colour of dates and raisins.

There is also scope for producing expensive oils such as cocoa butter from commodity oils using enzyme systems.

In waste treatment economically viable processes using cellulase are not widely available. Environcon in Canada have recently opened a $1.5 million pilot process using *Chaetonium cellulolyticum*, which appears to break down lignocellulose, to convert pulp and paper wastes to animal feed. This process is based on a development by Waterloo University, Centre for Process Development, Canada, which has been researching this area for many years. MIT have patented a toxic waste treatment system based on peroxidase, which they claim can remove phenols, cresols, naphthol compounds and aniline from waste streams. Peroxidase acts by producing free radicals which diffuse into solution and club together with the toxic substances. This causes the formation of large molecules which precipitate out of solution, facilitating their removal.

All these opportunities rely on bulk supplies of low-value enzymes. In other product areas, such as fine chemicals and health care, there will be opportunities for small quantities of high-value enzymes. Information Research Ltd (USA) has predicted that by 1992 many of the inborn errors of metabolism could be cured by the induction of missing enzymes. Other enzymes such as streptokinase, staphylokinase and urokinase are being looked at for treating blood clots. Urokinase is currently extracted from urine — a difficult and expensive process. Potential exists for collagenase as a wound debrider and hyaluronidase for treating burns and arthritic joints. Recombinant urokinase, superoxide dismutase and digestive enzymes are all entering clinical trials with health care applications in mind. In Japan, researchers at Idemitsu Kosan Co. Ltd have cloned and expressed a cellulase gene from *Clostridium thermocellum* into *C. thermosaccharolyticum* to allow butanol production from cellulase.

In the future it may be possible to produce designer enzymes either by modifying natural enzymes or *de novo*. Workers at Chicago University have modified hydrolytic papain into an enzyme that catalyses oxidation reactions by attaching a flavin group to the molecule. A molecule based on cyclodextrin which has a hydrolysis rate comparable to that of chymotrypsin has been designed at Columbia University. So far, synthetic enzyme systems fail to match up to all three natural parameters: rate, specificity and selectivity. Ultimately, genetic engineering should enable modifications at the gene level to alter the characteristics of an enzyme to those desired.

BIOSENSOR ENZYMES

Biosensors based on enzymes are a promising new development. The University of Pittsburgh Medical School are developing a potentiometric glucose electrode. This reliable glucose sensor is aimed at diabetics, to constantly monitor blood glucose concentrations and regulate insulin flow as required. The monitoring enzyme glucose oxidase is immobilised onto the ion-selective electrode in a cellulose acetate membrane on the platinum electrode. The resulting electric current is proportional to the amount of glucose oxidised by the enzyme. There is scope here for improving glucose oxidase, because its present stability is variable, ranging from a few hours to several months. This is thought to be related to the enzyme's heat stability at body temperatures. Once the technical hurdles have been overcome, several years of clinical testing will be required to meet regulatory approval for such an implantable device.

Many commercial applications will call for biosensors to analyse up to 12 variables at once. The most obvious applications are to monitor reactions inside the human body, where the minute size is a real advantage, but other 'real-time' measurements are needed in industrial fermentations, such as flow rate, product formation and contamination. It will also be possible to design cofactor electrodes.

Further developments in the field of enzyme biosensors are enzyme thermistors, which are based on the heat of the reaction, which is a linear function of the concentration of the enzyme and the substrate. Likely applications include immunological assays (such as measurement of antigens), clinical analysis, industrial fermentation control and detection of hazardous materials. For instance, mercury has an inhibitory action on immobilised urease down to 0.2 ppb. Rhodanase reacts with cyanide at concentrations as low as 10^{-5}M. Another interesting development in this field is the use of enzyme alanine aminotransferase to detect sweet-tasting amino acids in foods and animal feeds instead of catfish alanine receptors. This type of example offers encouraging support to research efforts aimed at replacing the need for animals to be used in product testing.

ENFETs

Recently a great deal of attention has been focused on FETs (field effect transistors), based on semiconductor technology. Opportunities for enzyme FETs (ENFETs) may also be found: when a specific chemical is passed through the membrane, the enzymatic reaction releases free electrons which cause current to flow between source and drain. The miniature size makes them best suited to monitoring and treatment in medical applications. ENFET sensitivity (10^{-4}M) is low compared with chemical FETs (10^{-6}), so that the main applications for ENFETs probably lie in product formation detection rather than reaction monitoring.

CONCLUSION

The second generation of industrial enzymes will be dependent both on the success of the enabling technologies such as genetic engineering and on the development of processes designed to utilise the full range of enzymes. For instance, many enzymes depend on cofactors for activation. These are unchanged at the end of the reaction. There appear to be three groups: metal ions; vitamin derivatives, which are intimately conjugated to the enzyme and form part of the active site; and small organic molecules, which are readily dissociated from the enzyme.

Although a large number of micro-organisms may produce the same enzyme, there are relatively few species used for commercial production. There is enormous scope here for surveying the whole range of enzymes and categorising their activities, pH and temperature tolerance, the conversions they catalyse, inhibitors and cofactors (if any). If there is no suitable enzyme already available to do the job required, then one can begin to think in terms of designing, or modifying an existing enzyme to do the job.

BIBLIOGRAPHY

1. *Chemistry and Engineering* (1983). 15 August, 27
2. *Chemical and Engineering News* (1979). 9 April, 26
3. *The Commercial Exploitation of Emerging Biotechnical Processes in Europe Over the Next 20 Years.* Barnum Communications, New York
4. *High Technology* (1983). November, 41–49
5. Godfrey, T. and Reichelt, J. (Eds.) (1983). *Industrial Enzymology*, Nature Press
6. *Nature* (1982). 23 September, 298–299
7. *New Scientist* (1982). 4 November, 298

17
A CASE STUDY IN ENZYMES: WASHING-POWDER ENZYMES

Z. Towalski and M. I. Boxer

INTRODUCTION

Enzymes have become one of the major industrial products of biotechnology. Growth of the enzyme market has been dramatic over the last 20 years and its expansion continues, new growth being often initiated from unusual and unexpected sources. This case study was selected because of this significance and because the SRC Enzyme Committee in 1966 implied that this development was an example of recent innovative activity and growth associated with the growth of enzyme science. Retrospective analysis of the technology has indicated that this technology had been around for years, certainly since 1913. It would appear, therefore, that certain factors had been operating on the diffusion of the technology which warrant further investigation.

THE DEVELOPMENT SCENE

The best way of introducing the study is by discussing aspects of washing clothes and the reasons for the use of soaps and detergents. Water alone is not efficient for cleaning clothes, unless a great deal of time and energy is spent on the process — for example, in ancient days, by thrashing the clothes against rocks in a stream and rubbing the clothes, a technique still used in many Third World countries today. This process is time-consuming and labour-intensive, and also damages the clothes by breaking down the fibres of the fabric. Large amounts of mechanical energy are required, because water is not particularly good at penetrating woven surfaces. Clean water has a high 'surface tension', so that it forms droplets on the surface of soiled clothes and does not easily penetrate the fabric. Soaps and detergents, however, can reduce the energy needed to loosen dirt, by coating the fibre structure of the fabric with a surface-active film. Surface-active agents, or surfactants, wet the surface of the clothes and allow the water to penetrate throughout the clothes, to dissolve or remove the dirt. Soaps are useful cleaning agents and reduce the surface tension of water considerably, but detergents are better because they reduce it even further. Surfactants also help by surrounding the dirt particles in the cleaning of clothes during the wash by forcing them up off the fabric and into the water. Agitation by hand or machine helps to loosen the dirt and causes droplets in suspension to break up still further. Once in the water, surfactants now help to keep dirt suspended by coating it with a layer of surfactant.

Soap and Detergents

'Soap' is one of the oldest surfactants known and has been around for at least 2000 years. The Celts made soap from animal fats and plant ashes. Later it came to be made from tallow or vegetable oil heated with caustic alkalis obtained from wood ashes of trees (potash) or maritime plants (barilla) to give sodium salts of fatty acids. Salt was then added to separate the soap from the liquid; hence the geographical association of soap manufacturers and brine springs and alkali works. Production of soap was a local activity supplying and meeting local market needs. William Lever, in 1823, changed this pattern by marketing soap in tablet form carefully wrapped for the individual, and created a market. Small soaperies such as Hudson's, Knight's and Pear's were bought out by Lever and helped to form the giant Lever Brothers. Likewise, a similar process of concentration in the USA produced the giant Procter and Gamble. In part, therefore, economic criteria such as the improvements in transport, the development of a mass market and the availability of energy to fuel mass production enabled the process of industrial concentration to gain momentum. This in turn led to a mass market and mass production with consequent loss of consumer sovereignty. In this oligopoly it came to mean accepting whatever the major manufacturers choose to sell. A policy of careful product matching between main rivals has also resulted in a limited choice of product type. 'Minority' markets are now considered to be uneconomical.

Soap technology advanced slowly until discoveries in organic chemistry led to the production of wetting agents that were superior to soap. These were first introduced in the 1930s, and the term 'detergent' is nowadays usually applied to these synthetic soapless agents derived from mineral oils. The composition of the household washing-powders is different in the USA and Europe, reflecting the temperature of the wash in Europe, in addition, usually contain 'builders' such as phosphates which help the washing action by softening the water, preventing the corrosion of metal parts in the washing machine, and help remove dirt by forming soluble complex ions which repel the fibre and loosen the dirt from its surface. Sodium perborate is added as a bleaching agent to remove stains by oxygenating the water, fluorescent materials to increase brightness and sodium carboxymethyl cellulose to help to prevent the dirt from being redeposited onto the garment and keep it suspended in the water. Cheap perfumes are also included to make the clothes smell fresh and clean. Other chemicals are added either to boost or to stabilise foam.

Enzymes in Washing Preparations

Although enzymes were making inroads in the food and leather industries, their application as washing agents was less spectacular in the first half of the twentieth century, and it was not until the 1960s that they gained widespread introduction. The factor which appears to have hastened their introduction was a way of introducing competition in a new dimension into an otherwise static market for soaps and detergents, since they coped with biological stains at the lower temperatures of a soak or wash and obviated the need to boil and bleach (two less popular domestic activities). When enzymes were introduced into detergent formulations, much was made of the fact that they were non-polluting, biodegradable and convenient. It was the enzymes' ability to catalyse reactions at relatively low temperature (10–65 °C) and normal atmospheric pressure which led to their initial success, but it was the enzyme producers' ability to meet the requirements of the large soapers and detergent manufacturers to supply enzymes which could work under the alkaline conditions of the typical domestic wash which further proved the importance of this innovation in the 1970s and 1980s.

Enzyme-utilising processes, however, have their own special problems, because most enzymes operate in conditions of mild pH (6–8). Washing systems are usually strongly alkaline solutions (pH 9–12). In this case the complex protein structure of enzymes, vital for their function, makes them susceptible to denaturation and inhibition, even by slight alterations of their physical environment. Their catalytic action is easily affected by mild forms of chemical change. Enzymes are also susceptible to a range of poisons, and therefore generally require high standards of purity from the substrate and clean operating conditions. Many enzymes require specific metal ions to activate the enzyme itself; some have a requirement for cofactors which need to be present as 'carriers' for the transfer of distinct chemical groups before the catalytic reaction proceeds to completion. These requirements vary from enzyme to enzyme and need to be determined for each enzyme in turn; thus, it is the more robust broader-acting extracellular enzymes (E.C. Enzyme Class I) which at present have found most industrial uses — in washing preparations in particular.

Dirt may be grouped into three types: (1) grease — resulting largely from the natural lubricants of the body, mixtures of hydrocarbons, fats, fatty acids, esters and glycerol, industrial fats and oils; (2) proteins — food, blood and body debris; (3) grime — made up largely of solid particles and polysaccharides such as mucus. They all prove difficult to remove from clothes washed only in water. Such biological materials are less frequently encountered as stains in the home but provide a major problem on protective wear used by some industries, such as food processing. Enzymes could be used to digest biological compounds such as grease, proteins or carbohydrates and help to loosen or digest them into more soluble products. In a laboratory a wide range of enzymes could be used to produce satisfactory results under carefully controlled conditions but not all lend themselves to being used as presoaking agents.

The development of washing-powder enzymes proved to be problematic, because industrial soaps and household detergent formulations are generally very alkaline, sufficiently so to inactivate potentially useful protein-digesting enzymes. Tailoring the supply of enzymes to meet the detergent producers' needs thus became the key to success. The innovation of incorporating enzyme preparations as cleaning agents shows three distinct phases of development, each with its distinctive commercial impact.

The First Phase

The use of enzymes as washing aids was initiated in Germany in 1913. Otto Rohm[16] developed the idea of making use of pancreatic enzymes to break down fat and protein in the presoak, on the basis of the premise that the main part of dirt in fabrics used by humans was composed of those two materials. According to the patent claim, the fabric could be cleaned in a shorter time, using less strength (these were the days of soap, the washboard and the clothes boiler) and at a temperature further below the boiling point of water than when enzymes were not used. In addition, harsh alkaline chemicals could be used in the main wash, less soap was needed and the fabric had a better appearance after washing. The Company Rohm and Haas of Darmstadt began to market this presoak washing powder as Burnus[17]. It enjoyed moderate commercial success and was sold for about 50 years in European countries, chiefly Germany, until the 1960s. The product consisted mostly of a mixture of sodium carbonate (washing soda) and small amounts of pancreatin, an extract of pancreatic glands, hence, a mixture of enzymes, including trypsin (a protease to digest protein) and emulsion (a lipase to digest fat). Unfortunately, sodium carbonate is very alkaline and in solution has a pH of 12–14. Such a pH adversely affected the stability and chemical activity of the enzymes; hence, doubt can be thrown on the benefits of incorporating enzymes in such a formulation. Nevertheless, Rohm's idea of using enzymes to digest away stains rather than to boil or bleach them out was a good one and was grounded in sound biological principles; consequently, the idea was picked up in 1934 by Dr E. Jaag, working for the firm of Gebrueder Schnyder, based at Biel, in Switzerland[3]. The use of other enzymes was also proposed by other enthusiasts — lipase by Altenberg[1] and amylase by Balentine and Sons[20] in the USA.

The Second Phase

A shortage of imported oils and fats in the war-torn Europe of the early 1940s led to a severe shortage of soaps and soap powders. This triggered a search for alternative solutions. Necessity once again became the mother of invention, and Zscharn[21] in 1940 suggested that enzymes could be used as an *ersatz seife*, or soap substitute. Enzymic presoaking agents gained increasing importance in Europe during the war as the severe shortage of fats and soaps continued. This was not the case in all countries. In Switzerland, however, pancreatic glands, a source of the enzyme trypsin, were used to produce insulin by a process which destroyed the gland's enzyme activity; thus, production of an enzyme-based washing-powder had to await the cessation of hostilities.

Jaag's investigations published in 1947[8] indicated that enzymes were of considerable value as washing aids but needed to be used carefully when washing protein-based fibres, such as silk and wool. After the war a pancreatic trypsin preparation with bile salts for fat emulsification was produced by the Swiss Ferment Company Ltd at Basle and marketed as Bio 38[3].

Jaag realised that, if enzymes were to have any future, they would have to be incorporated and used under normal wash conditions. After the war soap powders began to be replaced by much more efficient surfactants — the detergents. By 1959 Jaag had succeeded in developing an improved product using a bacterial neutral proteinase which was more effective as a presoak agent. This presoak enzyme detergent was called Bio 40 and was produced by Gebrueder Schnyder in co-operation with the Swiss Ferment

Company Ltd of Basle. (This company is now part of Novo Industri A/S of Denmark.)

The Third Phase

Independently of these developments, Novo Industri A/S, of Copenhagen, Denmark, began investigating a laundering problem: blood- and mucus-stained protective clothes from the local meat and fish processing industries. In 1958 they began a programme of research which lasted for 3 years and resulted in the introduction of an alkaline bacterial protease. Novo researchers developed a six-stage washing protocol that involved an enzyme treatment of the garments at Stage III. This method was technically successful but too complicated for the laundry to use and, hence, failed to be adopted[3]. Undeterred, Novo researchers identified the reason for failure as the complex washing procedure and looked around for alternatives. Knowing that they had a better enzyme to offer than the neutral protease produced by the Swiss Ferment Company Ltd, they entered into a co-operative agreement with Gebrueder Schnyder, who introduced Novo's alkaline protease, trade-named Alcalase, in their Bio 40 detergent preparation and improved its efficiency[19].

However, Gebrueder Schnyder began to suffer from local competition from the large soapers. Consequently, Novo and Grebrueder Schnyder entered into technical co-operation agreements with two small detergent producers — Kortman and Schulte, based in Rotterdam, Holland, and Blumller, in Odense, Denmark. In 1963 Korman and Schulte began to market a presoaking agent, Biotex, in Holland and in a short time captured 20 per cent of the Dutch detergent market[3]. The detergent market had been more or less static since the introduction of optical brightness in the 1950s. A loss of 20 per cent to a newcomer was considered to be serious and the two big soapers, Procter and Gamble and Lever Brothers (a Unilever operating company), began to look at enzymes with renewed interest.

The main aim of business is to make profit. One way of increasing profit is to increase the share of the market of the goods produced. Thus, the detergent manufacturer had to consider ways of at least maintaining and preferably increasing the market share of each product line. With just two manufacturers producing between them over 90 per cent of the washing powders available in the UK, the competition between the two companies (Levers and Procter and Gamble) is intense. Each brand of powder has a 'brand image' which aims the product at a particular section of the community. As a rule, rival brands are matched by image, although the overall 'company image' is also important in determining volume of sales.

THE MODERN ENZYME INDUSTRY

The Production of Washing-powder Enzymes

Most of the enzymes used by the detergent manufacturers are produced by one or other of the two major producers. Maxatase is marketed by the Dutch company Gist-Brocades and Alcalase by the Danish company Novo: of the two, Novo is the larger, being the major European and world producer of enzymes. Both companies' washing-powder enzymes are serine proteases, bacterial proteases which contain the amino acid serine in their active centre and which reach their highest activity under alkaline conditions.

The production of extracellular enzymes is theoretically simple; it relies on the culture of micro-organisms in a suitable medium. Enzymes produced by the micro-organisms are then recovered from the medium or the cells. Nevertheless, to make this process commercially viable, with respect to price, activity and quantity, and to satisfy stipulated standards of safety and hygiene, a variety of complicated techniques and complex scientific changes have had to be adopted. To illustrate this with the use of proteolytic enzymes, the procedures which have been used with alkaline protease are a valuable example.

The production of alkaline protease is based on the common soil-based organism *Bacillus licheniformis* or *Bacillus subtilis* in a process which was developed in the late 1950s and early 1960s. Large fermenters culture the organism at temperatures between 30 °C and 40 °C and this produces large quantities of protease. By holding the medium at pH 7, self-destruction is prevented, since the enzyme has an optimum pH of about 9–10, so at pH 7 it is relatively inactive. The enzyme extracted from this microbe is found to be stable in the presence of sequestering agents such as sodium tripolyphosphate and ethylenediaminetetra-acetic acid. It is also stable in the presence of mild oxidising agents such as sodium perborate. The wide-ranging catalytic activity of this protease is illustrated by its ability to break-down between 30 per cent and 35 per cent of the milk protein casein by hydrolysis into small fragments ranging from short peptide chains to individual amino acids. The amount of activity (in this case, hydrolysis) is determined spectrophotometrically using UV absorbance or colorimetrically using trinitrobenzene sulphonic acid.

Processing of the culture begins with the separation of the culture medium from the cells by centrifugation. The liquid is dewatered by either vacuum evaporation or ultrafiltration, or both. Remaining bacteria are filtered out. The enzyme is then separated out from the filtrate by salt or solvent precipitation. In the Novo process the enzyme is made into a paste granulation of the enzyme by mixing it with sodium chloride (common salt), which acts as a preservative, carboxymethyl cellulose, which acts also as a filler binder, and water. Originally the paste was then extruded and passed to a marumeriser, for shaping into spheres. The dried spheres are then coated with a non-ionic wax material. This wax coat is relatively biologically inert material which protects the biologically highly active enzyme and is thus less liable to form a dust. Shrouding of the enzyme in this matter is designed to protect both the workforce and the consumer from contact with the enzyme through inhalation. Recent developments in the wax-coating and enzyme-particle-producing technology have replaced the marumerisers with processes that make spherical enzyme agglomerates to a higher degree of uniformity. This further reduces the risk of producing fine enzyme-containing dusts and lowers the risk to the workforce or consumer of contact with the enzyme through inhalation.

Initial Market Growth

Many different procedures are used to wash clothes. Novo's experience with complex procedures suggested that if enzymes were incorporated in washing-powders in a one-step method, market resistance experienced with a complex washing procedure accompanying their introduction would be reduced. The introduction of enzymes into washing-powder formulations differs between Europe and the USA because of differences in clothes-washing procedures. In the United States method clothes are washed at about 50 °C for 10 min and then a bleaching agent is added separately. In Europe the temperature of the hot wash is raised slowly to just below boiling and held for 30–40 min. The bleaching action takes place in the second part of the wash. European washing-powders contain sodium perborate, which releases its oxygen at temperatures about 50 °C. Table 1 summarises European and United States washing-powder compositions.

Table 1 A comparison between European and US detergent washing-powder formulas. Source: Ref. 11

European enzyme detergent composition:

Constituent	*Composition(%)*
Sodium LAS[a] (40%)	25.0
Alkylphenol ethoxylate (11 EO)[b]	3.0
Soap, high-titre (80%)	3.0
CMC[c] (59%)	1.6
Sodium tripolyphosphate (STP)	38.0
Sodium perborate tetrahydrate	25.0
Sodium metasilicate	1.0
Alcalase 1.5 M	0.9
Sodium sulphate	*c*.2.5

(continued on next page)

United States enzyme detergent formulas:

Constituent	*Composition(%)*	
	Phos-phate	*Non-phosphate*
Sodium LAS[a]	10–20	10–22
Tallow alcohol sulphate	0–10	
Tallow ether sulphate	0–30	0–12
Sodium tripolyphosphate	up to 34	
Sodium carbonate		20
Sodium silicate	up to 12	up to 20
Sodium carboxymethyl cellulose (CMC)	0–0.5	0–1.0
Fluorescent whiteners	0.1–0.5	0.1–0.5
Alkaline protease	1.0	1.25
Sodium sulphate and perfume	qs	qs

[a] Linear alkane sulphonate surfactant.
[b] Average of 11 ethylene oxide units (see 'surfactants').
[c] Carboxymethyl cellulose, sodium salt.

By 1966 enzyme washing-powders had assumed a 2–3 per cent market share of the European laundry detergent market. In 1966 enzyme washing-powders first appeared in test marketing exercises in the United States[3]. Growth in the market share of the estimated enzyme washing-powder market reached 50 per cent in Europe by 1969 and was almost 45 per cent

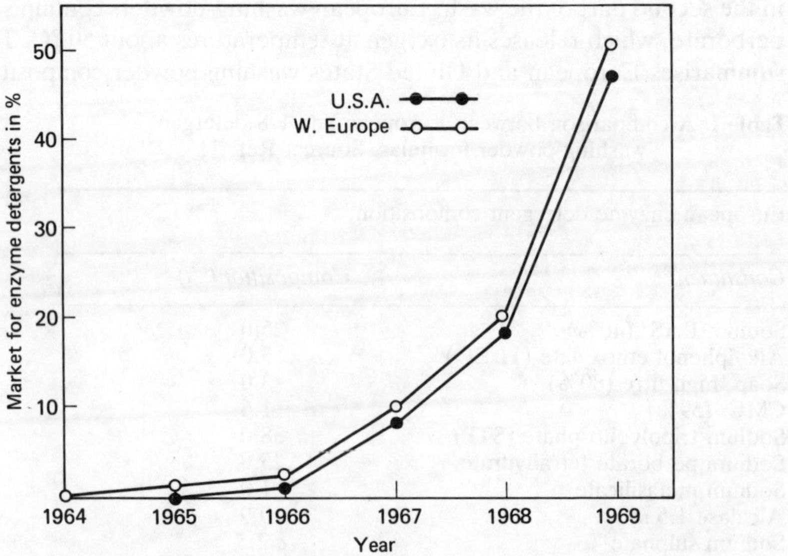

Figure 1 The market entry rate of enzyme detergents in the United States and Western Europe on the basis of the proportion of market (t). Source: Ref. 3

in the United States (see Figure 1). Growth in the market share was largely attributable to the demand for enzyme washing-powders by the housewife, who was now able to remove stubborn stains produced by sweat, mucus, blood, wine, etc., in the normal household wash.

The Social Dimension

Enzymes are proteins and this can cause problems in their usability. A source of difficulty arises in the way the human body interacts with enzyme glycoproteins present in the environment. Enzymes may have a direct effect on the body as specific chemical catalysts; indeed, proteolytic enzymes may be used to clean dirty wounds from dead cells and pus in topic dressings or to lyse blood clots[9]. Risks of damage by digestion are quite low compared with the allergic effects that may be triggered in certain people initiated by the enzyme glycoprotein being recognised as 'non-self' and the setting in motion of an immune response towards it. The immune response is a complex chain of events during the initial sensitising period. Sensitisation is 'dose-related', and it may take a matter of hours or years or may never reach the level to trigger a clinical response. The response is also enhanced by the presence of other allergy reaction agents, a process called potentiation. During the period of sensitisation, antibodies are produced in response to the enzyme, which now also acts as an allergen, and these become attached to mast cells within the connective tissues of the body. Once the individual has been sensitised, if there is further exposure to the same substance, an allergy response will result. Renewed exposure to the enzyme now causes the mast cell to explosively release mediators such as histamine and bradykinin, chemicals which act on smooth muscles, mucus glands and blood vessels in the body. If a massive release of mediators takes place, anaphylaxis ensues, which may result in a range of symptoms including skin rashes and eczema, swelling of the mucous membranes of the nose and throat, vomiting, vascular collapse and bronchospasm. Hay-fever-like symptoms carry with them severe discomfort but severe asthma-type attacks may prove to be fatal. Asthma is now recognised as an industrial disease if the patient has been exposed to the proteolytic enzymes used in detergent manufacture. People who suffer from allergies are called atopic[2].

Screening of atopic individuals was recommended by the 1971 Soap Detergent Industry Association (SDIA) Report[15], according to the following procedures.

'Skin testing should be done by the prick method using an agreed enzyme reagent, grass pollen, *Aspergillus fumigatus, Dermatophagoides culinae* and a control.' A questionnaire is completed, base-line assessment of lung function carried out, chest X-rays performed as well as the skin testing using the prick method and an agreed enzyme reagent, some other known allergens and a control. 'Those with significant findings as a result of any of the above procedures should not be recommended for work with enzyme powders.'

'The ventral surface of the forearm is used — one drop of the skin test solution is placed on a marked area of the arm. The tip of a number 20 hypodermic needle (26 swg) is passed through the drop and inserted into the skin at an angle so that the skin may be lifted on the bevel. The needle is withdrawn and the excess solution is wiped off immediately. A positive reaction is indicated by a weal larger than the control and should be read at 15 min. A positive reading is either a weal of 1 mm diameter or larger, or a weal greater than that which may be produced by the control.'

Dr Flindt[4], a former Unilever doctor, described the symptoms of workers exposed to enzyme dusts in the early days as breathlessness, usually of acute onset and sometimes very severe. At its worst peak, patients were unable to get out of bed and thought they were dying[4(a)]. This clinical picture has been confirmed by other investigators in the UK, the USA, Switzerland and Italy[5,6,13,14,18]. Juniper summarised some of the points as follows.

'Alcalase is an allergen; inhalation could sensitise the individual; subsequent inhalation could affect nose and lungs; effects could last a few minutes or up to 24 h; removal of patient from Alcalase could effect a cure, whereas failure to do so could mean increasingly severe episodes: repeated episodes could lead to incomplete recovery and even long-term impairment of health.'

The Industrial Situation

During this initial period of incorporating enzymes into powders, as more became known about the health hazards of enzymes, more precautions were taken such as exhaust ventilation, filter masks, specially designed overalls and health checks[10]. The safety precautions were those recommended in the SDIA Report of 1971[12], in which great emphasis is laid on the screening of atopic workers. Such workers are not recruited for employment. Thus, those people who are likely to develop a sensitivity to enzymes are screened out. The SDIA Report states that perhaps 30 per cent of the population 'may have atopic characteristics' but even non-atopics develop sensitivity; the ratio is believed to be about 2:1 atopic to non-atopic.

The detergent industry spent a great deal of money in protecting the workforce from hazards. The USDAW branch secretary at one site said,

'In the past we had occasion to eliminate silica from Vim. It had been used for years before it was identified with silicosis — as far as I am concerned, Lever Brothers did everything in their power on safety. It was a difference of medical opinion. I am concerned about those I represent and the evidence. NSDs were made without any noticeable problems but when the "Bios" were introduced, the same workers were then affected. In 1975 Lever Brothers opted out of the enzyme detergent market for commercial (not medical) reasons. The price was too high. It is regrettable that we did not make progress nationally. I hope I am wrong, but people in Procter and Gamble might develop chest pains they have not had before. The NSD department here is a lot happier now.'*

The decision to pull out of enzymes was also possibly prompted by the public's disenchantment with the product and the unfavourable publicity attributed to enzyme dusts and continued reports of allergic reactions. Radiant, in particular, was receiving a lot of adverse publicity. Such effects on the sale of microbial enzymes world-wide was dramatic as the market share for enzyme detergents fell to pre-1966 levels (see Figure 2). The

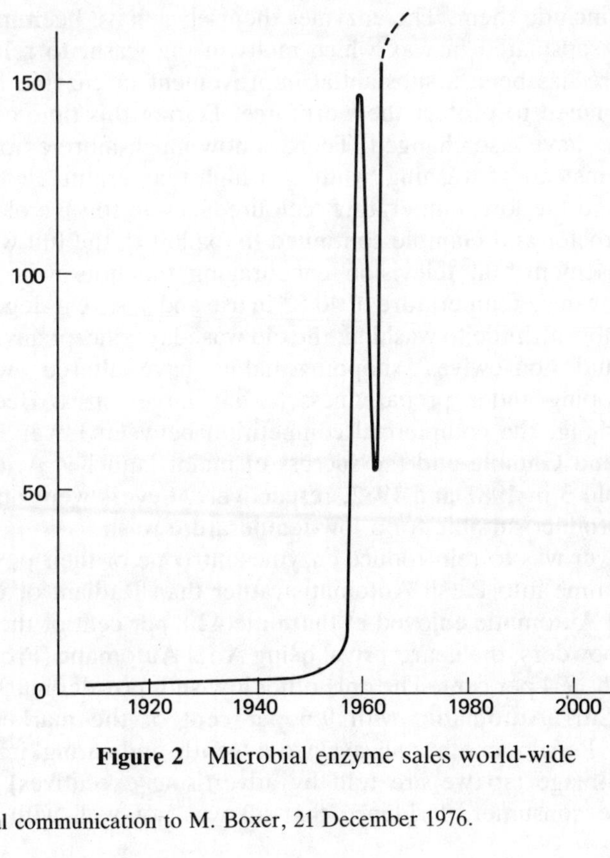

Figure 2 Microbial enzyme sales world-wide

*Personal communication to M. Boxer, 21 December 1976.

withdrawal of enzymes by Lever Brothers in 1975 was done relatively quietly after considerable discussion with the union at the factory involved. The union (USDAW) were concerned that there was a medical problem, as incidents convinced them that enzymes were the cause of some workers becoming 'distressed with asthmatic attacks'. Initially the enzymes were regarded just as a new ingredient, but after a while, when the dust hazard was highlighted, the enzymes were encapsulated (shrouded) by fixing the enzyme onto granular sodium bipolyphosphate and rendering the resulting granules substantially dust-free by spraying non-ionic material on; in the Novo process the enzyme was also crisped and screened to remove lumps, which made them heavier and more likely to fall to the floor instead of their being blown about.

The Current Situation

More than a decade has now passed since Lever Brothers withdrew enzymes from powders. During this period the other manufacturers have continued to include them. The enzymes themselves have been modified, now being encapsulated in wax which melts in the wash, to release the enzyme. There has been a substantial improvement in enzyme handling and plant designed to protect the workforce. During this time consumer washing habits have also changed. There is now much more emphasis on economy; so instead of washing 'whites' at high temperature, consumers are switching to the low-temperature 'coloureds' cycle to save electricity. This mood Procter and Gamble continued to exploit to the full with their Ariel advertisements on television encouraging the housewife to economise by showing a temperature of 40 °C in use and a 'save it' logo. There is a more flexible attitude to washing, the old washday concept having been abandoned and housewives' shopping habits have altered, with less frequent shopping and a preparedness to buy larger sizes. Because of product matching, the commercial competition between Lever Brothers and Procter and Gamble and the success of major launches Ariel Automatic and Bold 3 in 1981 and 1982, respectively, Levers were obliged to introduce a product suitable for a low-temperature wash.

Their answer was to reintroduce enzymes into one of their powders in 1983 but this time into Persil Automatic rather than Radiant or Omo, as before. Persil Automatic enjoyed at that time 42.8 per cent of the market for low-suds powders, the nearest rival being Ariel Automatic (Procter and Gamble), with 17.4 per cent. The only other low-suds powder marketed by Levers was Surf Automatic, with 9.6 per cent of the market share. However, the Persil name has always had a 'gentle and caring' image and this product image (so we are told by advertising executives) is what influences the consumer. Evidence that all was not well with enzyme washing-powders was accumulating from the trades unions, where certain groups of the workforce were not atopically screened and whose members came into contact with enzyme powders. Such was the case in the departments dealing with 'money-off coupons' cut from the packets of powder. Certain sorters experienced a variety of allergy reactions while handling those coupons from powders containing enzymes.

Other evidence of difficulties began to trickle in from groups such as the National Eczema Society, whose members reported that clothes washed in enzyme-containing detergents were associated with skin problems. Until the introduction of enzymes into Persil Automatic, it had been the one powder recommended by the National Eczema Society for those who experienced discomfort from other powders. It was the powder with the fewest complaints of skin problems as a result of its use. Following the introduction of enzymes, this advantageous situation was lost. A variety of articles appeared in the popular press and popular scientific journals[7]. It was recognised that Lever Brothers' decision to change Persil Automatic into an enzyme powder effectively reduced consumer choice of low-lather, enzyme-free detergents and now excluded a section of the population, suffering from eczema, from using its favourite detergent formulation[2].

Evidence of causality is always difficult to obtain without experimentation. Lever Brothers carried out exhaustive trials before launching the new product, and independent research by two leading dermatologists. The former revealed no problems but the latter showed that indeed some people do suffer adverse effects when using a New System Persil Automatic, or any other enzyme-containing washing-powder. It is interesting to note that Procter and Gamble say publicly that the tests they have carried out have proved negative[2]. Perhaps the tests they ran were not sensitive enough to detect these effects. Both companies state that the majority of those using enzyme-containing washing-powders do not suffer from any side effects. However, on 22 June 1983 Lever Brothers stated publicly that they would reintroduce the old-formula Persil Automatic within 2 months and the non-enzyme powder has been available since.

The soap and detergent market has long been regarded as having lost consumer sovereignty through being dominated by a few large companies, sale of brand name products being skilfully manipulated through brand image advertising. Competition in the market is restricted to a choice between rival brands. The opinions as to consumer need were ascertained by market research and marketing functions. Recent activities have shown how tenuous and fragile are the brand images which are so carefully and so expensively created. This policy reversal by Lever Brothers, which also included the relaunch of their other low-suds powder, Surf Automatic, as a low-temperature powder *without* enzymes, is a modest victory for the consumer and it took a product of biotechnology to bring it about. Surf Automatic has since (1986) been relaunched *with* enzymes. The packet is clearly labelled as such and has not attracted the publicity that was encountered by Persil. The launch of the biological powder Breeze (a product title previously reserved for certain overseas markets) as an 'automatic' powder confirms Lever Brothers' confidence in enzymes in the wash. It is, as yet, too soon to gauge consumer reaction.

The lesson for other biologically active products must be clear: innovators who develop new products or processes that interact with people must consider the *social dimension* of the market, in particular 'the artificially created brand image' and the expectations it engenders within the consumers, as well as the market criteria of price, demand and supply.

REFERENCES

1. Altenberg, J. (1930). German Patent
2. Boxer, M. and Towalski, Z. (1984). Can clean clothes damage your health? *New Scientist*, 19 January
3. Dambmann, C., Holm, P., Jensen, V. and Nielsen, M. H. (1971). How enzymes got into detergents. *Developments in Industrial Microbiology*, **12**, 11–23
4. Flindt, M. L. H. (a) *Lancet*, **1**, 1177 (1969); (b) *Lancet*, **1**, 430 (1978)
5. Gaffun, E. (1970). Respiratory effects of biological detergents. *British Medical Journal*, **4**, 52
6. Greenburg, M., Milne, J. F. and Watt, A. (1970). Survey of workers exposed to dusts containing derivatives of *Bacillus subtilis*. *British Medical Journal*, **2**, 629–633
7. *Guardian*, 9 November 1983
8. Jaag, E. (1949). Untersuchungen uber die Einwirkung von Fermenten auf verschiedene Fasermaterialien und deren Verunreinigungen. *Chemia*, **1**(3), 57
9. Johnson, A. J. and McCarty, W. R. (1959). *Journal of Clinical Investigation*, **38**, 1627
10. Juniper, C. (1978). Health and the factory environment: The medical contribution. *Journal of American Oil Chemists Society*, **55**, 185
11. *Kirk-Othmer Encyclopedia of Chemical Technology*, 3rd edn (1980)
12. *New Scientist*, **44**, No. 671 (1969)
13. Newhouse, M. L., Tagg, B., Pocock, S. J. and McEvan, A. C. (1970). An epidemiological study of workers producing enzyme washing powders. *Lancet*, **1**, 681–693
14. Pepys, J., Hargreave, F. E., Longbottom, J. L. and Faux, J. (1969). Allergic reactions of the lungs to enzyme of *Bacillus subtilis*. *Lancet*, **1**, 1181–1184
15. Recommended operating procedures for UK factories handling enzyme materials. SIDA and *Annals of Occupational Hygiene*, **14**, 71–87 (1971)
16. Rohm, O. (1913). Verfahren zum Reineryen von Waschestucker aller Art. German Patent 283,923
17. Rohm und Haas, company information
18. Shapiro, R. S. and Eisenberg, B. C. (1971). Sensitivity to proteolyte enzymes in laundry detergents. *Journal of Allergy*, **47**, 76–79

19. Starace and Barfoed (1980). Enzyme detergents. In *Kirk-Othmer Encyclopedia of Chemical Technology* 3rd edn
20. Tauber, H. (1949). *The Chemistry and Technology of Enzymes.* Wiley, New York
21. Zscharn, A. (1940). Enzyme sparen Seife. *Fette und Seifen,* **47**, 463

18
ENZYMES FOR INDUSTRY IN THE 1990s

Tony Godfrey

WHY ENZYMES?

There are many ways to make the case for the use of enzymes in industrial processing, most of them very acceptable in their own right. The enthusiastic inclusion of enzyme-catalysed steps in many natural product-modifying industrial processes has confirmed that there exists a commercial benefit from their use. 'Natural' in all its current contexts is now the 'in' consideration, and the properties of nature's own catalysts are receiving enormous attention at the biochemical and molecular biology level. Evidence is now emerging that it is possible to create synthetic enzyme structures exhibiting true catalysis, although the activities are not very impressive. There are few successes yet, but the target has been set by the critical view of the process engineer or chemist who does not appreciate the limitations of performance, stability, tolerance to other chemicals and storage quality which are present with today's enzyme products. The advent of immobilisation of enzymes heralded a major new generation of catalysts that have not achieved the significance expected of them, in terms of the wide range of processes they were predicted to influence. Those that have been scaled up are impressive for their efficiency and economy. The performance and the system configuration of an immobilised enzyme reactor conform more readily to conventional chemical processing and catalysis, but the really difficult concept to achieve is that of a system that has its own conditions, constraints, productivity, etc., but is based on natural chemistry.

Enzymes do it the way nature does it (well, most of the time, anyway): they have the inherent natural recognition of the optical and stereospecific variants of other natural molecules, and generally act on only one species; they function under very modest conditions; and they remain biodegradable when their catalytic potency is finished. When combinations of these characters are considered there are attractive prospects for enzyme-mediated processing.

WHAT ENZYMES?

The development of the industrial enzymes (Table 1), has been largely through the exploitation of the hydrolases acting to depolymerise natural materials such as starch, plant structural materials and proteins. The traditional biotechnical industries using such raw materials have accepted the enzymes as the natural way to create modifications. The newer applications have been built out of a widening appreciation of these same enzymes, either singly or in selected combinations, for their ability to modify substances other than those in the environment from which they have been extracted. Thus, we see conventional enzymes working in unconventional circumstances to create new and specific chemistry. This is well illustrated by the use of lipolytic enzymes in the modification of synthetic fats for targeted molecule arrangements.

Table 1 The current commercial enzyme profile

Enzyme type	*Industrial*	*Kit/research*
Oxidoreductases	2	65
Transferases	–	74
Hydrolases	45	130
Lyases	4	53
Isomerases	1	16
Ligases	–	26
	52	364

It has long been considered that enzymes require large amounts of water in order to function. There are now many examples of activity of value occurring in non-micro aqueous conditions, and biochemists also recognise that many enzymes function in the virtual absence of water in the cell environment.

The industrial hydrolases now cover almost all the known types and are available in bulk for full-scale processing. The subtle differences in characteristics of the same enzyme type from different sources are beginning to be exploited, to widen the scope of their action and to focus the process even more sharply.

In addition to the hydrolases, industrial quantities of several other enzymes have become available. The use of 'glucose isomerase' for the conversion of part of the glucose in starch-derived syrup to fructose has been available for some fifteen years now. Similarly, the use of glucose oxidase for the removal of glucose or oxygen from food systems has been applied on an industrial scale.

Biocatalysts Ltd was created to develop and exploit the growing demand for detailed expertise in the formulation and application of the already available enzyme types, and to isolate and prepare new enzymes for the industrial markets. The requirement to improve the processing of botanical raw materials, other than traditional products such as starch, has stimulated the creation of tailored systems. Several enzymes, working in concert, but present in both appropriate type and quantity, can be introduced to conventional processing and contribute significantly to the yield and purity of target products. The extraction of natural colours and flavours, chemicals of medical significance and new raw materials for biodegradable product creation have all been aided in this way. Similarly, the increasing understanding of the action of proteases and lipases has made it possible to create enzyme systems that make accurate and specified modifications of proteins and fats. The flavourings for pet foods and the generation of enhanced flavour internal to foods are examples.

WHAT NEW ENZYMES TO COME?

Within a very dynamic industry, the changes come quickly, and enzyme applications are no exception. Biotechnology has already offered the opportunity for the bulk production of enzymes from one species by another. To exploit this skill, we have to consider the current cost of such methods, and also to include a careful evaluation of the present and likely future regulatory status for the resulting enzymes. In terms of costs, we see that genetically engineered enzymes may have some way to go to compete with conventional ones. Prices of one to five orders of magnitude greater than current bulk enzyme costs are commonly indicated, and the first outcome will almost certainly be low-volume products with high value, such as pharmaceutically relevant catalysts. There is the one area of greatest prospect — the preparation of enzymes from accepted species of organism using the genetic information from a pathogenic or otherwise not accepted species. The exploitation of the proteases of the pathogens may well add to our resources to advantage.

Great advances in the development of plant-tissue culture have been made with speciality secondary metabolites as the targets, the alkaloids being the most notable of these. It is expected that a major shift in the balance of enzyme supply will derive from recognition of the many that can be recovered from these fermentation systems without climatic and seasonal intervention.

Protein engineering methods are also being applied to the modification of the character of enzymes, and we can expect to have a new range of catalysts with enhanced features. Tolerance of elevated temperatures giving higher productivity per unit time is the one area receiving great attention, but it is perhaps necessary to caution a little by asking about the means to regulate and/or inhibit activity at the conclusion of desired action. The creation of sensitivity to provide acceptable means for inactivation would also seem a suitable objective.

The great wealth of enzyme knowledge that is emerging from medical and general biochemical analytical research will provide a new sector for the development of industrial-scale processing. Consideration of the numbers of enzymes already identified and characterised by research shows that in excess of 2500 different catalysts could be available. Of these, some 300 are readily purchased through fine-chemical catalogues, and about 50 can now be obtained in sufficient quantity for industrial-scale work. Most of those awaiting development for industry require cochemicals of equally delicate and biological natures. We can expect to see the production of these substances, and the enzymes matching them, to increase in quantity and decrease in cost in the next few years. Biosynthetic routes to target specific compounds depend upon this development, and the list of desirable oxidations, reductions, saturations and unsaturations, and molecular rearrangements is growing. The acceleration of molecular biological research in the past two decades is leading to an awareness of the diversity of chemical selectivity within the living cell which attracts the industrial chemical mind very strongly.

HOW MUCH ENZYME BUSINESS?

The development of the industrial enzymes market has been dramatic and closely related to the explosive expansion of biochemical and molecular biological knowledge since the 1960s.

From 1930 to 1950 there was a slow growth, mainly in traditional industries such as leather, baking, starch hydrolysis and fruit and wine processing. The expansion came with the investment in bulk fermentation for antibiotic production which provided the technology and skills resource for other fermentations. The value of the market has reflected this development, which allowed economic pricing of bulk enzymes, many of which have now become commodity products, especially for starch processing, textile desizing and detergents (Table 2).

Table 2 Past and predicted world enzyme values compared with antibiotics and amino acids. Data based on US Trade Commission Report 1984, with extrapolation for world estimates: Values: Million $ (1980)

	1960	*1970*	*1980*	*1985*	*1990*
Amino acids	35	120	400	580	800
Enzymes	185	340	432	1200	2000
Antibiotics	1680	2840	5345	7960	15800

While the continued growth of the enzymes industry is confidently expected, it is instructive to consider it in context with other 'biotechnical' industrial products such as antibiotics and amino acids.

Research into biological systems is now a major feature of all the biotechnology programmes reported around the world. It is to be expected that efficiency of scale-up, economy of materials and purification procedures, and minimal capital plant and energy investment will create a strong drive for a shift from whole-organism-mediated operations to enzyme mediation in the next decade.

BIBLIOGRAPHY

1. *Biochemical Engineering and Biotechnology Handbook* (Ed. B. Atkinson and F. Mavituna). Macmillan, Nature Press, London (1983)
2. *Enzyme Technology – Recent Advances* (Ed. S. Torrey). Noyes Data Corp., USA (1983)
3. *High Technology Profiles and Outlooks — Biotechnology*. US Dep. Commerce (1985)
4. *Industrial Enzymology* (Ed. T. Godfrey and J. Reichelt). Macmillan, Nature Press, London (1983)
5. US International Trade Commission Report on Developments in Biotechnology and Possible Impact on Certain Sections of the US Chemical Industry, Washington, 1984

19
MONOCLONAL ANTIBODIES — THE BIOLOGICAL BASIS

J. J. Rimmer

INTRODUCTION

Monoclonal antibodies represent one of the success stories of biotechnology. In less than a decade they have progressed from the research laboratory to the market place and are already making a considerable economic impact. In scientific terms they are making substantial contributions to the study of human biology and disease, and the future prospects for applications in diagnostics and therapy are only now being more fully appreciated. In human medicine alone, gross sales rose from \$11 million in 1981 to \$28 million in 1983 to \$555 million in 1985. Future projected sales of \$1.7 billion in 1990 are confidently predicted[2]. The present chapter will briefly review the biological basis of monoclonal antibody production before turning to an assessment of some present and proposed future applications.

ANTIBODY MOLECULES AND THE IMMUNE SYSTEM

The immune system is a complex interacting network of cells, tissues and molecules which serves to protect us against potentially life-threatening attacks by viruses, bacteria and other parasites. This capacity for self-defence reaches its most sophisticated level in man and other mammals, and the immune responses of men and rodents have been extensively studied. The basic functional unit of the immune system is the lymphocyte and this cell has been implicated in all specific immune reactions. In adult mammals, lymphocytes are produced and mature in two distinct anatomical sites: the bone marrow and the thymus gland. Following a period of maturation in these organs, lymphocytes migrate into the circulation and come to rest in lymphoid organs throughout the body. These lymphoid organs (e.g. the spleen and lymph nodes) form important garrison posts in the war against invading pathogens, and their function is to capture foreign material (e.g. bacteria and viruses) and render it ineffectual. This process is achieved in large part by the synthesis and production of antibody molecules. The manufacture and release of antibody molecules is carried out by 'B' lymphocytes (i.e. those lymphocytes which matured in the bone marrow prior to emigration to the lymphoid organs). Antibodies are globular proteins made up of two 'light' and two 'heavy' polypeptide chains. These are arranged in such a way that the overall shape of the molecule resembles a letter 'Y' or 'T'. At the end of each of the two 'arms' of the antibody molecule are small grooves or clefts which vary in shape from one molecule to another. The basis of this variability lies in the amino acid sequences within the polypeptide chains, and differences in these sequences give rise to a large range of different-shaped grooves or clefts. It is these molecular grooves which allow the antibody molecule to bind to foreign material in a specific manner. The general term for foreign material which slots into this groove is 'antigen', and, hence, the groove is termed the antigen-binding site of the molecule. An antigen may be defined as a particular molecular pattern or configuration which is found as part of a virus, bacterium or foreign tissue cell (e.g. as in a surgical transplant) or as part of a larger complex molecule such as a foreign protein or polysaccharide. Under normal circumstances, antibody protects the body against pathogens or their products by binding onto them and promoting their disposal (e.g. by phagocytosis). The function of antibody need not be restricted to disposing of pathogens, however, since antibody molecules can be generated against almost any foreign substance, provided that it falls within a certain molecular size range. The immune system is capable of producing an almost inexhaustible range of antibody patterns or antigen-binding sites. Furthermore, each antigen-binding site is specific for its own particular antigen, and one antibody will only combine with its 'own' antigen. The precise nature of this relationship between antibody and antigen provides a biological system which is capable of a superb degree of discrimination. This capacity has been recognised for many years. However, the problem has been to harness this potential effectively in order to produce a consistent product in large amounts and at a commercially acceptable cost.

POLYCLONAL ANTISERA

The usual way to produce large amounts of antibody is to inject animals (e.g. goats or sheep) with the appropriate antigen and then collect the resultant antibody from the blood serum. In some instances (e.g. for medical uses) human volunteers are necessary and this may cause additional ethical problems. The major drawback with these methods of antibody production is one of consistency, since even if the same individual is used on two separate occasions to produce antibody against the same antigen, the overall specificity of the antibody can vary quite markedly. This is because large molecules and cells are in themselves a collection of different antigens each of which may elicit antibody production. Studies over the last 15–20 years have established that among the total population of lymphocytes within the body, each lymphocyte is only capable of synthesising and secreting antibody molecules of a single specificity — i.e. all antibody molecules produced by a single lymphocyte have exactly the same antigen-binding site. Furthermore each lymphocyte carries its own particular brand of antibody molecule on its surface membrane, where it acts as a receptor molecule for antigen. Antigen triggers off the process of antibody production by selecting the B lymphocyte which bears the appropriate cell-surface antibody. When antigen combines with this antibody it activates the B lymphocyte into mitosis, and as a result the original cell produces a clone of daughter cells (plasma cells) each of which synthesises and releases antibody molecules with exactly the same antigen-binding sites as the original parent cell. Clearly, a large, complex molecule or foreign cell carries many antigens on its surface, and as a consequence will trigger off the production of many different clones of antibody-producing cells each of which corresponds to one antigen. This makes standardisation difficult, since a conventional (polyclonal) antiserum is really a collection of different antibody molecules which bind to different fractions of the molecule or cell which elicited them. Each antibody may be produced in differing amounts, and they may bind to their corresponding antigen with differing degrees of affinity. Strict product control is therefore very difficult when employing polyclonal antiserum. For many purposes such antisera are perfectly adequate — indeed they have been in common use for years — but for applications which necessitate large amounts of specific, homogeneous reagents, monoclonal antibody provides a more satisfactory alternative.

MONOCLONAL ANTIBODIES

In 1975 Georges Kohler and Cesar Milstein published their now classic paper, in which they described details of a method for manufacturing monoclonal antibodies by a process of cell fusion and culture[3]. Since its initial description the technique has been improved by technical advances and by the availability of more suitable cell lines; but in essence the basic protocol has not diverged significantly from their original method. The process hinges upon fusing an antibody-secreting cell (plasma cell) with a myeloma cell from a continuously dividing cell line. First of all, a mouse is injected with the appropriate antigen in order to provoke an immune response and generate antibody-secreting cells *in vivo*. When the immune response is sufficiently developed, the mouse spleen is removed and dissociated into a single-cell suspension containing antibody-producing cells. These are then introduced to myeloma cells in culture in the presence of a chemical fusing agent called polyethylene glycol (PEG). It is usual to employ one of several myeloma cell lines which lack the ability to synthesise the purine salvage enzyme hypoxanthine phosphoribosyl transferase (HPRT). If these enzyme-deficient cells are cultured in a selective culture medium containing hypoxanthine, aminopterin and thymidine (HAT medium), they will die. If, however, such a myeloma cell fuses with an antibody-producing cell (which *can* synthesise HPRT), a hybrid is formed which is able to survive in HAT medium. Non-fused myeloma cells will, of course, die, as will unfused spleen cells, which have a limited life-span in culture, anyway. In this way, the HAT medium only allows survival of fused cells or hybridomas. Such hybridomas retain the characteristics of both parent cell types: they divide freely in culture and secrete antibody of a single antigen-binding specificity. This antibody, the product of a single immortal clone, is called monoclonal antibody. Following the production of hybridomas as described above, the cultures are subdivided into multiple small cultures. Individual subcultures which show synthesis of the required antibody are further diluted in order to identify and isolate the required hybridoma clones. By carefully controlling the cell concentration in soft agar cultures, it is possible to observe and remove single colonies. Clones secreting antibody of interest are selected and propagated in bulk tissue cultures or as ascites tumours in laboratory animals. The monoclonal antibody can then be harvested.

ADVANTAGES OF MONOCLONALS

The properties of monoclonal antibodies which have made them so useful to medicine and research are threefold. First, they exhibit far greater antigen-specificity than do conventional polyclonal antisera. Second, since the technique permits synthesis of a homogeneous product, monoclonal antibodies are therefore uniform, with definable, specific binding qualities. The third advantage is one of availability. In circumstances where the antigen of interest is rare, expensive or difficult to purify, large amounts of monoclonal antibody can be produced by obtaining enough antigen to immunise a laboratory mouse. The requirement for antigen purification is also reduced, since even the injection of a complex antigen mixture will give rise to a range of individual antibody-secreting clones from which the required one can be readily identified, selected and propagated. These properties of monoclonal antibody represent a major advance over conventional antisera. Antibody can now be regarded as a standard laboratory reagent.

As outlined above, monoclonal antibodies exhibit a number of major advantages over conventionally raised polyclonal antisera. Foremost among these are their greater antigen-specificity, homogeneity and availability. It is these three qualities which have proved so valuable in tackling some basic problems in biomedical and clinical research as well as in diagnostic and therapeutic work. In concluding this chapter a brief survey and review of some of the many applications of monoclonal antibodies is attempted. It is by no means comprehensive, but will illustrate some of the fields in which monoclonal antibodies are making significant contributions.

STUDYING CELL-SURFACE MOLECULES

One continuing source of interest to scientists and clinicians alike is the cell surface. Mammalian cells carry on their outer surface hundreds of different molecules which permit the cell to function appropriately in response to the demands of its environment. These cell-surface molecules perform a wide range of biological functions (for example, structural proteins, differentiation antigens, hormone receptors, enzymes, etc.) which are of interest to an equally wide range of scientists. Many of these molecules, when introduced into a foreign host, will trigger the immune response of the recipient into antibody production (i.e. they will act as antigens). It is therefore possible to investigate the nature of the cell surface by utilising specific antibodies against these cell-surface molecules. Thus, by tagging a fluorescent or radioactive label onto a specific antibody, it is possible to construct an image of the density and distribution of particular molecules on the cell surface.

Unfortunately, many cell-surface molecules are difficult to isolate and purify, and even then may only be available in minute amounts. This makes it difficult to raise conventional antisera against them. Hybridoma technology, however, has established a means of producing specific monoclonal antibodies to many of the elusive molecules on the cell surface. In such instances where the antigen is rare or difficult to purify, large amounts of monoclonal antibody can be generated by obtaining enough antigen to immunise a laboratory mouse. The need for antigen purification is also reduced, since even the injection of a complex mixture of antigens will ultimately give rise to a series of individual antibody-secreting clones from which the appropriate one can be identified, selected and cultured.

MAJOR HISTOCOMPATIBILITY COMPLEX

In illustrating this theme it would be instructive to consider some specific examples. One particularly important group of cell-surface molecules are those coded for by genes of the so-called major histocompatibility complex (MHC). This gene complex has been the subject of intensive study in both mice and men, where it has been discovered that, among other activities, it plays a major role in the rejection of foreign transplants. Genes within the MHC code for cell-surface molecules are commonly referred to as transplantation antigens. These glycoprotein molecules are present in varying amounts on the surface of most cells in the body. The MHC is highly polymorphic, and as a consequence each of us possesses our own set of transplantation antigens, which will differ to a greater or lesser extent from those of other individuals. In the context of transplantation surgery, these antigens continue to present the most significant barrier to successful organ and tissue transfer. They have been likened to 'flags of identity' on the cell surface, and when an organ or tissue is transplanted into a foreign host, the histocompatibility antigens are recognised by the recipient's immune response and the graft is destroyed with all the vigour that is normally reserved for life-threatening pathogens. To the immune system, potentially beneficial transplants and dangerous pathogens are both 'foreign' and they are disposed of with equal prejudice. In the population as a whole, however, there are individuals who are very similar with respect to the MHC antigens they possess. Organs or tissues exchanged between such individuals stand a much greater chance of survival, since the antigens on the cell surface of the donor's cells do not differ sufficiently from those on the host's cells to be recognised as foreign by the immune system. Thus, a graft exchanged between identical (homozygous) twins, for example, is never rejected immunologically. If the degree of similarity between donor and host MHC antigens can be maximised, then the chances of rejection are significantly reduced.

Information about MHC antigens on donor and host cells can be provided by studying the binding behaviour of a panel of specific antibodies to the different MHC antigens. This process is called tissue-typing and until recently depended entirely on the collection of antibodies from a number of sources. These include specifically immunised volunteers; women who have had a number of pregnancies and have thus been exposed to paternal antigens on several fetuses; transplant and blood transfusion recipients who have already manufactured antibodies against previously encountered foreign cells; and immunised primates. Clearly, such antisera are difficult and time-consuming to collect and test for use as standard tissue-typing reagents. Coupled with this is the problem that there are already many different antigens to test for and new ones are being discovered all the time, each of which is difficult to isolate and purify and requires a specific antibody to identify it.

Clearly, monoclonal technology provides a way to surmount these problems, and considerable effort is being channelled towards the large-scale isolation and characterisation of MHC antigens. In the near future it should be possible to produce a complete range of monoclonal antibodies which will identify the full spectrum of MHC antigens and lead to more efficient and universally standardised tissue reagents.

TUMOUR DIAGNOSIS AND THERAPY

Another group of cell-surface antigens which are of particular significance to biological research and medicine are those which occur on the surface of tumour cells. Following exposure to carcinogenic chemicals or certain viruses, laboratory animals will develop tumours. When a chemically induced mutation occurs in a cell, causing it to transform into a malignant cell, one consequence may be the production of altered molecules on the cell surface. Similarly, in virally infected cells the virus may incorporate itself into the host genome and code for novel cell-surface proteins on the infected cell. Many tumours in experimental animals are known to express novel molecules on the cell surface and these may be recognised by the immune system as being foreign. Often, however, these tumour-specific antigens are not capable of stimulating a vigorous immune response.

The status of tumour-specific antigens in man is still the subject of some debate, but the use of monoclonal antibodies will help to resolve this issue. They have already been used to detect what may be tumour-specific antigens on the surface of some leukaemic cells, colorectal carcinoma cells and melanoma cells, and work is currently under way to build up a 'dictionary of tumour antigens' using monoclonal antibodies. Such antibodies will be invaluable for tumour diagnosis and therapy. Because these antibodies are highly specific, they should bind only to tumour cells, and, by attaching radioactive tracers or therapeutic drugs, it should be possible to detect and destroy tumours hidden deep within the body.

One approach to detecting deep-seated tumours has been to use I-labelled monoclonal antibodies as tracer molecules which bind onto tumours and are subsequently located by external photoscanning or a relatively recent technique called single photon emission computerised topography (SPECT)[4]. In addition to diagnostic uses, monoclonal antibodies are being employed as drug delivery systems which will bind onto tumour cells and concentrate highly toxic drugs such as ricin in the tumour itself while leaving the rest of the body free from attack[1].

Some success has also been achieved by employing an *in vitro* approach to treating leukaemia. Work in Boston, Massachusetts, has provided a technique for eliminating leukaemic cells while leaving normal cells intact[5]. This is possible because it was discovered that in certain leukaemias the transformed cells carry on their surface a unique molecule called common acute lymphoblastic leukaemia antigen (CALLA). A highly specific monoclonal antibody has been produced against CALLA and will bind only to leukaemic cells bearing this molecule. Ninety-nine per cent of normal bone marrow cells do not carry CALLA on their surface and therefore go unrecognised by the antibody. The leukaemic patient's bone marrow cells are collected in a relatively simple operation and cultured in the presence of the specific monoclonal antibody against CALLA. A serum-derived component called 'complement' is added to the culture. ('Complement' is the term used for a group of naturally occurring proteins which, in conjunction with specific antibody, will kill cells by the formation of lesions in the cell membrane.) Since only CALLA-bearing leukaemic cells will bind the monoclonal antibody to their surface, these are the only cells which are destroyed by the action of complement. Normal bone marrow cells, which lack CALLA molecules, are left unharmed by the treatment and may now be cryopreserved pending the next phase of treatment. The blood and other tissues of the patient will still contain large numbers of leukaemic cells and these must be destroyed. The patient therefore receives a course of chemotherapy and irradiation to destroy the residual leukaemic cells in the body. The cryopreserved bone marrow cells, now free of leukaemic contaminants, are then removed from storage and reinjected into the patient, where they set about restoring the damaged haematopoietic system.

MISCELLANEOUS USES OF MONOCLONAL ANTIBODIES

The use of monoclonal antibodies for the study, characterisation and treatment of disease is also proving to be of great clinical value. Studies on a number of viruses have already been advanced by the ready availability of monoclonal antibodies against viral antigens. Already monoclonal antibodies to a wide range of viruses are available (e.g. rabies, herpes simplex, influenza and Epstein–Barr virus), and will contribute towards further understanding the mechanisms by which viruses infect cells. Antiviral antibodies produced by hybridoma technology also promise much for the future of vaccination. Monoclonal antibodies may also be used to isolate and purify harmless components from the surface of many parasites which can then be used as antigens in vaccination programmes to stimulate the host's immune response against infection. Promising results have already been obtained by producing monoclonal antibodies against trypanosome and schistosome parasites, and monoclonal antibodies are also available for a wide range of surface antigens from sporozoites of many malarial species[6]. There is a growing awareness of the enormous burden placed by parasites upon man, his livestock and other species, and monoclonal antibodies seem set to make major contributions in this field, as in so many others.

Despite the apparent promise of monoclonal antibodies outlined above, it would be short-sighted to ignore some of the constraints associated with their usage. Many of these constraints were highlighted at a meeting, 'Immunotechnology and Industry', organised by the British Society for Immunology[2]. It is apparent that much basic research is still required before monoclonal antibodies can be widely applied to human disease. The production of human rather than mouse monoclonal antibodies may circumvent some of the problems associated with treating patients with 'foreign' antibodies. In animal husbandry, where constraints on usage may be less rigorous than for humans, the question of cost-effectiveness must be considered. It is clear that large-scale industrial use of monoclonals as product purification reagents must await cheaper *in vitro* production.

The meeting mentioned above also considered broader topics such as government backing of 'basic' research through universities. The decline of the dual funding system was held by some to signal a gloomy future for academic research in the UK. Attention was also drawn to the fact that although monoclonal antibodies were 'invented' in the UK, they are currently being exploited elsewhere and the UK has failed to match its research effort with appropriate commercial backing. Despite these rather gloomy prognostications for the UK, it is clear that in terms of both science and the economy, monoclonal antibodies represent one of the major success stories of biotechnology.

REFERENCES

1. Anon. (1982). Antibody carriers of drugs and toxins in tumor therapy. *Immunological Reviews,* **62**
2. Jefferis, R. (1983). Monoclonal antibodies and British Industry. Too little... and too late? *Immunology Today,* **4**(5), 128–130
3. Kohler, G. and Milstein, C. (1975). Continuous cultures of fused cells secreting antibody of predefined specificity. *Nature, Lond.,* **256**, 495–497
4. Mach, J. P. *et al.* (1981). Use of radiolabelled monoclonal anti-CEA antibodies for the detection of human carcinomas by external photoscanning and tomoscintography. *Immunology Today,* **2** (12), 239–249

5. Ritz, J. *et al*. (1982). Autologous bone-marrow transplantation in CALLA-positive acute lymphoblastic leukaemia after *in-vitro* treatment with J5 monoclonal antibody and complement. *Lancet,* **2**, 60–63
6. Wakelin, D. (1983). Antigens yield to new technology. *Immunology Today,* **4**(11), 310–311

Part II
Applications of Biotechnology

20
MONOCLONAL ANTIBODIES — BIOTECHNOLOGY AND MEDICAL APPLICATIONS

A. D. Blann

INTRODUCTION

The past 20 years have witnessed the development of a vast repertoire of laboratory techniques that are slowly extending our knowledge. One of these, cell fusion, is providing new perspectives about the workings of the cell, its regulations, development and fate. The ability to grow cells outside the body (*in vitro*) makes available manipulations which are not possible with cells growing naturally inside an animal (*in vivo*). Cell fusion has applications because two cell types of known characteristics, and often widely differing backgrounds, can be fused to form a single cell — the daughter cell. This 'daughter' cell can be analysed for its own characteristics, some or all of which it will have inherited from its 'parents'. This kind of manipulation is virtually unheard of in nature, the major exception being the fusion of non-somatic gametes in sexual reproduction. Lesser organisms such as bacteria moulds and yeast can also fuse, but this facility is not as readily exploitable as is mammalian somatic cell fusion.

The phenomenon of a cell with more than one nucleus was observed by pathologists long before it became a research tool during the 1950s, when the practical seeds of cell fusion were sown. Many observations reported giant multinucleate cells associated with virus infections (such as mumps). The not unreasonable deduction was that the viruses were responsible for the multi- or fused cells. This property of certain virus groups was developed in the 1960s into a potent new tool for the induction of fusion between cells of varied backgrounds. However, the virus as a fusion agent has been succeeded by other methods, the latest being the use of pulses of electricity.

THE BIOCHEMISTRY OF SELECTIVE SURVIVAL OF FUSED CELLS

Several complicated mechanisms have to be employed to ensure the purity of a hybrid cell population. If we attempt to fuse two cancer cell lines, then not all the cells will fuse and some may fuse with the wrong partner (i.e. A × A hybrids, B × B hybrids) and there will be leftover As and Bs instead of only the desired A × B hybrids. Specific enzyme deficiencies have to be introduced into each cell line before the fusion event. This can be facilitated by adding to the tissue culture medium cytotoxic drugs, such as 8-azaguanine (8AG). This cytotoxic agent kills by replacing the normal constituents of DNA in the chromosomes and certain enzymes are required to do this. Hence, possession of the enzyme will lead to the incorporation of the poison, while absence of the enzyme will allow the cell to survive. The enzyme that is required to use 8AG is hypoxanthine-guanine phosphoribosyl transferase (HGPRT).

The cell has two different ways to make the DNA needed for its chromosomes: a *de novo* pathway, where the nucleic acids are slowly built up from simple amino acids and sugars, and a salvage pathway that utilises similar structures and reconverts them into the correct nucleic acids. The *de novo* pathway can be paralysed by a metabolic poison, aminopterin, but the salvage pathway is not affected by this drug. However, HGPRT is an essential member of the salvage pathway, so that culture in aminopterin of a cell line, A, selected for growth in 8AG (i.e. lacking HGPRT) will lead to its death because both of the methods of making DNA are inoperative; one is poisoned by aminopterin, the other is missing a particular enzyme. Cell hybrids A × B will survive because parent B has contributed one normal gene for the enzyme HGPRT and so can bypass the metabolic poisons. As a result of the extra demand placed on the salvage pathway (because the *de novo* pathway is blocked by aminopterin), extra hypoxanthine and thymidine are added. Hence, the cocktail is called HAT medium (hypoxanthine, aminopterin and thymidine). However, B × B cell lines will also survive in this medium. So how does selection work?

The situation becomes clearer when we consider the case of a fusion between a cancer cell and a normal untransformed cell. One other property of transformed cell lines *in vitro* is their immortality. Normal body cells will eventually die in tissue culture, as they have a limited life-span which can be as short as two or three weeks. So normal × normal cell hybrids will eventually die. Unfused cancer cells selected for HGPRT deficiency by growth in 8AG will die because they lack the enzyme, and so will be killed by the HAT medium. Only the fused hybrid normal/cancer cell will survive, because the normal cell provides a functioning HGPRT gene and the cancer cell provides the ability to live in a tissue culture environment which normal cells cannot.

Although initially far behind, plant fusion technology is rapidly catching up and has been able to side-step several zoological problems. Given that the world 'biomass' is overwhelmingly botanical, the potential is enormous in terms of food, pharmaceuticals, etc. The combination of fusions and genetic engineering techniques could, for example, improve yields and increase food values of seeds and fruit — yet astute farmers have been doing just this for centuries.

MONOCLONAL ANTIBODIES

Application of the principles of cell fusion and selection has permitted the development of a totally new method of making antibodies for use in research, diagnosis and treatment alike, which supersedes the conventional production of antibodies in rabbit, goat and sheep. The key is the myeloma cell, which is fused to a normal B lymphocyte. A myeloma is a type of cancer which is associated with an abnormal production of irregular antibody. All antibody molecules are made by B cells and each cell makes only one type of antibody, which is highly specific for its antigen; often more specific than an enzyme and its substrate. The disease, myeloma, is caused by one or even two or three single antibody-producing B cells that have lost their normal controls. The immune response is such that B cells are controlled (directly or indirectly) by the amount of antigen that they are to respond to: a large amount of antigen switches them on to make antibody and to divide, and when this antibody has removed the antigen, the falling levels of antigen will signal the B cells themselves to switch off into a dormant state. Myeloma cells, on the other hand, are B cells that have continued to divide and make unnecessary antibody. The dangerous characteristic of these cells is that their reproductive capacity is often high: they quickly multiply and invade other areas of the body, leaving their original home, which is generally a lymph node, the spleen or the bone marrow. However, the myeloma in the mouse has been turned to advantage in cell fusion. A myeloma can be induced with mineral oil and the active cells themselves isolated and grown up in tissue culture (which they will do because of their malignant character). The next step is to induce a deficiency in the enzyme HGPRT by using the poison 8-

azaguanine and then to isolate a mutant that fails to secrete its abnormal antibody molecules but still retains the machinery to do so. Immunologists are interested only in this defective genetic machinery inside the cell and not in its product. The correct information for making the antibody comes from the normal parent of the cell fusion and this is a B lymphocyte from an animal immunised with a specific antigen. An interesting illustrative analogy is to consider the myeloma cell as a motor car factory whose assembly line is not in operation because of a lack of direction. The B cell provides all the necessary plans, blueprints and instructions for the workers and robots to construct the product, which is assembled and exported.

In the construction of monoclonal antibodies, much preliminary work is necessary to work out the kinetics of the immune response. Armed with the immunising regimen, a high antibody titre to the antigen of interest can be obtained which should yield the most useful B cells. The spleen is removed from the mouse, as it is this organ that is most easily exploited (lymph nodes contain fewer B cells and are harder to remove), and the spleen cells are gently removed from inside the capsule. The myeloma cancer cells are generally grown in tissue culture flasks and reproduce quickly, so that large numbers can be obtained quite rapidly.

Although a virus (the Sendai virus) was originally used to fuse cells together, this has been replaced by another 'fusogen', polyethylene glycol (PEG), which is much safer and easier to handle. The fusion event consists simply of mixing the myeloma cells and the spleen cells in PEG, which takes only two or three minutes, as too much PEG can kill the cells. The fusion mixture is then seeded out into tissue culture vessels and biochemical selection begins with HAT medium being added. After 10–14 days or so, surviving hybrids will become apparent as the only live cells in the culture vessel, the HAT killing the myeloma parent and the normal spleen cells dying 'naturally' outside the body. The following 2 weeks are critical, as not all the growing cells will be useful — some will fail to make any antibodies at all, so their growth will contaminate useful hybrids. As a result, it is vital to constantly test each culture of hybrids to look for useful cells, and eliminate those that are unproductive.

Those cell populations that are making useful antibodies should be subcultured into smaller vessels and some of them frozen so that a catastrophe, such as a heavy bacterial contamination, does not destroy months of work. So now we have established a hopefully immortal series of antibody-producing cells that have retained their malignancy. The term 'hybridoma' can be employed for these cells; the suffix '-oma' is borrowed from other cancer terminology (e.g. carcinoma). However, all these hybridoma cells are making different types of antibody (because the spleen extract contained many different types of B cell, in addition to the one stimulated by the specific immunisation regimen). Therefore, we next have to sort out each useful cell type. The term 'monoclonal' derives from the knowledge that *each cell in a culture is derived ultimately from just one cell.* This is done by seeding out crude populations of hybridoma cells into agar jelly that is only just solid. By seeding at a low enough density we can follow the progress of a single hybridoma cell as it multiplies to form a clone. The clone can then be picked out of the agar with a Pasteur pipette and grown up in ever-increasing culture volume. Thus, the antibody produced will be of monoclonal origin, because all the cells making the antibody are derived from the same single cell. We are assured of this because the agar jelly prevents migration of cells between growing clones. The clone of hybridoma cells can also be grown inside the peritoneal cavity of a mouse (of a suitable strain), which makes the antibody much more concentrated.

It is difficult to understate just how important this kind of technology is to immunologists, biochemists and a host of other workers from many diverse fields. Just one advantage is the confidence with which investigators can make claims about their work. Using conventionally raised antiserum from a goat, there is little similarity between batches raised to the same antigen. Monoclonal antibodies contain no cross-reacting or contaminating antibodies, because they are all the product of a single cell line. Since the cell is a cancer cell, it should be able to live indefinitely *in vitro*, so that fresh stocks are always available and guaranteed to recognise the same antigen each time. However, not everything about monoclonal antibodies is wonderful. In some cases the antibody may not be very specific at all, in which case multi-antibody antiserum from a sheep may be more useful, as it would recognize several sites on the antigen. As the monoclonal antibody 'sees' only one site per antigen, that may not be enough. The whole method is also very expensive in terms of man-hours and reagents, and it is possible to work on a single fusion for months without the hint of a useful product. These cells also have the infuriating habit of dying without any apparent cause (although this may be associated with the loss of certain vital chromosomes), but something may be recovered if sufficient cells were stored frozen in liquid nitrogen.

The system described above is that of the mouse, yet it is possible to construct human and rat monoclonal antibodies, as the relevant myeloma cells have been developed in tissue culture. Rat fusions work on the same principle as the mouse, and rat hybridoma cells can be grown in the peritoneal cavity of a suitable rat. The human system is less well established, although significant advances are being made, but, of course, the system carries several major insurmountable problems. It is simply not possible to manipulate one's fellows as one can experimental animals. Hybridoma cells cannot be grown inside someone's peritoneal cavity; neither could the immunisation of a volunteer with a dangerous virus or a putative cancer antigen be sanctioned. Hence, human monoclonal antibodies are grown only in tissue culture, without the concentrating advantages of growth *in vivo*. This need not be too much of a problem, as antibodies can be concentrated by other means in the laboratory, without loss of potency. Neither rat nor human monoclonal antibodies have made much impact on the scientific/medical world, but mouse monoclonals have reached a stage of sophistication which has aroused considerable interest from commercial and drug companies.

APPLICATIONS OF MONOCLONAL ANTIBODIES

In spite of the above reservations, monoclonal antibodies can be used in place of conventional antisera in virtually any case, both as laboratory reagents and as therapeutic agents. Their unique properties make them ideal agents for the purification of small items such as hormones, interleukins and other peptides and glycoproteins. An example is the cell's own antiviral agent, interferon, once optimistically proposed as a cure for cancer. This chemical is extremely active and so is produced in the body in very small amounts, making purification difficult; indeed, the raising of a monoclonal antibody to it in the first instance was a major success. Armed with the antibody, interferon can be purified by affinity chromatography up to 500 or even 1000 times its initial concentration. Hence, large batches can be made of requisite purity, and the monoclonal antibody can be used time and time again and will remain constant. This is also important, as there are at least three types of interferon, so that an antibody to α-interferon will not collect types β or γ, unlike some conventional polyclonal antisera. The same monoclonal antibody can also be used in the estimation of interferon levels via a modified radioimmunoassay or immunoradiometric assay. The scientist can be sure that only α-interferon is being counted at each assay. The product was the first to be marketed by the UK's first major biotechnology company, Celltech.

Based in Slough, Celltech has interests in the recombinant DNA field, it has been quick to market several monoclonal antibodies such as that to α-interferon mentioned, and has another to γ-interferon which can also be used for purifying and assaying. Collaborating with Boots, a diagnostic division produces kits for use in the medical world to detect such blood proteins as thyroid-stimulating hormone and alpha-fetoprotein. The former may be used for the differentiation of hypothyroid, euthyroid and hyperthyroid conditions. The alpha-fetoprotein kit is used to test the maternal serum of pregnant women, and can be used to aid diagnosis of spina bifida and other fetal abnormalities at an early stage. Both these kits use monoclonal antibodies and have great potential, but not as great as Celltech's monoclonal antibodies to blood group antigens.

In partnership with Ortho Diagnostics, a current world market leader for the supply of conventional blood grouping reagents, Celltech will be entering a market worth in excess of £7.5 million. They are required to produce multi-kilogram quantities of monoclonal antibodies to the blood group substances A and B and have developed the world's largest fermenter to do this — it has a capacity of 1000 litres. The advantages of monoclonal blood grouping reagents are many, not least that they are free of pathogenic viruses, such as those which cause AIDS and hepatitis, and that a much improved quality control system can be applied. They will free many thousands of blood donations which would otherwise be used to make conventional grouping reagents, and so may instead be used for the preparation of blood products such as Factor VIII, required by haemophiliacs. The Blood Group Reference Laboratory at Oxford has now accepted the anti-A and anti-B monoclonals as substitutes for human-derived materials for clinical use in UK hospitals and blood transfusion centres. The products have been granted a licence by America's Food and Drug Administration, so that a much larger market is now available.

Monoclonal antibodies are the basis of a multitude of commercial kits used in diagnosis, such as that used in the detection of the hormone human chorionic gonadotrophin, identified with pregnancy. However, a useful illustration of the value of these reagents is amply demonstrated by their use in the study and treatment of diseases of the blood.

MONOCLONAL ANTIBODIES IN TREATMENT OF HAEMATOLOGICAL DISEASE AND TRANSPLANTATION

As therapeutic agents, monoclonal antibodies are in use to help patients with leukaemias, lymphomas, etc., and in bone marrow transplantation. In such cases the object is to remove those cells (often lymphocytes) whose aberrant behaviour is responsible for the illness, yet leave all the normal, healthy blood cells unaffected. This can be done by removing the patient's bone marrow, 'purging' to kill the cancerous cells and replacing the cleaned-up population. In some cases it is necessary to replace someone's bone marrow with that from another person — i.e. bone marrow transplantation. Here there will be problems with rejection, since it is the immunologically active cells in both the donor marrow and the patient that can react with each other with fatal outcome. Monoclonal antibodies can be used to suppress the action of the immune cells and so allow the new transplanted marrow in the recipient to succeed.

The same principle uses the same monoclonal antibodies as immunosuppression agents to prevent the rejection of a transplanted kidney. In this way the success of such a graft is much increased. Use of mouse monoclonal antibodies in man may be said to lead to allergic responses of the serum sickness kind (related to the responses of sufferers of hayfever, asthma, etc.), but there have been few reports of major adverse reactions.

This is an important benefit of monoclonal antibodies, because even if conventional antisera could be obtained of suitable specificity, such a large dose would be needed that unacceptable side-effects would follow. A novel application involves conjugation of the dangerous plant toxin ricin to a monoclonal antibody which is specific for cancer cells. Thus, the antibody targets the toxin directly on to the cell to be destroyed, avoiding normal cells. Many of these functions implicate the T lymphocyte, and the use of monoclonal antibodies has been essential for the elucidation of the various workings of these cells. The antibodies have been used to probe one of the major unknowns in immunology — that of the T lymphocyte receptor for antigen (the B lymphocyte receptor for antigen is in its antibody).

The use of monoclonal antibodies extends also to basic research in the mouse, where fundamental work on the induction of tolerance is being done. The use of the reagents to treat humans is only possible because of experimental work in rodents, which directly preceded the transfer of the method to the hospital ward. Moving away from the bedside, monoclonal antibodies are becoming reagents of choice in haematology laboratories where not even conventional antisera have been used. In these and other cases they are used to test patients' blood to find which of the various types of leukaemias may be present, or to discern whether an apparent leukaemic picture is actually a normal reaction, as in the case of infectious mononucleosis (glandular fever).

These antibodies are available world-wide and experiments can be directly compared between different groups. Hence, information about these cells can be quickly distributed and everyone can understand them, as they are based on the same set of reagents. Therefore, with widespread understanding there can be rapid agreement or resolution of problems, so that new facts are more quickly accepted or rejected. It is most unlikely that the frontiers of this branch of science would have been resolved so quickly without the use of monoclonal antibodies. The result of all this is that clinicians now have a much clearer idea of the workings of the body, so that they will be able to treat disease much more effectively, to prevent death and alleviate suffering.

BIBLIOGRAPHY

1. Acuto, O. and Reinherz, E. L. (1985). The human T cell receptor. *New England Journal of Medicine,* **312**, 1100–1111
2. Benjamin, R. J. and Waldmann, H. (1986). Induction of tolerance by monoclonal antibody therapy. *Nature, Lond.,* **320**, 449–451
3. Blann, A. D. (1986). Immunogold cell labelling system for T lymphocyte enumeration: an evaluation. *Med. Lab. Sciences,* **43**, 152–156
4. Erber, W. N., Pinching, A. J. and Mason, D. Y. (1984). Immunocytochemical detection of T and B cell populations in routine blood smears. *Lancet,* **i**, 1042–1046

5. Lennox, E. S. (Ed.) (1984). *Clinical Applications of Monoclonal Antibodies.* Churchill Livingstone, Edinburgh
6. McMichael, A. J. and Fabre, J. W. (Eds) (1982). *Monoclonal Antibodies in Clinical Medicine.* Academic Press, London

21
HAEMOPHILIA — A SHORT REVIEW OF THE DISORDER AND ITS TREATMENT

P. N. Robinson

THE DISORDER

Classical haemophilia (haemophilia A) is a bleeding condition caused by a congenital deficiency of Factor VIII coagulant activity in the bloodstream.

As the Factor VIII gene is located on the X chromosome, the condition is sex-linked, affecting 10–20 per 100 000 males in a population. The disease is found world-wide and has been reported in most racial groups[5].

The severity of bleeding in haemophiliacs varies from patient to patient and is related to the amount of Factor VIII present in the blood. Less severely affected individuals, with more than 2–3 per cent of Factor VIII, bleed after trauma but are normally spared spontaneous painful and crippling haemorrhages into joint muscles.

Severe haemophiliacs with no detectable Factor VIII bleed most commonly into joints, particularly knees, ankles, elbows and shoulders. The resulting haemarthrosis causes extreme pain and if untreated may take weeks or months to resolve. There may be irrevocable change to the membranes and cartilage and a resulting predisposition to further bleeds, and development of arthritic problems.

Bleeding into muscle tissue (haematoma) is also common and may be spontaneous or follow an injury such as a strain or direct blow. A severe haematoma may result in permanent limb deformities.

Intracranial haemorrhage is perhaps the commonest cause of death in haemophilia[4] and normally follows a known injury.

Finally, we need to consider surgery and dental extractions. These will obviously involve varying degrees of bleeding, and such operations could not be considered without the potent preparations of Factor VIII now available to give blood levels of Factor VIII adequate to control bleeding during and after surgery.

TREATMENT OF HAEMOPHILIA

The treatment of haemophilia is based on replacement therapy, which aims, by means of intravenous transfusion, to raise the circulatory plasma level of Factor VIII to that which will bring about cessation of bleeding (haemostasis) and to maintain this for the time necessary to allow the wound to heal. The cold precipitated fraction of plasma (cryoprecipitate) is rich in Factor VIII and solutions of cryoprecipitate prepared from single donors have been used for many years to treat bleeding episodes[16, 17]. While this method is still widely in use, it is being superseded by the use of freeze-dried preparations made from large pools of donor plasma.

In the UK freeze-dried human Factor VIII is prepared from freely donated blood at the Blood Products Laboratory, Elstree, and the Protein Fractionation Centre in Edinburgh.

There are also a number of commercial concentrates available, particularly from the United States.

Such freeze-dried preparations, which are reconstituted with sterile distilled water before use, provide a convenient and well-controlled source of material for the management of the haemophiliac.

The most important drawback associated with the use of freeze-dried concentrates prepared from large pools of plasma is the increased risk of transmission of hepatitis and, of greater current interest, Acquired Immune Deficiency Syndrome (AIDS) — see Chapter 21.

HAEMOPHILIA COMPLICATED BY FACTOR VIII ANTIBODIES

The development of antibodies to human Factor VIII clotting activity occurs in approximately 5–15 per cent of congenital haemophiliacs who receive Factor VIII replacement therapy[19]. Acquired haemophilia A — the spontaneous development of anti-Factor VIII antibodies in a patient with no previous haemostatic defect — occurs even less frequently, but together such patients represent a major challenge to the clinician's resources.

In haemorrhagic disorders prompt and effective treatment is vital in order to prevent irreversible damage to the patient. It is thus of great importance that an effective alternative form of treatment should be immediately available in cases of life- or limb-threatening bleeding in patients with Factor VIII inhibitors who have previously proved resistant to conventional haemophilia therapy.

There are two main approaches to the treatment of patients with Factor VIII inhibitors. The first is to attempt to either temporarily remove or permanently eradicate the inhibitor, and the second is to try to achieve haemostasis when the necessity arises using one of a variety of therapeutic approaches.

The former line of attack has been attempted using plasma exchange[23], affinity absorption of antibodies from plasma[14], immunosuppressive drugs[13], intravenous infusions of gammaglobulin concentrates[15] and immune tolerance regimens using either high- or low-dosage human Factor VIII[6, 18].

Although all of these methods have had reported success in certain patients, none has been found to be universally acceptable.

A number of clotting factor concentrates have been reported to be effective to some extent in securing haemostasis in those patients whose inhibitor titre against human Factor VIII is sufficient to preclude conventional Factor VIII concentrate therapy.

Both regular and 'activated' prothrombin complex concentrates have been evaluated in comparative studies and have been found to have an efficacy of between 50 and 65 per cent[11, 12, 20].

However, the mode of action of such materials is as yet unknown and there is no reliable means of predicting their likely effect or monitoring their efficacy by laboratory methods. The absence of a suitable laboratory assay also leads to problems in gauging correct dosage levels and frequency of treatment, and may preclude their use as cover for surgical procedures.

Another therapeutic option is the use of an animal-derived Factor VIII concentrate. Bovine and porcine Factor VIII preparations for therapeutic use were first prepared by Bidwell[1, 2]. Although used successfully during a period when human Factor VIII was in insufficient supply, there were important side-effects of allergic and pyrogenic character. Thrombocytopenia also occurred, which was thought to be due to an activity, associated with the animal Factor VIII preparation, causing aggregation of human platelets *in vivo* (so-called platelet aggregating factor). Because of these problems the use of such material was generally reserved for the treatment of potentially lethal haemorrhages and for life saving surgery[3]. The only animal concentrate available today is a porcine concentrate (Hyate C), manufactured by Speywood Laboratories Limited, Wrexham, Clwyd. This material shows few of the problems associated with the original preparations[9] and is a result of 10 years of development work aimed at reducing the adverse side-effects while retaining beneficial therapeutic properties.

The key to this success has been the development of a fractionation procedure, based upon a novel ion-exchange resin, developed from methodology originally described by Johnson *et al.*[8]. The result has been the production of a purer product, giving fewer allergic side-effects and notably a reduction in the level of platelet aggregating factor with concomitant fall in the incidence of thrombocytopenia.

The rationale behind the use of this concentrate in patients with an antibody against human Factor VIII is that while such antibodies are rarely absolutely specific for human Factor VIII, in the majority of cases the activity against porcine Factor VIII is considerably weaker.

This form of treatment for inhibitor patients, where antibody levels are sufficiently high to make treatment with human Factor VIII difficult or impossible, has the advantage that the result is a measurable level of Factor VIII in the patient's bloodstream, enabling dosage regimens to be evaluated and clinical effects predicted[10]. The use of porcine Factor VIII to cover surgical procedures is also facilitated.

FUTURE DEVELOPMENTS

The current concern over the association of AIDS with blood products may lead to the increasing use of porcine Factor VIII in patients where human Factor VIII would normally be used, notably those with a low antibody titre. This development will depend upon the success, both practically and economically, of heat treatment programmes now being carried out on both NHS and commercial human blood products aimed at the prevention of transmission of AIDS.

More importantly in the longer term, the recent announcement of the complete characterisation of the human Factor VIII gene[7], its structure[22] and cloning[21, 24] heralds the next generation of genetically engineered products which will provide a safe treatment for haemophiliacs with pure Factor VIII free from virus diseases such as AIDS and hepatitis.

REFERENCES

1. Bidwell, E. (1955). *Br. J. Haematol.*, **1**, 35–45
2. Bidwell, E. (1955). *Br. J. Haematol.*, **1**, 386–9
3. Biggs, R. (1976). In *Human Blood Coagulation, Haemostasis and Thrombosis*, p. 371 (Biggs, R. and Rizza, C. R., Eds.), Blackwell Scientific, Oxford
4. Biggs, R. (1977). *Br. J. Haematol.*, **35**, 487–504
5. Biggs, R. (1984). In *Human Blood Coagulation, Haemostasis and Thrombosis*, pp. 92–118 (Biggs, R. and Rizza, C. R., Eds.), Blackwell Scientific, Oxford
6. Brackmann, H. H. (1984). In *Factor VIII Inhibitors*, pp. 181–195 (Hoyer, L. W., Ed.), A. R. Liss
7. Gitschier, J. *et al.* (1984). *Nature*, **312**, 326–330

8. Johnson, A. J. *et al*. (1978). *J. Lab. Clin. Med.*, **92**, 194
9. Kernoff, P. B. A. and Tuddenham. E. G. D. (1981). *Br. Med. J.*, **283**, 381–382
10. Kernoff, P. B. A. *et al*. (1984). *Blood*, **63**, 31–41
11. Lusher, J. M. *et al*. (1980). *N. Engl. J. Med.*, **303**, 401
12. Lusher, J. M. *et al*. (1983). *Blood*, **62**, 1135
13. Nilsson, I. M. and Hedner, U. (1976). *Scand. J. Haematol.*, **16**, 369
14. Nilsson, I. M. *et al*. (1981). *Blood*, **58**, 38
15. Nilsson, I. M. *et al*. (1983). *Scand. J. Haematol.*, **30**, 458
16. Poole, J. G., Hershgold, E. J. and Paperhagen, A. R. (1964). *Nature*, **203**, 312
17. Poole, J. G. and Shannon, A. E. (1965). *N. Engl. J. Med.*, **273**, 1443–1447
18. Rizza, C. R. and Mathews, J. M. (1982). *Br. J. Haematol.*, **52**, 13
19. Shapiro, S. S. (1984). In *Factor VIII Inhibitors*, pp. 44–55 (Hoyer, L. W., Ed.), A. R. Liss
20. Sjamosoedin, L. J. *et al*. (1981). *N. Engl. J. Med.*, **305**, 717
21. Toole, J. J. *et al*. (1984). *Nature*, **312**, 342–346
22. Vehar, G. A. *et al*. (1984). *Nature*, **312**, 337–342
23. Wensley, R. T. *et al*. (1980). *Br. Med. J.* **281**, 1388
24. Wood, W. I. *et al*. (1984). *Nature*, **312**, 330–337

22
RECENT DEVELOPMENTS IN THE TREATMENT OF DIABETES

Cliff J. Bailey and Peter R. Flatt

INTRODUCTION

Diabetes mellitus is a growing health problem throughout the world. In the United Kingdom alone there are probably more than one million sufferers, many of whom rely on daily injections of insulin. There have been several important developments recently in the treatment of diabetes. Particularly significant has been the coming of age of human insulin.

HUMAN INSULIN

Since the introduction of insulin therapy over 60 years ago, commercial insulins have been obtained by large-scale extraction of pig and cattle pancreas. For several decades the preparations contained impurities which caused allergic responses, including the production of anti-insulin antibodies. With the more recent use of highly purified (so-called *mono-component*) insulins, these problems have largely disappeared and smaller doses of insulin have been successfully prescribed. Pork and beef insulins differ from human insulins by only 1 and 3 amino acid residues, respectively, and mono-component preparations of these insulins have proved most satisfactory in clinical practice. However, there has long been a feeling that purified human insulin would be most desirable. Also, the increasing prevalence of diabetes has prompted predictions of an imminent world shortage of extractable animal insulins.

To meet this challenge, Eli Lilly & Co collaborated with Genentech to produce biosynthetic human insulin (BHI) using a recombinant DNA technique. While Eli Lilly was scaling-up for commercial production, the Danish company Novo Industri prepared human insulin by chemical modification of pork insulin and the resulting product was introduced commercially in the UK in June 1982. Trials using human insulin prepared by both companies have shown that biological activity is generally the same as that of pork insulin, although preliminary results suggest that human insulin may be absorbed and inactivated more rapidly than monocomponent pork preparations.

INSULIN INFUSION PUMPS

A pressing need exists for more physiological methods of insulin administration. Daily injections of insulin eliminate the immediate life-threatening effects of Type 1 diabetes, but they constitute an inconvenient and crude therapeutic tool which lacks the regulatory finesse of a healthy pancreas. Even a meticulous patient may not achieve the constant control over blood glucose that is required to minimise the long-term complications of diabetes, such as failing sight, and cardiovascular, kidney and neural disorders.

A practical approach to this problem has been the introduction of small portable pumps with an in-dwelling cannula to enable continuous subcutaneous insulin infusion (CSII). Reports of several clinical trials indicate that patient-adjustable CSII regimens can provide a level of blood glucose control close to the physiological norm. Rapid improvements in the design and capabilities of the pumps have been encouraging, but more fundamental is the continuing research towards an autonomous and fully implantable system. This is envisaged to combine a blood glucose sensor, power source, microcomputer and rechargable insulin reservoir with a pump which releases insulin at precisely the required rate.

U100 INSULIN

Insulin users have recently experienced a change in the concentration of commercial insulin preparations which started in March 1983, when the existing 20, 40 and 80 units/ml preparations were replaced by 100 units/ml preparations (U100 insulin). Hopefully, the new uniformity will prevent the confusion which sometimes arose with the previous array of concentrations.

HOME GLUCOSE MEASUREMENT

Several reports have noted that regular blood-glucose monitoring by the patient is conducive to better control. Considerable success has been achieved with miniature battery-operated meters which accurately measure the concentration of glucose in a drop of blood from a finger prick. However, most patients continue to use the visually read test strips. These strips provide a similar but less accurate procedure based on a change in the colour of the strip when activated by the glucose in a drop of blood. Although Boehringer Mannheim has led the field with a two-tone visual strip, a cheaper strip has recently been introduced by the Ames Division of Miles Laboratories.

GLYCOSYLATED HAEMOGLOBIN

It is customary to check the blood glucose of a diabetic patient at each visit to the out-patient clinic. This can sometimes give a misleading impression, since patients are often more conscientious immediately prior to the visit. The non-enzymatic attachment of glucose to the protein moiety of haemoglobin (glycosylated haemoglobin) has been exploited as a measure of long-term blood glucose control. Recent improvements in methodology have made the measurement of glycosylated haemoglobin a routine parameter of choice in many diabetic clinics. Analysis of glycosylated haemoglobin identifies patients who have not achieved good control throughout the months between out-patient visits. Treatment can then be adjusted to accommodate the more typical needs of the individual.

NEW DIETARY RECOMMENDATIONS

The traditional emphasis on a low-carbohydrate diet for diabetics has been re-examined in recent years. In November 1982 the British Diabetic Association gave its approval to a revised approach to the diabetic diet. Foremost, there remains the key objective of regulating total energy intake to achieve and maintain an ideal body weight. The new recommendations also continue to stress the danger of eating highly refined sugars such as sucrose and glucose. However, it is now suggested that 50 per cent or more of total calories may be taken as carbohydrate, provided that this is in the form of complex polysaccharides. Especially favoured are the fibre-rich sources, including bread, potatoes, cereals and beans. The rationale here is to slow the rate of intestinal glucose absorption. This improves glycaemic control by reducing postprandial peaks and interprandial troughs of blood glucose. To counter the increased risk of atherosclerosis and hypertension among diabetics, the new recommendations suggest a low intake of saturated fats (less than 35 per cent of total calories) and a low intake of salt.

FUTURE DEVELOPMENTS

Recent years have seen promising advances in the therapeutic control and management of patients with insulin-dependent diabetes. For the future treatment of these patients it may become feasible to implant insulin-secreting cells from cultured or cryo-preserved tissue banks. At present the major obstacle to transplantation is the rejection of allografted and xenografted tissue. However, several devices to segregate the transplanted tissue from antibodies and cellular components of the host's immune system are now receiving investigation. The application of gene therapy is a further possibility, in which genetic material coding for the biosynthesis and secretion of insulin could be introduced into selected cells of the patient. Also, encouraging research in the area of diagnosis points towards the future identification of those at high risk of developing insulin-dependent diabetes.

ANTIDIABETIC DRUGS: NEW DEVELOPMENTS IN THE TREATMENT OF DIABETES

Diabetes mellitus is a complex and heterogeneous disorder of intermediary metabolism characterised by an increased concentration of glucose in the blood, termed *hyperglycaemia*. Insulin, a polypeptide hormone secreted into the blood by the B-cells of the pancreatic islets of Langerhans, normally prevents hyperglycaemia. The hormone stimulates the cellular uptake and metabolism of glucose and inhibits excess production of glucose by the liver. In diabetes there is either a lack of insulin, an impaired ability of cells to respond to insulin (insulin resistance), a surfeit of factors which oppose the action of insulin, or a combination of these defects.

Characteristics of the main types of diabetes are documented in Table 1. In Type 1 diabetes destruction of the pancreatic B-cells results in a complete (or almost complete) lack of insulin, rendering an acute life-threatening condition in which the patient is dependent upon insulin therapy for survival. The less-critical nature of Type 2 diabetes does not necessitate insulin therapy and patients are usually treated by dietary adjustments and oral hypoglycaemic drugs where required[2, 3]. Protracted hyperglycaemia, even at modest levels, can hasten the onset and increase the severity of an array of chronic morbid complications, including

Table 1 Features of Type 1 and Type 2 diabetes mellitus

	Pancreatic B-cell population	*Insulin response to glucose*	*Insulin resistance*	*Hyperglycaemia*	*Possible causes*	*Treatment*
Normal	Fully functional	Rapid	Absent	Absent	—	—
Type 1	All or most B-cells destroyed	No response or negligible response	Often present	Very severe, culminates in fatal coma if untreated	B-cell attack by virus toxins or autoantibodies	Diet plus insulin
Type 2 (mild)	Approximately normal	Delayed response, normal or increased in magnitude	Present	Modest, mainly after meals	Over-eating, obesity, chronic infections, pregnancy, other metabolic stresses	Diet alone or plus oral hypoglycaemic drugs
Type 2 (severe)	Approximately normal or reduced	Delayed response, often reduced in magnitude	Present, often very severe	Modest to severe	Deterioration of mild Type 2 condition	Diet plus oral hypoglycaemic drugs

cardiovascular disease, kidney failure, neural disorders and blindness[4]. Thus, the rigorous control of blood glucose concentrations in diabetic patients is a major therapeutic objective[5].

Oral Hypoglycaemic Drugs

This section concentrates specifically on the use of oral drugs in the treatment of Type 2 diabetes. This type of diabetes accounts for 80–90 per cent of known diabetic patients, about half of whom receive oral drugs. The disorder is now estimated to rank as the third most common chronic condition after cancer and cardiovascular disease in several western societies. Accurate prevalence figures are not available, since the disorder often passes undiagnosed for many years until it is brought to attention by the appearance of one or more chronic morbid complications. Available evidence suggests that in Britain, Type 2 diabetes may affect 3–5 per cent of persons aged over fifty and the present usage of oral hypoglycaemic drugs accounts for an annual expenditure in excess of £12 000 000.

Oral hypoglycaemic drugs were introduced during the 1950s. Those most widely prescribed in Britain are the sulphonylureas, with 12–15 per cent of the market accounting for the biguanide metformin (Figure 1). The mechanisms through which sulphonylureas and metformin influence blood glucose concentrations are illustrated in Figures 1 and 2[3, 7–9].

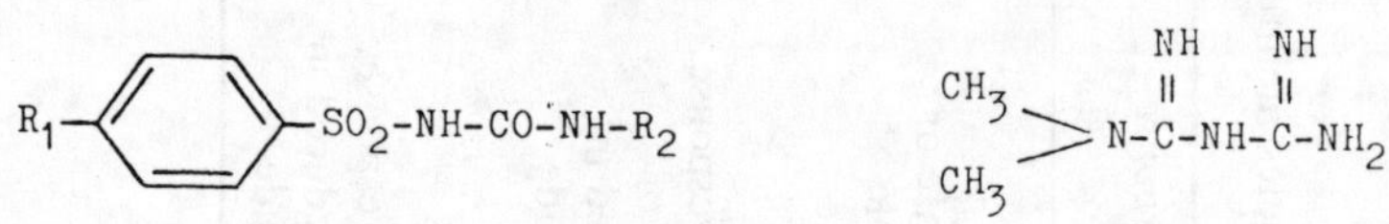

Figure 1 Sulphonylureas and metformin

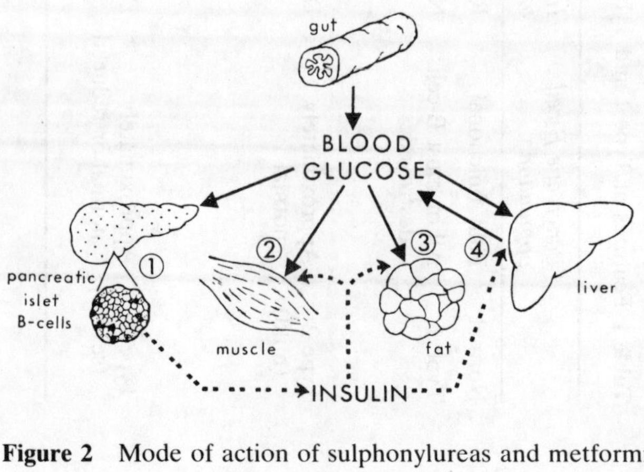

Figure 2 Mode of action of sulphonylureas and metformin

Sulphonylureas
This group of drugs, which derives from sulfonamide antibiotics, acts initially to lower blood glucose concentations by evoking a rapid, transient release of insulin. After several months of treatment the pancreatic B-cells often lose their responsiveness to sulphonylureas, but the hypoglycaemic efficacy is usually maintained. This results from an effect on other tissues, such as liver, fat and muscle, where the drugs enhance the action of insulin. This effect is also prone to disappear slowly and sulphonylurea treatment ceases to benefit Type 2 diabetic patients at a rate of about 10 per cent per annum. If a combination of the sulphonylurea with metformin is not successful, the patient's condition may deteriorate to the extent that insulin treatment is required.

The range of available sulphonylurea preparations offers a diversity of pharmacokinetic properties. In general, sulphonylureas are rapidly absorbed and rapidly cleared, with hypoglycaemic actions lasting from 6 hours to more than 1 day. Sulphonylureas are at least partly degraded in the liver and excreted mostly in the urine. Hence, their hypoglycaemic activity may be altered if hepatic or renal function is impaired and excessive amounts of the drugs can precipitate dangerous episodes of severe hypoglycaemia. Since the insulin-releasing effect of sulphonylureas is not accompanied by increased insulin biosynthesis, it is possible that the long-term preservation of B-cell function is not assisted. Associated with their enhancement of insulin-mediated glucose uptake by fat, sulphonylureas are prone to induce weight gain, which is not a favoured action in a large proportion of Type 2 patients who are already over-weight. Although sulphonylureas interact with many other medicines and often interfere with renal fluid handling, they are generally well tolerated and are widely used.

Metformin
Metformin is the only antidiabetic biguanide used in Britain. The development of biguanide drugs stems from the identification of guanidine in *Galega officinalis* (goats' rue), a traditional herbal treatment for diabetes in Europe. Metformin does not affect insulin secretion but appears to increase the actions of insulin on muscle and liver. Its effect usually lasts for 4–5 hours. Since the drug has little effect in subjects with normal glucose concentrations and does not lower glucose concentrations below normal in diabetic patients, its action may be more aptly described as 'antihyperglycaemic' rather than 'hypoglycaemic'. Thus metformin does not incite the danger of severe hypoglycaemia. Metformin is rapidly absorbed and cleared. It is excreted in the urine apparently unchanged and for this reason it is not suited to patients with impaired renal function. Metformin can increase lactate production and if prescribed inappropriately, this may engender a risk of lactic acidosis. At the onset of metformin

therapy and if the drug is taken without food, the patient may experience gastro-intestinal incommode, while in the long term it can impair the absorption of some vitamins. However, metformin does not induce weight gain and does not interact unduly with other medicines, which favours its use in over-weight patients and patients receiving other treatments.

The Need for New Oral Hypoglycaemic Agents

While sulphonylureas and metformin have proven very valuable treatments for hyperglycaemia of Type 2 diabetes, it is evident that they are not ideal. Although they may alleviate severe hyperglycaemia, these drugs are often unable to reinstate and permanently maintain glucose concentrations within the normal range. Hence, the risk of chronic complications arising from modest or intermittent hyperglycaemia is not eliminated. Clearly there exists a pressing need for new types of orally active hypoglycaemic agents.

Insulin resistance, excess hepatic glucose production and a delayed insulin response to glucose are the main defects responsible for hyperglycaemia in Type 2 diabetes[2]. A new hypoglycaemic agent should therefore combat one or more of these conditions. Insulin resistance is probably the most invidious defect, since it directly impedes the usage of glucose by a diversity of cell types in most tissues and organs. The defect is initially countered by increased secretion of insulin, the extent being dependent upon the adaptive capacity of the pancreatic B-cells. When the demand for insulin exceeds the supply and the B-cells suffer degenerative changes, the severity of diabetes escalates rapidly. Ironically, conditions of insulin resistance can be exacerbated by insulin treatment[10]. Moreover, the large doses of insulin required to overcome pronounced insulin resistance can have undesirable effects on other metabolic processes, such as abnormal lipoprotein synthesis which is linked with the development of cardiovascular complications. Given also the inconvenience of insulin injections, it is apparent why insulin therapy is not a treatment of choice in Type 2 diabetes.

The problematic relationship between insulin and insulin resistance highlights the need for agents targeted directly against insulin resistance.

The relationship also indicates that agents acting to increase insulin secretion cannot be expected to offer more than a short-term respite for the Type 2 patient. Agents which reduce insulin resistance might also reduce hepatic glucose production, since hepatic gluconeogenesis is normally suppressed by insulin.

New Oral Hypoglycaemic Drugs Under Consideration

Without labouring the specifications of an ideal new hypoglycaemic agent, it would be desirable if the efficacy could be maintained for many years to meet with the on-going nature of the diabetic condition. A slowly acting agent would be acceptable if the risks of an over-effect, resulting in severe hypoglycaemia, could be safeguarded against. A highly potent hypoglycaemic principle would more appropriately possess pharmacokinetic properties which would allow its action to be rapidly induced and rapidly terminated, giving least opportunity for an over-effect while optimising compatibility with the normal daily fluctuations in the physiological control of glucose homoeostasis. Safety aspects, as always, stand foremost and oral route of administration would be convenient and facilitate patient compliance.

Many substances with apparent hypoglycaemic activity have been reported and patented, but surprisingly few have received detailed attention, almost certainly owing to toxicity or adverse side effects at the dose required to lower glucose concentrations in human diabetes. Also, the immense resources and expenditure required to develop a new drug must have curbed the speculative indulgence of the pharmaceutical industry.

Compounds under consideration as possible new oral hypoglycaemic drugs are documented in Table 2. Published evidence concerning their efficacy derives mainly from studies in normal and diabetic laboratory animals. Only where a compound is already in clinical use for another purpose has its effect been studied in Type 2 human patients. Since diabetes in animals is a heterogeneous disorder, no single animal model of diabetes precisely duplicates every feature of any particular presentation of diabetes in man. Thus, it is informative to study the activity of a prospective hypoglycaemic agent in a range of different animal models, although an ability to counter any of the animal syndromes of diabetes does not guarantee a suitable antidiabetic effect in man[11]. Animal studies can valuably indicate the mode of action of a hypoglycaemic agent and it is notable that Table 2 is not over-abundant in new compounds with a demonstrated ability to increase insulin action and thereby reverse insulin resistance. An on-going search for new types of hypoglycaemic substances is therefore indicated.

Traditional Treatments for Diabetes

Traditional herbal remedies for diabetes continue to receive widespread use, particularly in societies which do not receive adequate supplies of conventional medicines[12]. The plant materials used in these remedies represent a potential source of new hypoglycaemic compounds. Informa-

Table 2 Compounds under consideration as possible new oral hypoglycaemic agents

Compound	*Hypoglycaemic activity observed in:*	*Proposed mode of action*	*Ref.*
AS-6	Normal and diabetic rats and mice	↑ Insulin secretion and action	17
B 807-27	Normal rats, mice, guinea-pigs and dogs	↓ Hepatic gluconeogenesis	
	Diabetic pigs and mice	↓ Fatty acid oxidation	18
BRL 26830	Normal and diabetic rats and mice	↑ Insulin secretion and thermogenesis	
		β-adrenoceptor agonist	19
Ciglitazone	Normal and diabetic rats, mice and hamsters	↑ Insulin synthesis and action	20
Fenfluramine	Type 2 diabetic patients	↑ Insulin action	21
HB 699	Normal rats, mice and dogs	↑ Insulin secretion	22
Hydrazonopropionic acids	Normal and diabetic rats, mice, hamsters and guinea-pigs	↓ Hepatic gluconeogenesis	
		↓ Rate intestinal glucose absorption	23
Linogliride	Normal rats, mice and human subjects	↑ Insulin secretion	24
Mebendazole	Type 2 diabetic patients	↑ Insulin secretion	25
McN 3716	Normal and diabetic rats and dogs	↓ Hepatic gluconeogenesis	
		↓ Fatty acid oxidation	26
McN 3802	Normal mice and guinea-pigs	↓ Hepatic gluconeogenesis	
	Diabetic mice	↓ Fatty acid oxidation	27
Piroglíride	Normal rats, mice, dogs and monkeys	↑ Insulin secretion	
	Diabetic rats and mice		28
TA-078	Normal and diabetic rats, mice and dogs	↑ Insulin secretion	29

tion concerning the efficacy of traditional remedies is extremely limited, deriving almost exclusively from unsubstantiated anecdotal accounts. Realising the unexplored possibilities of this natural pharmacy, the World Health Organization has recommended that a screening programme should be undertaken[5]. It might be envisaged that traditional herbal remedies should not gain acceptance if they produce adverse effects. However, a recent example concerning the traditional use of unripe ackee fruits (*Blighia sapida*) in Jamaica and central America serves as a notable contradiction to this view. The fruits yielded a potent hypoglycaemic extract (hypoglycin), which proved unsuitable for further development owing to an unacceptable toxic effect on hepatic metabolism[13]. Some other traditional antidiabetic treatments showing hypoglycaemic activity are listed in Table 3. The nature of their active principles and their toxicity are now under investigation.

Examples of treatments for Type 2 diabetes which have recently been developed from the study of traditional plant remedies are the unabsorbable gel-forming polysaccharides guar and konjac. These substances are added to the diet to reduce the rate of intestinal glucose absorption. The galactomannan guar, from seeds of the Indian cluster bean (*Cyamopsis tetragonolobus*) has long been used as a stabilising agent in the food and cosmetics industry[14] and a glucomannan, konjac, was discovered in roots of the Japanese konjac plant *Amorphophallus konjac*[15]. Another polysaccharide, acarbose, isolated from cultured strains of actinomycetes, also provides a dietary supplement for the treatment of Type 2 diabetes. It inhibits the action of intestinal α-glucosidase enzymes, thereby delaying the rate of carbohydrate digestion[16].

Table 3 Some traditional antidiabetic plant treatments which have proved to produce a hypoglycaemic effect

Plant	*Traditional antidiabetic treatment in:*	*Hypoglycaemic activity observed in:*	*Ref.*
Momordica charantia	Asia/W. Indies	Normal and diabetic rats and mice Type 2 diabetic patients	30
Coccinia indica	Asia	Type 2 diabetic patients	31
Kuwaiti herbal preparation	Middle East	Normal and diabetic rats	32
Teucrium oliverianum	Middle East	Diabetic mice	33
Trigonella foenum-graecum	Africa, Asia, Europe	Normal and diabetic rats, rabbits and dogs	34

CONCLUSION

Presently available sulphonylureas and metformin are useful oral hypoglycaemic drugs for the control of hyperglycaemia in Type 2 diabetic patients. However, these drugs have limitations, particularly with regard to duration of efficacy and reinstatement of a physiologically normal pattern of blood glucose homeostasis. New formulations of sulphonylureas are introduced with regularity but entirely novel compounds, to counter insulin resistance, reduce hepatic glucose production and improve insulin biosynthesis and secretion, are long overdue.

REFERENCES

1. National Diabetes Data Group (1979). Classification and diagnosis of diabetes mellitus and other categories of glucose intolerance. *Diabetes,* **28**, 1039–1057
2. DeForonzo, R. A., Ferrannini, E. and Koivisto, Y. (1983). New concepts in the pathogenesis and treatment of non-insulin dependent diabetes mellitus. *Am. J. Med.,* **74**, Suppl. 1, 52–81
3. Lebovitz, H. E. and Feinglos, M. N. (1983). The oral hypoglycaemic agents. In *Diabetes Mellitus, Theory and Practice*, 3rd edn (Eds. Ellenberg, M. and Rifkin, H.), pp.591–610, New York, Medical Examination Publishing Co.
4. Keen, H. and Jarrett, J. (Eds.). *Complications of Diabetes*, 2nd edn, London, Edward Arnold
5. World Health Organization Expert Committee on Diabetes mellitus. Second Report, Technical Report Series No. 646, Geneva, World Health Organization, 1980
6. British National Formulary, No. 10. London: British Medical Association and The Pharmaecutical Society of Great Britain, 1985
7. Lockwood, D. H., Gerich, J. E. and Goldfine, I. (Eds.) (1984). Symposium on effects of oral hypoglycaemic agents on receptor and postreceptor actions of insulin. *Diabetes Care,* **7**, Suppl. 1, 1–125
8. Hermann, L. S. (1979). Metformin: a review of its pharmacological properties and therapeutic use. *Diab. Metab.,* **5**, 233–245
9. Bailey, C. J. (1985). The antihyperglycaemic action of metformin. In *Diabetes and Metformin*, Krans, H. M. J. (Ed.), Royal Society of Medicine Int. Congr. Series No. 79, pp.17–26, London, Royal Society of Medicine
10. Bailey, C. J., Lord, J. M. and Atkins, T. W. (1985). The insulin receptor and diabetes. In *Recent Advances in Diabetes*, Vol. 1, Nattrass, M. and Santiago, J. V. (Eds.), pp.27–44, Edinburgh, Churchill Livingstone
11. Bailey, C. J. and Flatt, P. R. (1986). Animal models of diabetes. In *Recent Advances in Diabetes*, Vol. 2, Nattrass, M. (Ed.), Edinburgh, Churchill Livingstone
12. Bailey, C. J., Day, C. and Leatherdale, B. A. (1986). Traditional plant remedies for diabetes. *Diabetic Med.,* **3**, 185–186
13. Kean, E. A. and Pogson, C. I. (1979). Inhibition of gluconeogenesis in isolated

rat liver cells by methylenecyclopropylpyruvate (ketohypoglycin). *Biochem. J.*, **182**, 789–796
14. Day, C. (1986). Guar gum. *International Industrial Biotechnology*, **6**, 3
15. Doi, K., Matsuura, M., Kawara, A. and Baba, S. (1979). Treatment of diabetes with glucomannan (Konjac mannan). *Lancet*, **i**, 987–988
16. Aubell, R., Boehme, K. and Berchtold, P. (1983). Blood glucose concentrations and glycosurea during and after one year of acarbose therapy. Arzneimittelforsch, *Drug Res.*, **33**, 1314–1318
17. Hosokawa, T., Ando, K. and Tamura, G. (1985). An ascochlorin derivative, AS-6, reduces insulin resistance in the genetically obese diabetic mouse, db/db. *Diabetes*, **34**, 267–274
18. Wolf, H. P. O., Eistetter, K. and Ludwig, G. (1982). Phenylalkyloxirane carboxylic acids, a new class of hypoglycaemic, and hypoketonaemic effects of sodium 2-5(4-chlorophenyl)-pentyl-oxirane-2-carboxylate (B 807-27) in fasted animals. *Diabetologia*, **22**, 456–463
19. Carroll, M. J., Lister, C. A., Sennitt, M. V., Strewart-Long, N. and Cawthorne, M. A. (1985). Improved glycaemic control in C57BL/KsJ (db/db) mice after treatment with the thermogenic β-adrenoceptor agonist, BRL 26830. *Diabetes*, **34**, 1198–1204
20. Diani, A. R., Peterson, T., Sawada, G. A., Wyse, B. M., Gilchrist, B. J., Hearron, A. E. and Chang, A. Y. (1984). Ciglitazone, a new hypoglycaemic agent. 4. Effect on pancreatic islets of C57BL/6J ob/ob and C57BL/KsJ db/db mice. *Diabetologia*, **27**, 225–234
21. Ribes, G., Trimble, E. K., Blayac, J. P., Wollheim, C. B., Puech, R. and Loubatieres-Mariani, M. M. (1981). Effect of a new hypoglycaemic agent (HB 699) on the *in vivo* secretion of pancreatic hormones in the dog. *Diabetologia*, **20**, 501–505
22. Verdy, M., Charbonneay, L., Verdy, I., Belonger, R., Bolte, E. and Chaisson, J. L. (1983). Fenfluramine in the treatment of non-insulin dependent diabetics: hypoglycaemic versus anorectic effect. *Int. J. Obesity*, **7**, 289–297
23. Oellerich, M. and Haeckel, R. (1980). Hydrazonopropionic acid, a new class of hypoglycaemic substances. *Horm. Metab. Res.*, **12**, 182–189
24. Tutwiler, G. F., Tuman, R. W., Wallace, N. H. and Zawalich, W. (1984). Studies in isolated rat islets with linogliride fumarate (LIN), a new orally effective hypoglycaemic agent. *Excerpta Medica Int. Congr.*, Series No. 652, p.1434, Amsterdam, Excerpta Medica
25. Caprio, S., Ray, T. K., Boden, G., Smith, R. H., Mozzoii, M. A., Hoeldtke, R. D., Shuman, C. R. and Owen, O. E. (1982). Mebendazole increases insulin secretion: improvement of insulin secretion in diabetic patients during mebendazole administration. *Excerpta Medica Int. Congr.*, Series No. 577, p.200, Amsterdam, Excerpta Medica
26. Tutwiler, G. F., Kirsch, T., Mohrbacher, R. J. and Ho, W. (1978). Pharmacologic profile of methyl-2-tetradecylglycidate (McN-3716), an orally effective hypoglycaemic agent. *Metabolism*, **27**, 1539–1556
27. Lee, S. M., Tutwiler, G., Bressler, R. and Kircher, C. H. (1982). Metabolic control and prevention of nephropathy by 2-tetradecylglycidate in the diabetic mouse (db/db). *Diabetes*, **31**, 12–18
28. Zawalick, W. S., Weill, V. A. and Matschinsky, F. M. (1980). Piroglíride: a

new oral hypoglycaemic drug which accelerates glucose usage and insulin secretion by islets of Langerhans. *Diabetes,* **29**, 410–412
29. Iwai, H., Inamasu, M., Totsuka, T., Shimazaki, T., Morita, T. and Takeyama, S. (1983). Hypoglycaemic activity of 1-α (3,4-dimethoxyphenethyl-aminomethyl)-2-hydroxybenzylalcohol-1,2-fumarate (TA-078) in the mouse, rat and dog. *Biochem. Pharmacol.,* **32**, 849–855
30. Bailey, C. J., Day, C., Turner, S. L. and Leatherdale, B. A. (1985). Cerasee, a traditional treatment for diabetes. Studies in normal and streptozotocin diabetic mice. *Diabetes Res.,* **2**, 81–84
31. Azad Khan, A. K., Akhtar, S. and Mahtab, H. (1980). Treatment of diabetes mellitus with *Coccinia indica. Brit. Med. J.,* **280**, 1044
32. Al-Awadi, F. M., Augelo-Khattar, M. and Gumaa, K. A. (1985). On the mechanism of the hypoglycaemic effect of a plant extract. *Diabetologia,* **28**, 432–434
33. Ajabnoor, M. A., Al-Yahya, M. A., Tariq, M. and Jayyab, A. A. (1984). Antidiabetic activity of *Teucrium oliverianum. Fitoterapia,* **55**, 227–230
34. Ribes, G., Sauvaire, Y., Baccou, J. C., Valette, G., Chenon, D., Trimble, E. R. and Loubatieres-Mariani, M. M. (1984). Effects of fenugreek seeds on endocrine pancreatic secretions in dogs. *Ann. Nutr. Metab.,* **28**, 37–43

BIBLIOGRAPHY

1. Alberti, K. G. M. M. and Krall, L. P. (Eds.) (1985). *The Diabetes Annual*, Vol. 1, Elsevier, Amsterdam
2. Clark, A. J. L. *et al.* (1982). Biosynthetic human insulin in the treatment of diabetes. *Lancet,* **ii**, 354–357
3. Goldstein, D. E. *et al.* (1982). Clinical application of glycosylated haemoglobin measurements. *Diabetes,* **31**, Suppl. 3, 71–78
4. Hansen, B. *et al.* (1982). New approaches to therapy and diagnosis of diabetes. *Diabetologia,* **22**, 61–67
5. Nattrass, M. and Santiago, J. V. (Eds.) (1984). *Recent Advances in Diabetes*, Vol. 1, Churchill Livingstone, Edinburgh
6. Pickup, J. (1986). Human insulin. *Brit. Med. J.*, **292**, 155–157

23
TOWARDS THE BIONIC MAN — CURRENT TRENDS IN THE DEVELOPMENT OF BIOMATERIALS

Brian J. Tighe

THE BIONIC MAN: FANTASY AND REALITY

The concept of the bionic man was most effectively brought to public attention in the 1970s by the American television series 'The Six Million Dollar Man'. The programme centred around Steve Austin, an experimental plane test pilot who was reduced to a near-lifeless mutilated shell in a horrific crash. With the cry 'we have the technology — we can rebuild him' a team of scientists got to work and created the Bionic Man. No problems of rejection or failure of materials for him! The end result was a superhuman creature whose synthetic and electronic replacement parts far outperformed Mother Nature's originals. The only hint of realism in the series came when the Bionic Man's (non-bionic) girl friend jumped from a plane without the aid of a parachute. The scientists reluctantly performed a second series of operations to create the Bionic Woman. The programme producers, presumably to heighten pathos and avoid undue competition with the star of the show, permitted a hint of reality to creep in. The polymeric material encapsulating the electronic controls implanted in the girl's brain caused a massive embolism and Jamie Summers died. Of course, we subsequently learned (when it was realised what hot commercial property the 'Six Million Dollar Woman' would be in her own television series) that she made an unexpected, miraculous recovery.

There, at a stroke, the only resemblance in the series to real life problems in biomedical materials research disappeared.

A more realistic view of progress and potential in the fascinating area of biomaterials research is obtained at the meetings of the European Society for Biomaterials. 1986 saw the Tenth Anniversary Meeting (held in Bologna, Italy[6]) of the society, which was formed, coincidentally, when Steve Austin was at his most famous!

Although neither the present nor the foreseeable future offers the hope of producing replacement devices which confer 'bionic man' performance, it would be a fallacy to believe that patients who benefit from the various biomedical supports are necessarily life-long invalids. It is not infrequently the case that some prosthetic device is needed for a patient to continue a professional career. The artificial finger joint is a case in point, where recipients have ranged from typists to a first division goalkeeper whose football career was saved by this simple, but effective, prosthesis.

In recent years a great deal of progress has been made in biomedical research. One of the major growth subjects has been the design and development of materials, particularly polymers, to fulfil the growing number of functions involved. This chapter highlights one or two such functions, showing the particular balance of properties required, against the general background of broad developments in biomedical applications of polymers.

A comprehensive overview of developments in the field of biomedical applications of polymers as reflected in the clinical and scientific literature from the mid-1970s to the early 1980s is contained in a series of review articles[17–19]. The fact that these reviews contain specific reference to over 700 papers selected from some 5000 publications in this period illustrates the level of activity in this fast-growing area. Some of the well-established examples of biomedical functions together with the types of polymer used and the particular problems associated with each function are shown in Table 1.

POLYMERS AND THEIR PROPERTIES

In order to appreciate why polymers have such a useful range of properties for this type of application, it is necessary to understand how their properties arise and may be modified. This is particularly important in appreciating the development of the so-called 'hydrogel' polymers, which are playing an increasingly important role in biomedicine. The single characteristic that unites both natural and synthetic polymers is the fact that, as the name (poly-mer) suggests, they are composed of many units linked together in long chains. The name of the polymer, in the case of synthetic materials, indicates the nature of the chemical entity that makes

Table 1 Examples of biomedical applications of polymers

	Function	*Polymer*	*Problems*
Bones and joints	Hip and knee-joint prosthesis	High-density polyethylene/stainless steel	Biocompatibility Wear Fatigue
	Finger-joint prosthesis	Polypropylene	
	Total elbow prosthesis	Silicone rubber	
Soft tissue	Maxillofacial prosthesis	Silicone rubber	Tissue compatibility Mechanical properties
	Ear prosthesis	Polyurethane	
	Frontal sinus prosthesis	Polyvinylchloride	
	Breast prosthesis	Silicone rubber	
Ducts and canals	Urinary tract prosthesis Urethral prosthesis Ureteral prosthesis Bladder prosthesis Total implantable lung	Silicone rubber Polytetrafluoroethylene Polyethylene terephthalate Polyvinylchloride composites	Biocompatibility Attachment Mechanical properties
Vascular and related prosthesis	Vascular grafts Arterial grafts Haemodialyser shunts	Woven or knitted polyethylene terephtalate Expanded polytetrafluoroethylene	Blood compatibility Dynamic mechanical properties (compliance)
	Heart valves	Silicone rubber Polyurethanes Polypropylene Polytetrafluoroethylene	Blood compatibility Elastic properties Mechanical properties
Tendons and ligaments	Tendon prosthesis Ligament prosthesis	Nylon cord Polyethylene terephthalate cord Silicone rubber rod	Biocompatibility Stress/strain behaviour Ease of anchorage
Permeability-related applications	Haemodialysis membranes Artificial liver support systems Drug release systems Contact lenses (extended wear)	Cellulose acetate and its derivatives New synthetic copolymers Hydrogels	Blood and/or tissue compatibility Permeability and permselectivity Mechanical properties

up the units, e.g. poly(ethylene), poly(vinylchloride) or poly(methyl methacrylate). In practice the brackets are often omitted. In order to get some idea of the unique structure of polymers, imagine a molecule of water or oxygen enlarged to the size of a tennis ball. A molecule of a typical synthetic polymer, such as those mentioned above, on the same scale would be of similar cross-sectional diameter but several hundred feet long. It is this great length of polymers, reflected in their alternative name, macromolecules, in relation to their cross-sectional diameter that gives them their unique properties.

Perhaps the best way of visualising the way in which the individual polymer molecules arrange themselves is by taking several pieces of string to represent a collection of molecules. The most usual arrangement will be a random one in which the pieces of string are loosely entangled. By changing the structure of the molecules, the regularity of packing, the resistance to deforming forces and the elastomeric nature of the material can be modified.

At the extremes of behaviour are encountered flexible, elastomeric properties and hard glassy properties; these differences can be explained in terms of differences in mobility in the constituent polymer chains. Thus, at very low temperatures all polymers are hard and glassy. As the temperature is raised, however, the thermal energy of the system increases and eventually the individual polymer chains will possess enough energy to begin to undergo rotation. A useful analogy is that, at this point, the tangled chains are like a plate of spaghetti which receives enough rotational energy to turn into a plate of worms. In terms of physical properties this corresponds to a change from glassy behaviour to rubbery or leathery behaviour. For this reason the temperature at which this change occurs is known as the glass-rubber transition temperature (T_g); the actual temperature at which the transition occurs will depend on the chemical nature of the polymer. Thus, for natural rubber it is below 0°C and for polymethyl methacrylate (perhaps best known as the commercial plastic Perspex) it is around 100°C. For this reason, at room temperature natural rubber is readily deformed, whereas polymethyl methacrylate is glassy.

As the normal human body temperature is around 37°C, it is important to design the polymer structure to be 'glassy' or 'rubbery' at that temperature — according to the functions to be fulfilled. It is also important to ensure that the glass transition temperature is not sufficiently near to 37°C for changes in mechanical behaviour to occur during use.

HYDROGELS: A UNIQUE FAMILY OF BIOMATERIALS

A well-known technique for converting glassy polymers into flexible materials is the incorporation of a 'plasticiser'. This is a mobile component, often a high-boiling organic fluid, which will act as an 'internal lubricant'. Its presence separates the polymer chains and allows them to move more freely. Its function is to lower the temperature at which the transition from glassy to flexible behaviour takes place. The best-known domestic example of this is polyvinyl chloride. In its unmodified state it is familiar as a rigid glassy material extensively used as corrugated covering for car ports and similar extensions. When a plasticiser is incorporated, the material is converted into the flexible material used (with the addition of suitable colouring material) as 'vinyl' seat coverings in cars and on chairs.

These principles have been used in an ingenious way in the class of polymers known as 'hydrogels'. These materials are making an important contribution in the biomedical and related fields. The simplest such polymer is poly(2-**h**ydroxy**e**thyl **m**eth**a**crylate) known as polyHEMA. Developed by Czechoslovakian chemist Otto Wichterle, its basic structure is very similar to that of poly**m**ethyl **m**eth**a**crylate (polyMMA). The most significant difference is that every unit of the polymer chain contains a hydroxyl (OH) group. The two structures are illustrated below.

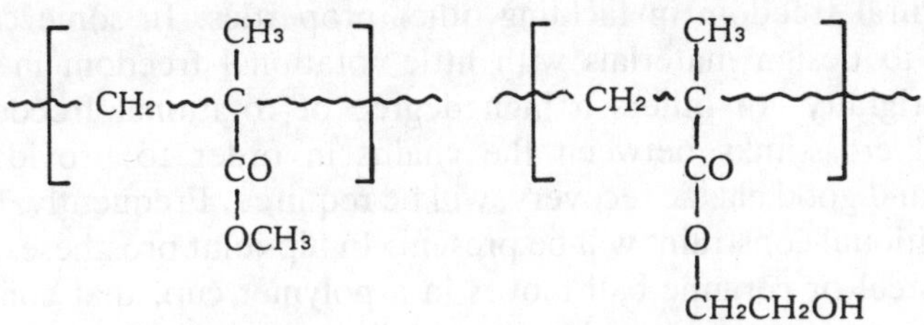

Poly MMA **Poly HEMA**

In the absence of water, both polymethyl methacrylate and polyHEMA are hard, clear and glassy. When placed in contact with water, however, the materials behave quite differently. Whereas polymethyl methacrylate remains hard and glassy, the hydroxyl groups in polyHEMA take up water, converting it to a clear flexible elastomeric gel. The water acts just like a conventional plasticiser. It behaves as an 'internal lubricant' and allows the chains to move freely — hence the flexible behaviour. More significantly, however, the water performs two other functions. It acts as something of a bridge between natural and synthetic systems, producing a very interesting enhancement in biocompatibility or, perhaps more properly, biotolerance. Additionally, it confers membrane properties on the hydrogel, allowing transport of oxygen and water-soluble metabolites through the polymer matrix.

POLYMERS IN BIOLOGICAL ENVIRONMENTS: THE MAJOR PROBLEMS

The types of polymers described in previous paragraphs provide a range of properties from hard and glassy plastics through tough hydrophobic rubbery materials to the soft water-containing hydrogels. This range of materials enables the requirements of prostheses for various body sites to be addressed. Table 1 gives some representative examples, together with the particular problems that each presents. It is apparent that the universal problem is that of compatibility, whether to blood, other body fluids or tissue. Mechanical properties, in various forms, also represent a continuous challenge, and of the remaining problems, that of permeability or permselectivity is of vital importance in some types of application.

The mechanical property requirements for a biomedical application are often fairly apparent. Indeed, it is frequently the case that if mechanical properties are not reasonably correct, little meaningful evaluation of other properties can take place. It will be obvious, then, that, as earlier discussion has shown, the interdependence of mechanical properties and polymer structure is such that mechanical property requirements restrict the structural freedom in tackling other properties. In some cases it is sufficient to design materials with little rotational freedom in order to promote rigidity. In others a high degree of rotational freedom, with occasional cross-links between the chains in order to provide 'elastic memory' and good elastic recovery, will be required. Frequently, however, some additional constraint will be present. In hip-joint prostheses, where a stainless steel or ceramic ball moves in a polymer cup, that constraint is wear. Whereas many materials have excellent dry wear characteristics, in the *in vivo* situation polyethylene performs better (in conjunction with stainless steel) than any of the possible alternatives. In the same way polypropylene, which has very high flex-fatigue resistance, is the most commonly used material for finger-joint prostheses.

It is usually the case that the physical nature of the material and its mechanical properties provide the short-term barrier to successful application, in the sense that they are necessary for a device to work. Having overcome this hurdle, however, the problem that limits the success of the material in the longer term is usually biocompatibility. As Table 1 illustrates, the fact that the body does not readily tolerate synthetic materials is an underlying problem in the great majority of biomedical applications.

BIOCOMPATIBILITY: THE ROLE OF PROTEIN DEPOSITION

Although vitally important, mechanical properties are arguably much less interesting than are the problems surrounding biocompatibility — in particular, blood compatibility. The underlying processes involved are common to many situations in which materials come into contact with biological solutions. Thus, the deposition of tear fluid debris on contact lenses, the clotting of blood at foreign surfaces, the formation of dental plaque and marine fouling all exemplify the same phenomenon — the adsorption of protein at solid surfaces. A great deal of research has centred on this problem from the point of view of both the events involved and their relationship to the surface properties of the material on which they occur. It is the question of blood contact that has attracted most attention — in particular, that of the thrombo-resistance of polymers. The adsorption of proteins has been found to have many analogies with the adsorption of gases at surfaces and the construction of adsorption isotherms to characterise the behaviour is commonplace. The adsorption of a protein monolayer occurs rapidly when a surface is brought into contact with a protein-containing solution, and, in natural systems, this primary adsorption process exerts a profound influence over subsequent events. Interest has centred, therefore, on the nature of that influence and on the possible ways of relating the primary adsorption process to the nature of the surface at which it occurs.

Although no single parameter has produced a satisfactory basis for correlation with thrombo-resistance, the surface energy concept has proved most useful. The most convenient way to describe this is in terms of the critical surface tension of the surface, which is defined as the surface tension of a liquid which will just spread on and wet a surface. This arises because surface forces are exerted by the constituent atoms of solids, just as in the case of fluids. A good illustration of surface forces is found in comparing the effect of scrambling eggs in a conventional metal saucepan(high-energy) and a non-stick pan (low-energy). Because of the strong interaction of the proteinaceous material in the former case, a good adhesive bond is formed and the pan is difficult to clean. The polytetrafluoroethylene lining of a non-stick pan is, on the other hand, easy to clean because the surface forces are very low. The difference between the surface tension of a fluid and the critical surface tension of the surface with which it is in contact gives rise to an important concept. This is the concept of 'interfacial tension', which takes into account the polar and non-polar contributions to the surface energy mismatch at the interface. It forms the basis of several attempts to quantify the compatibility of the liquid and solid phases.

CRITERIA FOR BLOOD COMPATIBILITY

Although it is quite acceptable to use the term 'biocompatibility' to describe the phenomenon associated with the interfacial activity between a synthetic and a biological material, no single material can be spoken of as 'biocompatible'. This is simply because different environments produce somewhat different requirements and no single material meets all of them. It is a case of 'horses for courses'. Much effort is expended upon attempts to match materials to substrate, especially in the case of applications involving blood contact.

Considerable conflict exists, however, as to what criteria should be used in defining biocompatible materials. They may seem puzzling, since it would be reasonable to assume that these criteria should be based on the natural substrate. Whereas many workers subscribe to the 'Minimum Interfacial Energy Hypothesis' (which suggests that for a material to exhibit maximum biocompatibility it should have a low interfacial tension with the biological system), there is considerable disagreement as to the actual values involved. The problems arise because characterisation of the natural surfaces inevitably involves some disturbance of their normal hydrated physiological state. It is the nature of these changes and the extent of their reversibility on rehydration that provides the basis for dispute.

One of the most interesting classes of polymer in this connection has been mentioned previously. These materials are the hydrogels. It is believed that the low interfacial tension between a hydrogel and an aqueous phase should reduce the tendency of plasma proteins to absorb and also reduce conformational changes in the secondary and tertiary structure of the globular proteins, when they so absorb.

This highlights the central point in the search for more biocompatible surfaces. It has been demonstrated that all non-physiological materials acquire a protein layer very rapidly on exposure to blood. There is no way of preventing this, but it does appear that in the competitive albumin–fibrinogen adsorption process primary albumin adsorption leads to greater compatibility than does fibrinogen adsorption. This in turn is attributed to the fact that albumin has a greater tendency to absorb in its native conformation than does fibrinogen. The synthesis of surfaces which promote initial albumin adsorption rather than fibrinogen appears to be one of the keys in the search for non-thrombogenic surfaces. This and related observations are used in attempts to develop laboratory procedures for determining a 'thrombogenic index' which would ideally grade materials and enable their performance in the body to be predicted.

TRENDS IN BIOCOMPATIBILITY RESEARCH

In addition to studies of protein and platelet adsorption on materials with well-characterised surfaces, in an attempt to quantify compatibility phenomena, other important approaches to the problem of biocompatibility are followed. One such approach is the attempt to improve compatibility by modification of surfaces with species found in the biological environment. An example is found in the binding of heparin and heparin-like polysaccharides to polymer surfaces[10, 13]. Variations on the conventional ionic binding techniques include initial grafting of chloromethylstyrene with gamma-rays, quaternisation with pyridine and finally exchange of heparin anions for chlorine by means of the sodium salt.

Although improved thrombo-resistance is invariably claimed, the mechanism of action of heparin complexes of this sort is still in doubt and under investigation[9]. A second group of techniques is associated particularly with vascular grafts and the need to establish a pseudo-intima of endothelial cells on the surface of the prosthesis. The existence and consequences of widely different degrees of maturity in this prosthetic pseudo-intima have been demonstrated[5]. Attempts to improve the rate of endothelial cell growth in this respect have included preliminary seeding of the polymer surface with endothelial cells prior to implantation or, alternatively, precoating with extracellular matrix. Both procedures were reported to be successful.

Studies of cell attachment, morphology and growth on the surfaces of synthetic polymers have much wider implications than those referred to above. The ability to control cell spreading has consequences in wound healing and scar tissue formation, for example. It appeared from studies[2] based on the work of Folkman and Moscona that it is possible to mediate cell growth on poly(2-hydroxyethyl methacrylate) by simply varying film thickness. This has been more properly interpreted in terms of the inhomogeneity of very thin evaporated coatings[12].

Despite this, a great deal of interest has centred recently on both the possibility of achieving such growth mediation and the use of tissue culture models of this sort to study the vital, early stages of 'biological interface conversion processes' that occur when a synthetic material is placed in a biological environment[11]. The fact that a session was devoted to the topic of tissue culture models at the 1986 ESB Congress[6] illustrates its current importance. An additional area of growing significance is the use of the eye as a body site for non-invasive studies of the effect of biological environments on polymers.

THE OCULAR ENVIRONMENT AND THE CONTACT LENS

Although drawings by Leonardo da Vinci have been interpreted as demonstrating the principle of the contact lens, it was not practically utilised until the end of the nineteenth century. Despite this it was perhaps the earliest environment to provide information on the interaction of a wide range of materials with a biological environment. It is an example of a device which illustrates extremely well the problems of providing this careful balance of properties in the extended-wear contact lens. The extended-wear lens which is left in position during successive day and night periods may be compared to an ocular bandage. Indeed, its value is not simply cosmetic. It is widely used in ocular surgery — for example, in cases of aphakia following cataract removal. The relative ease of carrying out clinical work in this field does tend to set it apart from other areas of biomedicine. The benefit of this fact has been obvious in the development of the previously mentioned hydrogel polymers. Far more innovative work has been carried out with such materials in this field than in any other. This is illustrated by the fact that well over 100 hydrogel patents relating to contact lenses have been published, involving vast numbers of chemical compositions. The number would undoubtedly have been higher had the US Food and Drugs Administration not decided to classify hydrogel (soft) contact lenses as drugs. This research, particularly in the light of the stringent requirements of the FDA, has provided information on a wide range of materials which will greatly assist future work on their use in other biomedical applications.

Although polymethyl methacrylate has been used for many years as the standard 'hard' contact lens material, its properties are far from ideal. It was not, however, until hydrogel polymers appeared on the scene that any serious competitor emerged. Because hydrogels were soft and much more comfortable, thought was given for the first time to the question of overnight wear. As will become apparent, this requires a careful balance of properties to accommodate the requirements of the three essential features of the ocular environment — namely the cornea, the eyelid and the tear fluid[16].

CONTACT LENS MATERIAL: MECHANICAL, TRANSPORT AND SURFACE PROPERTIES

The cornea governs the permeability requirements of the material. Since it is avascular, it needs to respire directly from the atmosphere, as this is the major source of the oxygen required to maintain its biochemical balance. In cases where the oxygen supply is curtailed, the Krebs cycle is disturbed and corneal oedema results. In the closed eye condition the oxygen concentration in the tears is maintained, although at a reduced level, from various peripheral blood vessels. Since the lens acts as a relatively immobile membrane at night, it is not difficult to calculate the level of permeability required. In hydrogels the transport processes are governed by the water content of the hydrogel, and, as a rule of thumb, a polymer containing 70 per cent water is required for lens wear during sleep. This, then, is the first design feature fixed.

The presence of this quantity of water in the polymer also affects the mechanical properties of the material. Here the balance of properties is governed by the eyelid. Although soft materials are, in general, more comfortable to the eyelid, there is a lower limit to the stiffness that can be tolerated. This is because the eyelid exerts a fairly considerable deforming force during the blink cycle. It is here that the polymer scientist needs to manipulate cross-link and backbone structure while maintaining the same water content in order to limit deformation under eyelid load to around 5 per cent with good elastic recovery.

Although it is easy to appreciate that the polymer must be sufficiently wettable to sustain a coherent layer of tear fluid, it might be thought that all problems associated with wettability would be overcome by the presence of water in the material. Unfortunately, this is not the case. The surface chemistry of the hydrogel governs the complex series of events that lead to a variety of possible deposits on the surface of the lenses during wear. Here, as in blood contact devices, the phenomenon of protein adsorption is an important first step, but the subsequent deposits are complex and contain calcium, mucin and mucopolysaccharides in addition to the major tear fluid proteins. Although this problem does not prevent extended wear of lenses, it does cause problems. Thus, ocular compatibility and blood compatibility are both problems in materials science that are far from being solved, although progress is continually being made.

Apart from the development of new contact lens materials, the detailed studies outlined above have benefits in two 'spin-off' areas. One has been referred to. It is the study of the biological interactions that occur between tears and polymers — information obtained here is of value in the wider context of biocompatibility studies. The second is in the area of intraocular lens design. These implanted lenses, used after cataract surgery, have traditionally been made of polyMMA. New materials, which have encouraged new device designs, have followed in the wake of the similar work on contact lens materials.

ARTIFICIAL ORGANS: KIDNEY, LIVER, HEART AND PANCREAS

It is apparent from Table 1 that several applications require the difficult combination of blood compatibility and permeability or permselectivity in order to function satisfactorily. One of the best-known, and indeed oldest, examples is that of haemodialysis. The pioneer work was done by Kolff in The Netherlands some 40 years ago. Remarkably, commenting on the investigation of this project, he said at a recent conference: 'We did not go to the movies for five years during the German occupation so we had nothing better to do than make artificial kidneys'. The early devices consisted of large rotating drums, which caused blood and dialysing fluid to pass through and over a 20 m length of Cellophane tubing, respectively.

This is an example of a device in which the first membrane material to be adopted is still in active use. There have been many refinements to the hardware and that, coupled with the critical dialysate control required, constitute the major problems. In recent years, progress has been made towards the wearable artificial kidney. There are many problems, not all of which centre round the essential feature, a solid adsorbent whose function is the continuous regeneration of the dialysate.

Another device that relies upon a solid adsorbent is the artificial liver support haemoperfusion system. The liver, unlike the kidney, is capable of regeneration following damage — for example, in cases of acute liver failure. In such cases, where causes range from drug overdose to viral hepatitis, the liver support system is designed to take over the toxin-removing function of the liver pending the restoration of its normal capability. Since blood is passed directly over the adsorbent in haemoperfusion, it is necessary to encapsulate it in a biocompatible membrane. The major problem here is not biocompatibility in itself but the fact that, in addition, the membrane must be permeable to the relatively large molecules (for example, bilirubin and cholic acid) which are normally dealt with by the liver.

Whereas interest in haemoperfusion for treatment of drug overdose continues to grow, clinical activity in its use as a basis for an artificial liver support system seems to be limited in Britain to the Williams group at Kings College Hospital. The position in artificial kidney research appears to be similar. The field has been reviewed[8], with some emphasis on future developments, but disappointingly little active progress (in portable systems, for example) has been reported. Questions of molecular weight selectivity and its importance, the relative ineffectiveness of heparin in preventing deposition at membrane surfaces and the dispersion of silica particles from dialysis (silicone rubber) tubing into various body organs still require research effort. Somewhat similar problems are raised by the question of an implantable artificial pancreas, which is frequently discussed[3, 15]. Undoubtedly the issue that attracts most public attention in this area is the total artificial heart. A useful collection of papers serves as a state of the art review and describes various aspects of its development and use[21]. In particular, various aspects of design and construction of the currently evaluated systems are discussed together with details of the animal work involved. Maximum survival periods still seem to be uniformly less than 1 year with any system.

TENDONS, LIGAMENTS, SUTURES AND ADHESIVES

This group of applications shares the problems associated with the need to achieve secure anchorage. In all four cases progress seems difficult to achieve.

An example[14] of the work required to take experimental prostheses to a satisfactory stage for general use is found in the case of a new polyethylene-based ligament (Polyflex). In a group of nine patients the ligament was so unsatisfactory as to require removal in a third of the patients, and half of the remaining group complained of pain. Despite the fact that the prosthesis had reached the stage of clinical use, it was concluded that many problems relating to the material and the operation remain to be resolved. A new technique for attachment to soft tissue in tendon and ligament replacement and repair has been recently developed[1]. This entails the use of a filamentous carbon–poly (lactic acid) composite as a tissue scaffold. Following implantation, soft tissue ingrowth into the carbon fibre network occurs. In a rabbit study the bond developed rapidly and was mechanically secure under physiological loading conditions. Strength tests after 12 weeks demonstrated that breaking strains equivalent to those of the natural system had been reached.

During the last decade the use of synthetic absorbable sutures has become widespread. There are two currently available materials, poly(glycollic acid) and a polylactic-glycollic copolymer. Because of their stiffness the polymers are normally produced as sutures in the form of braided filaments, except in very fine diameters. A new flexible monofilament polymer (PDS) produced by ring opening of *p*-dioxanone has recently been developed, and several reports of its properties and performance have now appeared.

One significant advance that has been possible with this material is the fabrication of biodegradable flexible fixation clips used in surgery. These have the advantage that, unlike metal clips, they are absorbed into tissue and metabolised. This is particularly useful in ligation procedures.

The field of surgical adhesives is widely recognised as being in need of a new generation of materials, but as yet there is little sign of innovative work. The potential advantages of 'super glue' in surgery are obvious and this group of cyanoacrylate adhesives are now becoming the subject of wider clinical evaluation, with varying degrees of success. Typical examples of use include middle ear and intracranial surgery, reattachment of tendons and reconstruction of the frontal sinuses.

LONG-TERM IMPLANTS: CAUSES OF FAILURE

The last decade has seen the opportunity for comparative studies on a variety of prostheses. In many cases failure or death has resulted in the premature removal of an implant. In other cases retrieval of the component has been part of a planned experimental protocol. In both situations valuable information relating to performances — in particular, the causes and mechanisms of failure — is obtained. Information is summarised here from the fields of joint prostheses, ducts and canals, and vascular and arterial grafts.

Although not without problems, hip-joint prostheses are obviously the most widely performed and arguably the most successful of joint prostheses. Similar emphasis is now being placed on total knee prostheses, in terms of both stress analysis of the various tibial component designs and material evaluation. In a review[4] of failed knee-joint prostheses one design was found to have a failure rate in 3 years of over 70 per cent. Failure was apparently largely due to either the generation of wear debris or the loosening of the polyester femoral component. This extremely unsatisfactory situation reflects the proliferation in the past of designs based on intuitive thinking rather than interdisciplinary research. Experimental techniques to assess deformation and wear of knee-joint components have been developed, and carbon fibre reinforced ultra-high molecular weight polyethylene appears to offer promise as a constructional material.

An interesting and thorough examination of several silicone rubber ureteral prostheses after implantation for periods of 4–6 years in dogs and in humans has been published[20]. This appears to be the longest implantation period so far studied, and the deposits appear to be predominantly lipoidal, with possible initiation sites for calcium-based incrustations. There are strong indications that progress is being made in understanding the initiation mechanisms involved in the calcification of elastomers during prolonged use in various body sites.

Early synthetic arterial prostheses with up to 15 years' use have been examined and a detailed report of some 500 presented, in which a significant proportion showed eventual fibre deterioration. The conclusions suggest that, in addition to manufacturing problems, factors related to storage and sterilisation, direct *in vivo* degradation is involved. A similar conclusion is reached in the case of a prosthesis[7] which failed after 7 years and in which hydrolytic degradation was found to have initiated the failure process. The most dramatic complication of reconstructive vascular surgery, even after long periods, is considered to be infection. Against the moderate success encountered with large-diameter grafts must be set the generally inadequate performance (essentially through occlusion resulting from compatibility problems) of small-diameter prostheses.

AN ECONOMIC OVERVIEW

What of the economic aspects? The investment interest in this area has been realised by many large companies. This is illustrated by the announcement recently of the activities of Sorin Biomedica, owned by the Fiat Motor Company. One of the company's areas of activities is artificial organs, and their contribution to the kidney dialysis field, the Orion, is apparently to be launched in Britain shortly.

It is no secret that many very sophisticated areas of surgery use materials which the polymer scientist would regard as rudimentary. The cost of the operation is so vast in relation to the cost of materials that the cost of improved materials represents no barriers to their use. Two simple examples have been mentioned. They are the use of polypropylene and polyethylene in finger-joint and hip-joint prostheses. The surgical adhesive field is even more susceptible to the introduction of new products.

Methyl methacylate dental cement is used to grout in the ball during hip-joint operations simply because there is no available alternative. It is far from ideal and a successor would be readily accepted. More sophisticated adhesives such as variations of super glue have a host of applications, ranging from soft-duct prostheses to neurosurgery. The relationship of the cost of material to the final cost of the product is typified by the case of contact lenses.

The material cost for polymethyl methacrylate used in hard lens manufacture is of the order of three or four pence per lens. The manufactured lens would be sold to the practitioner at something of the order of £5 and the final cost to the patient something of the order of £100 a pair. A recently introduced alternative hard lens material sells at approximately £2 per lens blank. The manufactured lens sells at around £8 and the final cost to the patient is something over £100 per pair. Obviously the final cost of the lens to the patient is largely related to the time and care expended by the practitioner and the overheads incurred in his particular practice. The significant fact, however, is that these new gas-permeable hard lens materials are fast outselling conventional polymethyl methacrylate even though the basic material cost is some 50 times greater. The increase in cost is of little importance compared with the final cost of the product, and that observation is reflected time and time again in biomedical materials research.

The general economic picture can be deduced from the following information. The base raw material (polymer) cost for the materials listed in Table 2. Even for more expensive, specialised polymers such as polyurethanes, silicone rubber and polytetrafluoroethylene, the cost is less than £10 per kg. The manufacturing cost of any device will be several times greater than the raw material cost, but these are minute compared with the cost of operations, let alone subsequent hospital care. It is obviously very difficult to obtain realistic costings for surgical operations, but figures of £7000 for complex surgery involving a team such as that involved in kidney transplantation have recently been quoted.

Table 2 Typical cost of biomedical polymers

Polyethylene	50p/kg
Polypropylene	65p/kg
Polyvinylchloride	60p/kg
Polyethylene terephthalate	£2/kg

A SUMMARY OF PRESENT ACHIEVEMENTS

Although the bioengineer's dream — Steve Austin — described at the beginning of this chapter still represents an unattainable goal, there has been considerable progress in this area during the last 10 years. The fundamental problem was described at the outset as biocompatibility, particularly blood compatibility. This remains the outstanding feature but considerable progress is being made — witness the development during the recent past of arterial grafts and prostheses. In addition, very real progress has been made in the development of artificial internal organs, a field of work which presents formidable problems. Although the more sophisticated of these devices have not yet reached human clinical trials, the implantable artificial pancreas has progressed well in animals, as has that most daunting of tasks, the total artificial heart. The 'record' for the maximum period of survival for an animal on this device is consistently, although slowly, advancing. At present, it stands at something under a year. Similar results were obtained with one of the human subjects, who, having no alternative hope, underwent a total heart prosthesis.

No reference has been made in this chapter to such activities as the growth in biosensors. A sign of the interest here is the appearance in 1985 of a new journal devoted to the topic. In one aspect alone, the development of field effect transistor (FET) based sensors with the capacity for multiple sensing on a tiny 'chip' coupled to feedback controls, may well revolutionise biomedical devices. Here again the requirement for appropriately permeable encapsulants having adequate compatibility with the appropriate biological environment presents a challenge for the biomaterials scientist.

A summary of the range of achievements in biomaterials science and bioengineering is shown in Figure 1. Although these are not all, by any means, routine prostheses, they give a useful broad overview of the present position.

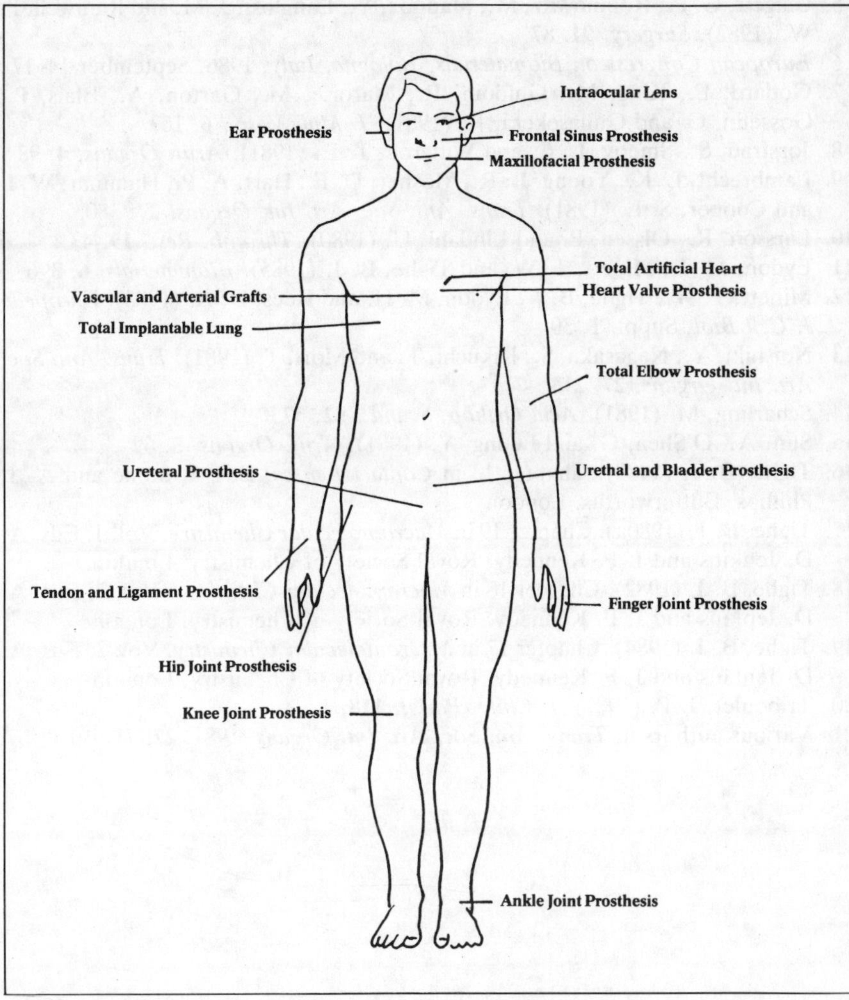

Figure 1 Examples of current prostheses

REFERENCES

1. Aragona, J., Parsons, J. R., Alexander, H. and Weiss, A. B. (1981). *Clin. Orthop.*, **160**, 268
2. Belmont, A. S., Kendall, F. M. and Nicolini, C. A. (1980). *Cell Biophys.*, **2**, 165
3. Bessman, S. P., Thomas, L. J., Kojima, H., Sayler, D. F. and Layne, E. C. (1981). *Trans. Am. Soc. Art. Int. Organs*, **27**, 7
4. Cameron, H. U. and Federkow, D. M. (1980). *Arch. Orthop. Trauma Surg.*, **97**, 87

5. Clagett, G. P., Robinowitz, M., Maddox, Y., Langloss, J. M. and Ramwell, P. W. (1982). *Surgery,* **91**, 87
6. *European Congress on Biomaterials, Bologna, Italy*, 1986, September 14–17
7. Godard, F., King, M., Guidoin, R., Maroise, M., Garton, A., Blais, P., Gosselin, C. and Gunasekera, K. (1981). *J. Mal. Vasc.,* **6**, 167
8. Jorstrad, S., Smeby, L. C. and Wideroe, T. E. (1981). *Artif. Organs,* **4**, 98
9. Lambrecht, L. K., Young, B. R., Mosher, D. F., Hart, A. P., Hammar, W. J. and Cooper, S. L. (1981). *Trans. Am. Soc. Art. Int. Organs,* **27**, 380
10. Larsson, R., Olssen, P. and Lindahl, U. (1980). *Thromb. Res.,* **19**, 43
11. Lydon, M. J., Minett, T. W. and Tighe, B. J. (1985). *Biomaterials,* **6**, 396
12. Minett, T. W., Tighe, B. J., Lydon, M. H. and Rees, D. A. (1983). *European J. Cell Biol*, Suppl. 1, 30
13. Noishiki, Y., Nagasaka, S., Kikuchi, T. and Mori, T. (1981). *Trans. Am. Soc. Art. Int. Organs,* **27**, 213
14. Scharling, M. (1981). *Acta Orthop. Scand.,* **52**, 575
15. Sun, A., O'Shea, G. and Leung, Y. (1981). *Artif. Organs,* **5**, 69
16. Tighe, B. J. (1980). Chapter 13 in *Contact Lenses*, Eds. J. Stone and A. J. Phillips, Butterworths, London
17. Tighe, B. J. (1980a). Chapter 19 in *Macromolecular Chemistry*, Vol. 1, Eds. A. D. Jenkins and J. F. Kennedy. Royal Society of Chemistry, London
18. Tighe, B. J. (1982). Chapter 18 in *Macromolecular Chemistry*, Vol. 2. Eds. A. D. Jenkins and J. F. Kennedy. Royal Society of Chemistry, London
19. Tighe, B. J. (1984). Chapter 17 in *Macromolecular Chemistry*, Vol. 3. Eds. A. D. Jenkins and J. F. Kennedy. Royal Society of Chemistry, London
20. Triboulet, J. P. (1981). *J. Chir. (Paris),* **118**, 1
21. Various authors in *Trans. Am. Soc. Art. Int. Organs*, 1981, **27**, 71–191

24
THE RELEVANCE OF THE WARNOCK REPORT ON HUMAN FERTILISATION AND EMBRYOLOGY TO BIOTECHNOLOGY

Edward J. Yoxen

To a hard-headed, industrial biotechnologist, preoccupied with fermentation yields, patent law and cash flow, the Report of the Committee of Inquiry into Human Fertilisation and Embryology (the Warnock Report) may not seem particularly relevant[1]. Three lines of argument weaken that view. First, 'test-tube babies' and 'genetic manipulation', of whatever kind, are connected in the public imagination. Controversy and uncertainty in one field may well influence the acceptability, appeal or public image of the other, despite the technical differences between them. Second, the actions of Lady Warnock's Committee are an excellent example of consensus building in British scientific and medical policy-making. It is not impossible that regulatory issues in biotechnology could at some point become contentious again. If that were to happen, attempts to manage such controversy would take a similar form. Moreover, the idea of a Licensing Authority, which is the proposed regulatory solution for *in vitro* fertilisation (IVF) owes something to the example of the Genetic Manipulation Advisory Group (GMAG), as it does also to the licensing of animal experiments by the Home Office. Third, although biotechnology and external human fertilisation are, in general, very different kinds of activity, there are nonetheless technical links between them. For example, some of the hormones used to induce ovulation prior to IVF have been made by recombinant bacteria. Similarly, the increase of medical interest in infertility, which produced the IVF procedures, is linked with increased concern with the diagnosis and control of sexually transmitted disease. DNA probes may have a role in the future diagnosis of such conditions, as may monoclonal antibody kits in pregnancy testing. The continued expansion of IVF could increase the market for some biotechnology products. These, incidentally, are my arguments: they do not appear in the Warnock Report itself, for all its breadth.

The Committee of Inquiry, chaired by Mary (now Lady) Warnock, was established in July 1982, essentially to advise the Secretary of State for Social Service, Norman Fowler, on policy for *in vitro* fertilisation. Its deliberations and the publication of a Report took 2 years. There were persistent rumours that it was going to be further delayed, because of disagreements within the Committee. Since its appearance, the volume of public discussion about IVF in the media has increased markedly, as has parliamentary interest. Both Houses of Parliament have debated the report[2]. Enoch Powell, MP, has presented a private members' bill to the House of Commons, which was given a second reading on 15 February 1985. This would ban all research on human embryos (contrary to the recommendations of the Report), and require notification of every attempt to implant a human embryo[3]. If enacted into law, it would severely restrict the practice of *in vitro* fertilisation in the UK. The controversiality of the field has also increased somewhat since the report appeared, because of a highly publicised case of surrogate motherhood, which involved controversial legal manoeuvres, although *in vitro* fertilisation had not been employed[4]. Lady Warnock's own concern that the principal ideas in the Report should not be lost in the theologically charged battles over the use of embryos or in the legal debates about surrogacy has led her to publish several articles urging Parliament to take legislative action, but neither too swiftly nor too slowly[5]. Government spokesmen, well aware that this whole area exposes conflicting strands in ideology, have indicated that legislation will be presented to Parliament in this session to regulate *in vitro* fertilisation, if problems in drafting it can be overcome and if unrest within their party can be held in check.

The Warnock Report has a rather distinctive style, which blends Civil Service administrative and legal concern with ramified questions with moral philosophy. This is not an unusual mixture. It is also slightly didactic. In press and TV interviews Lady Warnock in talking of 'seminars on why the moment at which life begins was not an admissible philosophical question', has let this aspect show rather more[6]. Throughout the Report, there is a real concern to show by what reasoning particular recommendations were reached. Within a framework that is held to be compatible with 'a tolerable society', much is made of the force of logic —

i.e. we have agreed to X; it is therefore only consistent to agree to Y.

It remains to be seen how persuasive this rationalist approach will prove, say, on the Tory backbenches. It is certainly possible that this rather high-minded approach was used to render down disagreements within the sixteen members of the Committee of Inquiry. At the same time one should not overlook the influence of Lady Warnock's attitudes to the nature of morality. In an interesting debate with a more pragmatic and utilitarian philosopher, she has argued for a moral code that is not based on a calculus of benefits and costs — i.e. strict utilitarianism — but on one that acknowledges deeply held moral sentiments that are contrary to the general view.

'I think that my position is very like (and owes something to) that adopted by Stuart Hampshire. Hampshire holds that the essence of morality is the existence of a set of not necessarily coherent or unified principles which constitute barriers against what is felt to be wrong-doing, a set of prohibitions. There are, in society, some things which must not be done, or, if done, must not be condoned. Those prohibitions operate within certain areas of life. They govern, in any society, attitudes towards human life and human death, towards the treatment of the dead, towards sexual relations, the relations of loyalty and honesty. Different societies will regard different principles as inviolable; and with a geographically single society, prohibitions will change from time to time. But if any society, at any given time, had **no** principles, this would be the end of morality for that society'[7].

She concludes her article as follows:

'What is needed is a principle, and an institution of legal restraints and surveillance, which will enable research to proceed and therapeutic techniques to be developed, but constantly watched, not merely by the medical profession and the research biologists, but by lay-people as well, the stupid, the prejudiced, the sentimental, the religious and the moralistic lay. Only so, I believe, shall we ensure that these sentiments out of which morality is composed will be respected[8]'.

That is a fairly remarkable characterisation of lay opinion, even if her point is that the essence of morality consists in acknowledging different notions of acceptability to find common ground.

The composition of the Committee was in itself quite interesting. Leaving aside the Civil Service 'observers' and the legal adviser, it was made up of seven women (including the chair) and nine men; not quite equality of representation by gender, but much closer than usual. Three people were legally qualified and two more were magistrates. Seven were medically qualified, four of them being professors, and one of them former President of the British Medical Association and a past Vice-Chairman of the Royal College of Obstetricians and Gynaecologists. None of the doctors was directly involved with IVF, although the biologist on the

Committee has a non-clinical interest in it. Two Committee members were social workers and one an Anglican theologian, and one, Lady Warnock, is a philosopher. Whether any of the Committee members had personal experience of infertility is unclear, although one of them has dealt professionally with couples seeking to adopt.

It is also fair to say that considerable efforts were made to canvass opinions, from researchers, from professional bodies, from churches and pressure groups. Certainly the impression given by the list of some 300 individuals and organisations submitting evidence to the Committee is of widespread consultation. It includes submissions from district, area and regional health authorities, from university departments, from Royal Colleges of Medicine and professional and learned societies. In addition, evidence was offered by organisations as diverse as Action for Lesbian Parents; Bharatiya Vidya Bharan — Institute of Indian Culture; the British Evangelical Council; Donors' Offspring; Magistrates Association — Inner Manchester Branch; Soroptomist Club — Knutsford and District; the Science Fiction Association; and the Yorkshire Pro-Life Co-ordinating Committee. At the same time it has been suggested that the committee failed to solicit the views of long-established women's organisations, such as the Women's Institutes[9].

The recommendations fell into two main categories: one concerned with new bodies to organise services and regulate the field, the other concerning specific legislative actions — for example, to modify the law relating to illegitimacy and the registration of births or to criminalise certain activities, such as setting up a surrogacy arrangement, with or without IVF.

The Report begins with infertility services, and notes that they are under-funded within the NHS, that they are in need of reorganisation and that very inadequate data are available on the scale of involuntary fertility, its treatment and causes. A working party is proposed to tackle this complex of questions and essentially to decide now what role IVF should have within NHS infertility services. The next question is artificial insemination by donor, and here two themes enter the discussion and continue right through the Report. The first is that of anonymity in donation and the second that of licensing. In a sense, then, the reform of the already established practice of artificial insemination by donor (AID) becomes a model for all subsequent discussion.

Under existing law children born after AID, where the donor is by definition not the husband of the biological mother, are technically illegitimate, even though, by falsifying the identity of the father on thc birth certificate in an act of perjury, this is often concealed. The Warnock Committee proposed, however, to legitimise the AID child, and in so doing to annul in law the rights and duties of the genetic father (i.e. the sperm donor). A further consequence would be that his identity would be concealed in perpetuity. This legal form, with appropriate modifications, is proposed later for the egg and embryo donation made possible by IVF. One may say that the effect of these proposed changes would be to separate in law donors much more decisively than is at present possible from the children their donations bring into being and to unite these children with their recipient parents without resort to minor illegality. The Report also proposes the setting up of a licensing authority to control AID, the task of which would include enforcing a limit of 10 donations per sperm donor and checking standards of organisation and care.

In Chapter 5 IVF is tackled in its own right for the first time. After some careful discussion, it is endorsed, and a licensing authority proposed here too. Essentially, the rest of the Report works through the ramifications of this crucial decision. Many of the issues are intended to be overseen by the proposed licensing authority, which might revise its policy from time to time. But some questions are held to require immediate legislation, both to enable the authority to act and to give greater force and political cogency to the regulatory framework. Thus, for example, following the precedent set with AID, it is proposed to redefine legitimacy so as to transfer legal parenthood to those people who receive the children born after embryo or egg donation, regardless of the genetic relation of either of them to the children. Surrogacy, in all its possible forms, with and without IVF, is considered at length and it is proposed to make surrogacy contracts unenforceable in law (i.e. to strengthen and formalise the attitudes of the courts at present). But the Committee, with several dissenting voices, proposed to go furher and effectively ban all but the most private and informal surrogacy by criminalising the activity of arranging a surrogacy contract. Had this proposal been on the statute book in late 1984, it would have made illegal the actions of Mrs Harriet Blankfeld of Surrogate Parenting Associates Inc. in establishing the contract for Mrs Cotton to act as a surrogate mother. But it would also apply to arrangements drawn up by doctors in their infertility practice and for this reason this recommendation has been criticised as too restrictive[10].

The question of embryo freezing was also discussed, since this has begun in Britain, although no child has been born here after having been stored as a frozen embryo[11]. The technical advantages of doing this are that several attempts can be made, under optimal hormonal conditions, to achieve implantation, without the need for invasive surgery to obtain fresh ova. Equally, couples may wish to store 'their' embryos for rather longer periods, or to make them available for donation when necessary or for research. Such procedures immediately raise serious questions of ownership, legal liability, custodial responsibility and inheritance. Here, if it is not apparent already, the truly extraordinary legal ramifications of this new technology appear. The Committee proposed that the future of frozen embryos be subject to five-year reviews by the biological parents and the storage authority. On the death of the parents within such a period, the rights of usage or disposal should be transferred to the hospital alone. This issue has already arisen in Australia, with the death in a plane crash of a couple whose embryos were in refrigerated storage[12]. Legislation should be passed that would ensure that there could be no property rights in embryos. It is similarly recommended that inheritance by 'children' not yet implanted, although conceived and in the storage medium, of the estate of their biological father would be disallowed in law; and the time of birth, not time of fertilisation, should be decisive in cases where the 'older' child inherits. Finally, the sale of embryos should be criminalised.

This leads on to the question of research, and here it is proposed that a limit of 14 days be set as the maximum time for which human embryos may be grown in culture for research purposes. This is 3 days less than that proposed by the Royal College of Obstetricians and Gynaecologists in 1984[13]. It is very much less than the 6 weeks proposed by the Australian philosopher Peter Singer and his co-author, Deane Wells, in their book *The Reproduction Revolution*[14]. Their argument is that only at this point can one detect electrical activity in the brainstem of a human embryo and this phenomenon is used to determine the moment of death at the other end of the life-cycle. This position clearly stands out from those advocated by research organisations and professional bodies in the UK, all of which have seen controlled experimentation as permissible, up to a limit that approaches the time of normal implantation after *in vitro* fertilisation, provided that the embryos on which experiments have been carried out are not themselves re-implanted[15]. It is also precisely this which Enoch Powell's Bill seeks to prevent. This issue was seen by the Committee as specifically requiring legislation. With this in mind, the Medical Research Council (MRC) issued a statement saying that an informal application of the Warnock criteria will be practised by a committee convened by the MRC, with public representation, to monitor and specify procedures for MRC-funded work in this area[16].

It would be claiming an unrealistic degree of prescience to say that most of the Warnock recommendations were to be expected. But with no major medical catastrophes known to have resulted from IVF, clear evidence of demand and the balance of public sentiment as revealed by public opinion surveys in its favour for the alleviation of infertility at least, it would have been rather remarkable had not the Warnock Committee cautiously endorsed it. Equally, the precedent for this kind of flexible regulation has been set in the UK by Genetic Manipulation Advisory Group, if not by older traditions of the licensing of research. Interestingly, what is proposed is not professional self-regulation, although consultants' autonomy to select patients is upheld, nor is it secret direction by an expert committee and a Civil Service secretariat. Lay participation and specific legislation are clearly thought necessary to increase the political legitimacy of the proposed arrangement. It may yet founder in legal difficulties and public controversy, though I doubt it. This kind of consultative exercise follows a tradition in Britain of soliciting opinions to form a consensus without full public participation in debate. In other political cultures, such as those in Australia or the United States, things are done differently. It will be interesting to see whether in other countries *in vitro* fertilisation will turn out to be more or less controversial.

NOTES

1. *Report of the Committee of Inquiry into Human Fertilisation and Embryology* (London: HMSO, 1984; Cmnd. 9314)
2. See D. McKie, Gingerly into the thickets of life, *Guardian* (1 November 1984), 2; Anon., Lords oppose embryo research, *New Scientist* (8 November 1984), 6; A. Veitch, Fowler deflects Tory embryo ambush, *Guardian* (24 November 1984), 1; Anon., Anxiety over potential embryo abuse, ibid., 4
3. C. Brown, Time factor works against success with anti-embryo experiments bill, *Guardian* (19 February 1985), 4; Anon., Warnock proposals in trouble [editorial], *Nature,* **313** (7 February 1985), 418
4. Anon., Surrogate mothers, *British Medical Journal,* **290** (26 January 1985), 308. This is a discussion of the legal status of actions taken with respect to the baby commissioned by unknown parents and borne to term by Mrs Kim Cotton as a surrogate mother; for reporting of some of the events in this case see S. Helm *et al.*, How to buy a baby, *Sunday Times* (13 January 1985), 15
5. M. Warnock, Meddling with the conception, *Sunday Times* (25 November 1984), 12; M. Warnock, The surrogacy scandal, *The Listener,* **113** (24 January 1984), 2–4; P. Chorlton, Warnock favours urgent law against surrogates, *Guardian* (8 January 1985), 4
6. Diverse Reports (Channel 4 TV, 7 November 1984); see also the article by Mary Warnock cited below
7. M. Warnock, *In vitro* fertilization, the ethical issues, II, *Philosophical Quarterly,* **33** (1983), 247
8. ibid., 249
9. Statement in the Diverse Reports programme; see note 6 above
10. Statement by Mr Robert Winston of the Hammersmith Hospital, London, in the *Panorama* Programme, 'Cash on Delivery' on BBC 1 (14 January 1985), devoted to the Cotton Case. Mr Winston, who runs an NHS infertility clinic, takes the view that occasionally surrogacy arrangements between sisters, or close relatives or friends, will be requested and should be allowed with counselling and medical supervision
11. This has, however, occurred in Australia, on 28 March 1984
12. C. Holden, Two fertilized eggs stir global furor, *Science,* **225** (6 July 1984), 35; J. Sellar, Surrogacy ban by Victoria State, *Nature,* **311** (27 September 1984), 288
13. *Report of the RCOG Ethic Committee on* in vitro *Fertilization and Embryo Replacement or Transfer* (London: RCOG, March 1983)

14. P. Singer and D. Wells, *The Reproductive Revolution: New Ways of Making Babies* (London: Oxford University Press, 1984)
15. T. Beardsley, MRC tests the climate, *Nature,* **300** (25 November 1982), 309; T. Beardsley, BMA reports on ethics, *Nature,* **303** (19 May 1983), 190; T. Beardsley, Human embryo experiments: societies urge a softer line, *Nature,* **302** (28 April 1983), 739
16. J. Turney, Baby research licence move, *The Times Higher Education Supplement* (25 January 1985); *Report of Inquiry Into Human Fertilization and Embryology: Medical Research Council Response* (London: MRC, 1985).

25

BIOTECHNOLOGICAL INVOLVEMENT AND ADVANCES IN WASTE-WATER TREATMENT

P. K. Robinson

INTRODUCTION

In rural areas the recycling of animal (including human) and plant waste has been practised by man for centuries. The progressive urbanisation of human populations, however, has necessitated the development of alternative waste disposal strategies capable of dealing with the large volumes of waste products generated locally in towns and cities. Historically, water has been used as a vector for the removal of a large proportion of urban wastes, and nowadays such waste-waters typically comprise domestic and industrial effluents, surface drainage and groundwater, containing both organic and inorganic matter in suspended, colloidal and soluble states. Such 'sewage' (as it is commonly called) is generally considered by the layman to be aesthetically objectionable, a risk to health and only worthy of disposal.

Conventional sewage treatment processes are designed to purify waste-water so that the final effluent may be suitable for discharge to a watercourse without causing any marked pollutional impact. The degree to which treatment is required is, therefore, dependent not only on the quality of the sewage itself, but also on the condition of the watercourse prior to effluent addition, the ability of the watercourse to self-purify the effluent and the further use to which the watercourse is to be put. In the UK, however, discharges to inland watercourses must usually conform to the 30:20 standard — i.e. contain no more than 30 mg/l of suspended solids and 20 mg/l of biochemical oxygen demand (BOD). Such conventional treatment involves physical, chemical and biological processes, the biological process, relying on the metabolic activity of a mixed culture of micro-organisms to degrade complex compounds, being of great importance. Indeed, the use of micro-organisms in waste-water treatment represents the largest applied use of micro-organisms by man. Smith[25] considers sewage treatment to be the largest of all biotechnological industries, yet the least recognised by lay people.

The first section of this chapter therefore sets out to describe conventional waste-water treatment processes with special reference to the biological involvement in these processes. Some of the new developments in plant design are also included in this section. Such conventional treatment processes are well able to purify waste-water so as to greatly reduce the pollutional impact of the waste on the environment. Indeed, the report of the Working Party on the Control of Pollution[31] concluded that little new technology is required to solve our water pollution problems, but that too often we are failing to apply biological treatment techniques which were well-known 50 years ago. The report suggests that the greatest need is for a more efficient administration and a system of financial incentives and controls to ensure that the best available technology is used. The rising costs of sewage disposal have, however, led to a greater readiness to consider waste-water as a potential resource, and much of the current biotechnological interest in sewage treatment involves reclamation rather than disposal. Such interest led the Joint Working Party from the Advisory Council for Applied Research and Development, the Advisory Board for Research Councils and the Royal Society[16] to include waste treatment as one of the six areas of biotechnology possessing the greatest potential for exploitation over the next 5–10 years. It may also be envisaged that two further areas identified by the Report — single-cell protein (SCP) and fuel production from biomass — may be integrated with waste treatment processes so as to utilise products and by-products.

The second section of this chapter deals with some of the more recent advances in waste-water treatment processes and in by-product utilisation involving the use of 'biotechnology'. The section may be broadly divided into two subsections, the first considering advances in secondary treatment processes and the second dealing with advances in sludge disposal and utilisation. Before considering either of these two topics, however, one interesting biotechnological development worthy of mention is the use of micro-organisms for sewer cleansing. Sewers which transport raw sewage to the treatment works frequently suffer from problems of grease accumulation; greasy materials in the sewage build up on the inner surfaces of the sewers and trap particulate matter, thereby reducing the effective capacity of the sewers to carry their designed flow. The introduction of grease-consuming bacteria into the sewage flow, however, acts so as to break down these accumulations, thereby cleaning the sewer system. In a comparison of the costs of such biological cleaning with conventional cleansing techniques, Baig and Grenning[2] estimated that the 3000 miles of sewers served by the Washington Suburban Sanitary Commission could be cleaned at a cost of $1.3 million/annum using grease-consuming bacteria, while the cost of using conventional cleaning techniques was $4 million/annum. In a service area of 3 million people, it was thus possible to make savings of $2.7 million/annum. Inoculum cultures for use in waste-stream and spill treatment are marketed by about 20 companies, and had a 1978 market value of $2–4 million (this figure, of course, includes the use of micro-organisms in treating oil and industrial spillages); one estimate of the total potential market is $200 million[11].

CONVENTIONAL WASTE-WATER TREATMENT PROCESSES

Figure 1 represents the conventional, multi-stage, waste-water treatment process. It seems best to consider each of the various stages in turn.

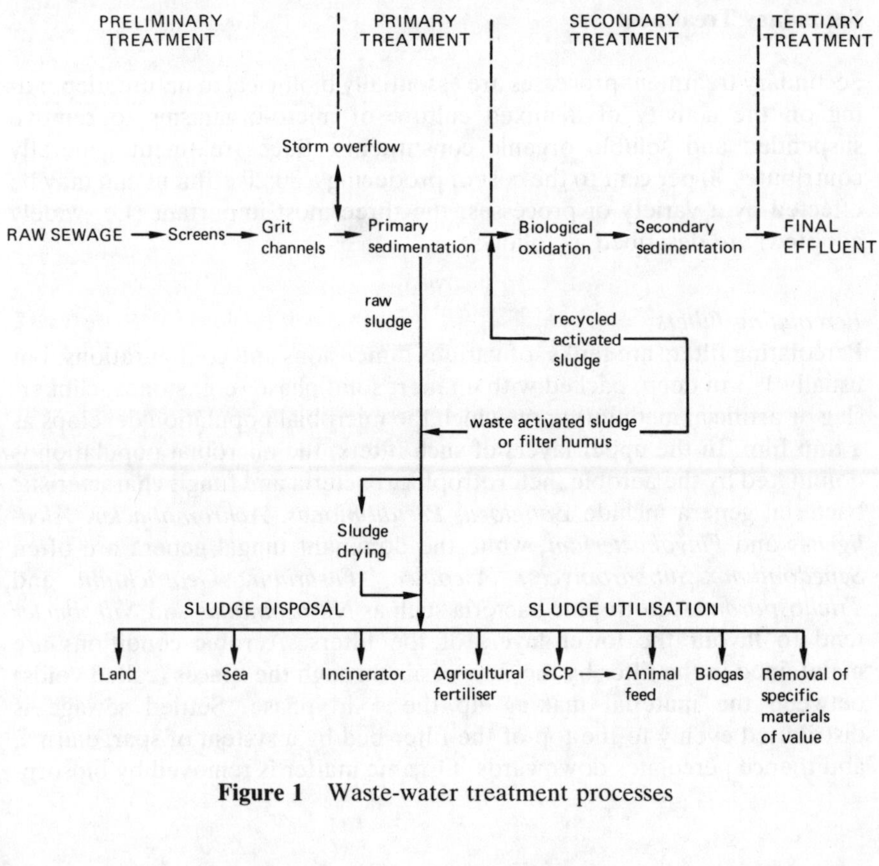

Figure 1 Waste-water treatment processes

Preliminary and Primary Treatment

Preliminary and primary treatment involves essentially physical processes. Influent raw sewage is first passed through screens to remove large items of solid matter and then passes through a series of grit channels, where the reduced flow rate encourages the settlement of grit and small stones. Screenings may be returned to the sewage flow after size reduction or may be buried or incinerated, while grit may be washed prior to use for landfill or disposal by burial. Primary treatment involves the further reduction of flow rate by passing the waste-water into primary sedimentation tanks. This facilitates the settlement of finer particulate matter, thereby forming a primary (or raw) sludge. Generally, such primary sedimentation will remove 65 per cent of sewage suspended solids and reduce BOD by 35–55 per cent. Owing to the simplicity of the processes involved, costs are low, preliminary treatment (screening and grit removal) contributing only 5 per cent to the total cost of producing a 30:20 effluent and sedimentation (both primary and secondary) contributing only 25 per cent.

Secondary Treatment

Secondary treatment processes are essentially biological in nature, depending on the activity of a mixed culture of micro-organisms to remove suspended and soluble organic constituents. Such treatment generally contributes 30 per cent to the cost of producing a 30:20 effluent and may be effected by a variety of processes; the three most important (i.e. widely adopted) are described as follows.

Percolating Filters

Percolating filters are tanks, of various dimensions and configurations, but usually 1–3 m deep, packed with an inert solid phase (e.g. stones, clinker, slag or artificial medium) upon which the microbial population develops as a thin film. In the upper layers of such filters, the microbial population is dominated by the aerobic, heterotrophic, bacteria and fungi; characteristic bacterial genera include *Zoogloea, Pseudomonas, Achromobacter, Alcaligenes* and *Flavobacterium*, while the dominant fungal genera are often *Sepedonium, Subbaromyces, Ascoidea, Fusarium, Geotrichium* and *Trichosporon*. Autotrophic bacteria such as *Nitrosomonas* and *Nitrobacter* tend to favour the lower layers of the filters. Aerobic conditions are maintained in the filter by the flow of air through the spaces (called voids) between the material making up the solid phase. Settled sewage is distributed evenly to the top of the filter bed by a system of sparge arms, and thence percolates downwards. Organic matter is removed by biosorption, coagulation and bio-oxidation by the heterotrophic organisms dominant in the upper layers of the bed, the energy and materials thus derived by the heterotrophes being used for their maintenance and the synthesis of new biomass (i.e. new film). In the lower layers of the bed the conversion of ammonia to nitrite and thence to nitrate is mediated by the autotrophic bacterial population. Various species of protozoa and algae also occur in the microbial film and, while the algae play little part in the purification process, the protozoa play a major role in the removal of bacteria freely suspended in the film, thereby effecting the clarification of the filter effluent. In a well-managed filter a grazing fauna including nematodes, rotifers, worms and larval flies acts so as to prevent excess microbial film developing, and it has been shown[29] that filters containing a macro-invertebrate population produce a higher-quality effluent than those with no such fauna. Thus, the ecology of percolating filters is complex and is more thoroughly reviewed by Hawkes[13].

The percolating filter may, therefore, be considered as a heterogeneous, two-phase, continuous culture involving a mixed culture of micro-organisms. Effluent leaves the filter containing a quantity of suspended solids, often dark brown in coloration, comprising sloughed microbial film and organic solids derived from the grazer community. This humus is removed by settlement in humus tanks, generally reducing BOD by 85–95 per cent and suspended solids by 70–90 per cent.

Filter efficiency (as usually measured by BOD removal) is primarily dependent on two factors: contact area and contact time. Contact area is dependent on both the size and shape of the solid phase. The introduction of plastic artificial media (e.g. Flocor and Biopac) with larger surface areas than conventional media and large void capacities has, however, been limited, as a result of the high capital cost of such media, and for the most part natural media provide an acceptable surface for microbial colonisation. Contact (or residence) time is inversely related to hydraulic loading. Consequently, lower hydraulic loadings allow for a longer residence time and a more complete degradation of the organic matter, and as hydraulic loading rises, effluent quality falls. Practically, however, it is found that recycling some of the final effluent, thereby increasing hydraulic loading but lowering sewage strength, prevents excess microbial film accumulating without adversely affecting effluent quality.

Temperature is known to affect the rate at which all biological reactions proceed. The effect of temperature on filter efficiency is, however, complex. Certainly the reduced activity of the grazer population at low temperatures (i.e. winter) allows a thicker microbial film to develop, which may impede percolation and cause a phenomenon known as ponding. But in a well-managed bed in which ponding is prevented it has been shown practically that reduced winter temperatures have little effect on filter efficiency.

Activated Sludge

Activated sludge is a homogeneous continuous culture process involving the aeration of settled sewage with an inoculum of micro-organisms for (conventionally) 5–15 h. During this time there is a rapid removal of organic matter by biosorption and flocculation, while oxidation and biosynthesis proceed at a slower rate. The effluent from the process is subsequently run into a secondary sedimentation tank, allowing settlement of the flocs, and a proportion of this sediment is returned to the aeration tank to act as a microbial inoculum for the incoming sewage (Figure 2a). Both BOD and suspended solids are reduced by 85–95 per cent in this conventional process.

The organisms responsible for the degradation of organic matter in an activated sludge plant are essentially similar to those in percolating filters.

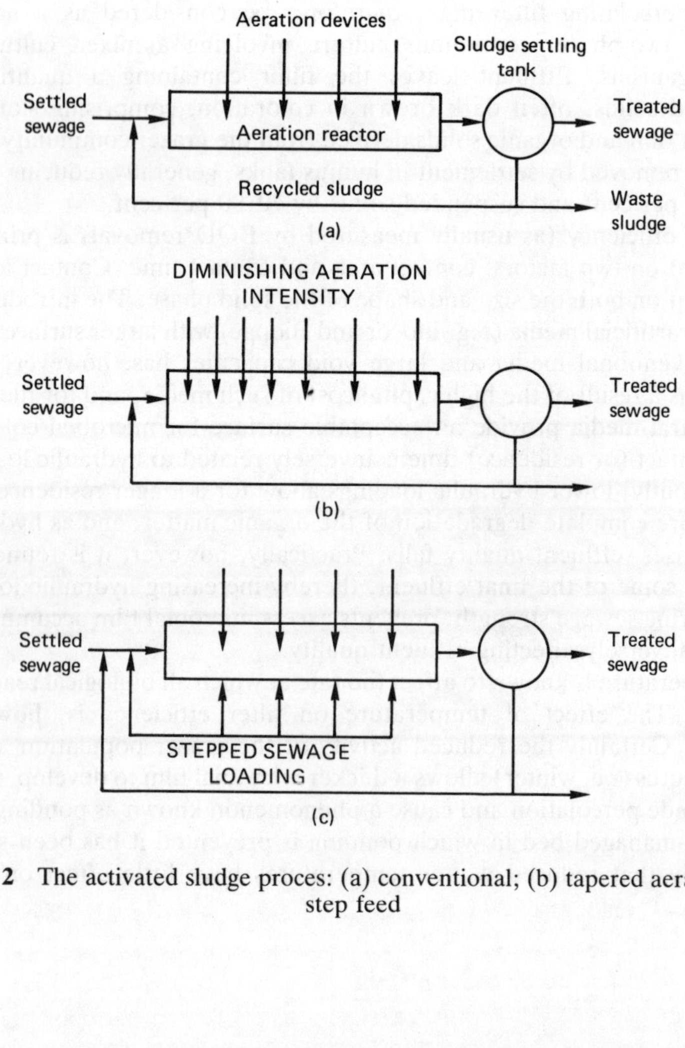

Figure 2 The activated sludge proces: (a) conventional; (b) tapered aeration; (c) step feed

In the former system, however, conditions are unsuitable for the macro-invertebrate grazer population. Consequently, activated sludge plants do not suffer from the fly nuisance common in plants employing percolating filters, and in urban areas this gives the former process a distinct advantage over the latter. Fungi are also less dominant in the activated sludge process than in percolating filters, although they remain important in that under conditions of inhibited bacterial growth, the growth of fungi may cause sludge bulking (this is further discussed later). Protozoans are abundant in activated sludge, Curds[10] listing 228 species associated with the process. Nematodes and rotifers are also present, though in small numbers.

A variety of modifications of the conventional activated sludge process (described above) have been adopted so as to provide better aeration and to achieve better sludge settlement. Some of these modifications will now be considered.

Aeration systems not only provide the oxygen requirements of the aerobic micro-organisms responsible for degradation of the raw sewage, but also provide the turbulence necessary to keep the microbial flocs in suspension (i.e. in contact with the substrate). Aeration may be provided by bubbling filtered compressed air through porous tiles or domes situated on the bottom of the tank (reactor), or by a variety of mechanical surface aeration devices. A major disadvantage with the conventional aeration process, however, is that the oxygen demand within the reactor is not uniform, being greater at the inlet end, where settled sewage enters. Oxygen supply may, therefore, be deficient at the inlet end of the reactor and superfluous at the outlet end. The provision for increased aeration at the inlet end of the reactor thus allows aeration to be tapered to the demands in the reactor. Generally, such tapered aeration involves the distribution of approximately 45 per cent of the total input to the inlet end (one-third of the reactor), 25 per cent of the input to the outlet end (one-third of the reactor) and 30 per cent of the total input to the middle one-third of the reactor. The addition of sewage at several points along the length of the reactor similarly prevents local deoxygenation from occurring at the inlet end of the tank. This is called step-feed, stepped loading, or, rather inaccurately, step-aeration (Figure 2c).

The duration of aeration is also of importance. Prolonged aeration of the order of 1–2 days and a high rate of sludge return allow for a very complete oxidation, and the low volumes of sludge produced by such extended aeration may be particularly desirable in plants whose location makes sludge disposal particularly expensive. On the other hand, high-rate activated sludge plants may use detention times of only 1–2 h, thereby producing effluents unable to conform to the 30:20 standard, but adequate for discharge to estuarial or coastal waters.

Sewage treatment works adopting the activated sludge system require only about 15 per cent of the land area demanded by plants operating percolating filters and are therefore often favoured in population-dense areas where land is scarce and (therefore) expensive.

Such plants are economically competitive with those using percolating filters, being cheaper to construct, but, having a higher energy requirement, being more costly to run. Actual costs will, of course, fluctuate with the cost of raw materials, labour and energy (i.e. electricity) and be dependent on plant size, sewage quality and the degree of treatment necessary. Tebbutt[26], however, considered that the capital cost of a medium-sized conventional plant (either activated sludge or percolating filter), would be in the region of £45 m^{-3} day^{-1} and that such a plant could produce a standard 30:20 effluent at a cost of 2 p/m^3.

In the conventional activated sludge treatment processes described previously, the upper limit for the rate of transfer of oxygen into solution is approximately 100 g O_2 $m^{-3}h^{-1}$ (Ref. 5) which, under conditions of high organic loading, may limit the rate at which treatment processes proceed. To increase this oxygen transfer rate using conventional aeration techniques requires additional power and therefore increased running costs. However, two new processes — the deep-shaft process and the use of commercial oxygen — are able to transfer oxygen into solution at a higher rate, with no corresponding increase in power consumption.

The deep-shaft process, pioneered by ICI, is capable of supplying oxygen at a rate of 2–3 kg O_2 $m^{-3}h^{-1}$ and involves the use of a shaft 50–300 m deep to generate a head of pressure under which the solubility of oxygen in water is increased. Similarly, the use of commercial oxygen (rather than air) in activated sludge aeration is able to produce a fivefold increase in oxygen transfer rate, and although the energy (and cost) required to produce the oxygen must be offset against the benefits gained from its usage, Sidwick *et al.*[24] claim that the process has a lower overall cost (cf. conventional activated sludge) for plants treating dry weather flows of 4500 m^3/day.

Sludge bulking is a condition in which the sludge loses its properties of rapid settlement, settling poorly in the secondary sedimentation tanks so that it tends to carry-over to contaminate the final effluent. Bulking sludges characteristically contain species of filamentous organisms — in particular, the filamentous bacterium *Sphaerotilus*. Although the phenomenon is still not completely understood, it appears to be encouraged by the combined effect of high C:N and C:P ratios, low dissolved oxygen concentrations or nitrogen deficiency, while low sludge age may also promote the phenomenon. Bulking is thus generally confined to smaller works where there may be considerable fluctuation in organic loading and/or inadequate technical supervision. Once established, a bulking site may be remedied by recirculating a proportion of the final effluent, adding finely divided inert materials, or by controlled chlorination. In extreme cases it may be necessary to discard the sludge altogether and allow a new one to develop.

The fluidised-bed process circumvents many of the problems of sludge settlement found in the conventional activated sludge process. This process employs beds of sand or other small media, 0.2–2 mm in diameter, upon which the microbial biomass develops. Settled sewage is forced up through the bed at such a rate as to keep the particles of medium suspended in free motion. The effluent leaving the reactor thus needs little or no secondary clarification, and the beds may be intermittently cleaned by air scouring, to produce thick sludges. Thus, whereas the upper limit for biomass concentration in the conventional activated sludge process is about 3–5 g/l, that achieved in fluidised-bed reactors is of the order of 15–40 g/l. This ability to produce thick sludges allows plants utilising the fluidised-bed process to be small, such that the land area required may be only 20 per cent of that needed for a conventional activated sludge plant. Cooper and Wheeldon[9] thus calculated that for carbonaceous oxidation the fluidised-bed process may be 25 per cent cheaper in terms of net present value (NPV) than conventional activated sludge processes, and Wheeldon and Bayley[28] concluded that for plants treating greater than 30 000 m^3/day the NPV for both processes differed by less than 15 per cent, while for smaller plants treating less than 30 000 m^3/day the fluidised-bed process seemed less economical than conventional treatment. The treatment of waste-waters using fluidised-bed reactors is thoroughly reviewed in Cooper and Atkinson[8].

Oxidation Ponds

Oxidation ponds are shallow basins usually 1 m deep which receive a continuous flow of settled or raw sewage. Detention times may range from a few days to several months. In such ponds microbial degradation of organic matter proceeds, thereby providing a eutrophic (nutrient-rich) environment. The minerals thus formed are then utilised by members of the producer population, of which the algae are the most important group. Aerobic conditions are maintained in such ponds by the release of oxygen (as a by-product of photosynthesis) by the producers, and this oxygen in turn is used by the aerobic heterotrophes. These processes are summarised in Figure 3.

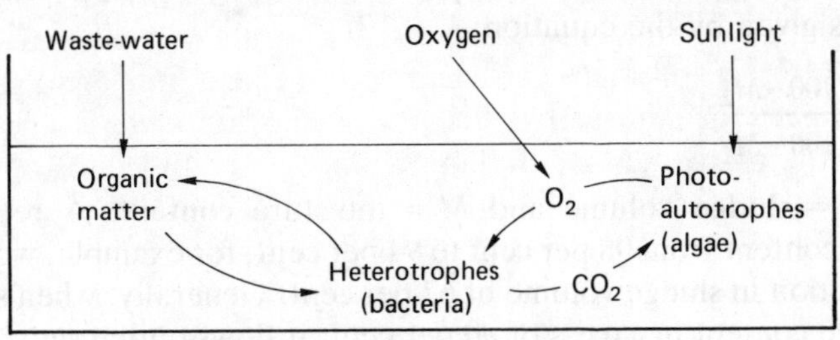

Figure 3 Oxidation pond processes

Oxidation ponds are cheap to construct and operate, but require large areas of land, are prone to develop odours, and provide a breeding ground for flies. Since the maintenance of aerobic conditions is so dependent on the activity of the producer population, the use of oxidation ponds is largely restricted to countries in which the climate is sunny, warm and dry — e.g. Australia, India and southern USA; in Britain the use of oxidation ponds is largely restricted to tertiary treatment.

Tertiary Treatment

Tertiary treatment processes are generally carried out to produce an effluent of better quality than the 30:20 standard. Such treatment processes include lagooning, and filtration, irrigation over grassland and microstraining. Effluents containing 10 mg/l suspended solids and 10 mg/l BOD (or better) are attainable when using such processes, while the cost of such treatments would add a further 10–20 per cent to the cost of conventional (i.e. preliminary, primary and secondary) treatment.

Sludge Treatment

The end result of conventional sewage treatment is purified water and a residual sludge from the primary and secondary sedimentation processes. The treatment and disposal of this sludge is a costly process, contributing 40 per cent to the cost of producing a 30:20 effluent. The population of England and Wales alone produce 1.2 million tons (dry weight) of sewage sludge annually, comprising 800 000 tons primary and 400 000 tons secondary sludge, and costing £85 million per annum to treat and dispose: US sludge production has been variously estimated at 3.65–7 million dry tons per annum.

Wet sludge contains 90–99.5 per cent water (moisture) and, since sludge volume is directly related to moisture content, it is desirable to reduce moisture content as much as possible so as to facilitate sludge handling. The general relationship between sludge volume and moisture content is given by the equation:

$$\frac{V_1}{V_2} = \frac{100-M_2}{100-M_1}$$

where V = sludge volume and M = moisture content. A reduction in moisture content from 98 per cent to 94 per cent, for example, would result in a reduction in sludge volume of 67 per cent. Generally, when sludge has a moisture content in excess of 90 per cent, it flows quite freely; when the moisture content approaches 85 per cent, a noticeable thickening occurs;

and when the moisture content is between 70 per cent and 80 per cent, the sludge no longer flows and is known as sludge cake.

Methods of dewatering sludge are both diverse and numerous. Drying beds are commonly employed in Britain; these are shallow basins into which sewage sludge may be run and allowed to settle for a period of time, after which supernatant water may be run off and further losses may occur through evaporation and drainage. Drying is thus highly dependent on climatic factors (i.e. rainfall and humidity), and may take up to several months in the wetter parts of Britain. Sludge dried by this method has a moisture content of 45–60 per cent. Alternatives to this method of sludge dewatering include mechanical processes such as filter pressing and vacuum filtration, which are able to produce sludge cakes of 50–65 per cent and 70–90 per cent moisture, respectively.

Sewage sludge is also commonly anaerobically digested. This process is schematically represented in Figure 4 and comprises three stages. Since, however, the organisms responsible for the first two stages are difficult to distinguish (since many are able to carry out both processes), the process is often considered to comprise only two phases, viz. the acid-forming stage and the methane-forming stage. The acid-forming stage is characterised by the solubilisation of complex organic molecules by extracellular enzymes and the subsequent formation of volatile fatty acids, alcohols, aldehydes and ketones by a variety of bacterial species, including members of the genera *Alcaligenes, Bacillus, Escherichia, Micrococcus, Proteus* and

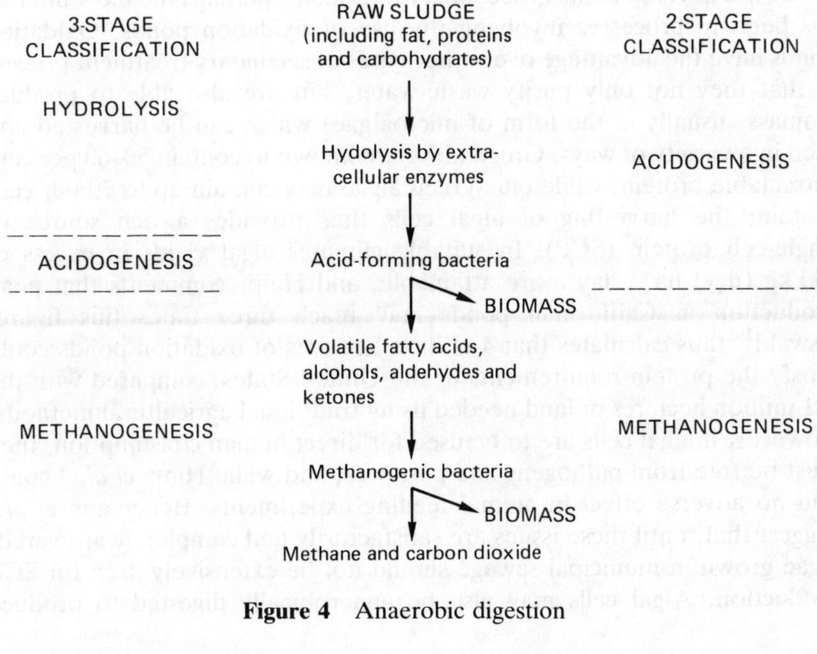

Figure 4 Anaerobic digestion

Pseudomonas. In the methane-forming stage, species of the genera *Methanobacterium, Methanobacillus, Methanococcus* and *Methanosarcina* convert these products to methane and carbon dioxide. During digestion about 40 per cent of the organic matter is converted into gaseous products and the biogas comprises about 70 per cent methane and 30 per cent carbon dioxide. In Britain mesophilic digestion (at about 30°C) is usually adopted, since it is easier to control, though slower than thermophilic digestion (at about 50°C). The sludge, after digestion, is reduced in volume by about 67 per cent. Biogas utilisation will be further considered below.

Disposal of sludge may be to land or sea, or by incineration. However, the expense and pollutional impact accompanying such disposal encourage the future development of sludge utilisation rather than disposal, and this will be further discussed in the next section.

RECENT ADVANCES IN WASTE-WATER TREATMENT PROCESSES

As described in the first section, secondary treatment processes involve the degradation of sewage mediated by a mixed microbial culture. Desirable metabolic characteristics could certainly be added to organisms responsible for this process by recombinant DNA technology. But it seems unlikely that such a course of action would be cost-effective as regards the treatment of municipal sewage, although for a works treating specific industrial wastes such a proposition may be economically more attractive.

Of the developments in secondary treatment, perhaps the most interest has been in processes involving the use of oxidation ponds. Oxidation ponds have the advantage over other types of secondary treatment process in that they not only purify waste-water, but are also able to produce biomass (usually in the form of microalgae) which can be harvested and used in a variety of ways. Green algae are known to contain 50–60 per cent extractable protein, while blue-green algae may contain up to 70 per cent protein; the harvesting of algal cells thus provides a rich source of single-cell protein (SCP). In suitable climates algal yields in excess of 100 kg (dry) ha^{-1} day^{-1} are attainable, and Hall[12] comments that peak production in Californian ponds may reach three times this figure. Oswald[19] thus calculates that 4 million hectares of oxidation ponds could satisfy the protein requirements of the United States, compared with the 121 million hectares of land needed using traditional agricultural methods. However, if algal cells are to be used for direct human consumption, they must be free from pathogens and parasites, and while Hintz *et al.*[14] could find no adverse effect in animal feeding experiments. Benemann *et al.*[3] suggest that, until these issues are satisfactorily and completely answered, algae grown in municipal sewage should not be extensively used for SCP production. Algal cells may also be anaerobically digested to produce methane (equivalent to 11.6 MJ/kg algae) and Hall[12] suggests that a net product of 520 GJ ha^{-1} $year^{-1}$ of methane seems feasible, the cost of methane thus produced in 1976 being $2.6–3.9/GJ. Together with the anaerobic digestion of sewage sludge this may supply local energy demands. Algal biomass may also be burned to produce electricity and an average yearly algal yield (50–60 dry tonnes/ha) would produce approximately 74 000 kWh of electricity. Unfortunately, while oxidation ponds are energetically efficient, the harvesting of microalgal cells is a difficult and costly process. Benemann *et al.*[3] estimate that the cost of harvesting may range from $5 to $140 per 10^6 l, depending on the separation technique adopted, and as yet the full-scale commercial exploitation of microalgal biomass has not been realised, although pilot-plant operation is well advanced.

Difficulties in biomass harvesting in oxidation ponds may be circumvented by the selective culture of macroscopic vascular plants (commonly called macrophytes), which may then be harvested either manually or by using simple mechanical harvester/shredders; one such development is the culture of the water hyacinth, *Eichhornia crassipes*. Wolverton and McDonald[30] describe work carried out by the US National Aeronautics and Space Administration (NASA) and the National Space Technology Laboratories (NSTL) near Bay St. Louis, Mississippi, in which *Eichhornia* was grown in both a primary lagoon receiving a light organic loading and a secondary lagoon receiving the effluent from a primary aerated lagoon. BOD was reduced by 95 per cent (from 110 mg/l to 5 mg/l) in the primary lagoon and by 70 per cent (from 50 mg/l to 15 mg/l) in the secondary lagoon, while suspended solids were reduced by 90 per cent (from 97 mg/l to 10 mg/l) in the primary lagoon and by 71 per cent (from 49 mg/l to 14 mg/l) in the secondary lagoon. Maximum annual productivity of *Eichhornia* was found to be 154 tonnes ha^{-1} $year^{-1}$. Harvested plants were composted (aerobically digested) to form a stable soil supplement, and since metal concentrations were found to remain low if grown in domestic sewage, the hyacinth was suggested to be useful as an animal feedstuff. If anaerobically digested, it was calculated that one hectare of pond could provide sufficient biomass to generate 58 400 m^3 biogas containing 35 100m^3 methane annually. Interestingly, it was found that evapotranspirative water vapour losses from the plant were substantial, and could be recovered by overlaying the hyacinth-covered area with greenhouse covers and drainage pipes. Using such apparatus, it was found that up to 40 per cent of the daily waste-water could be recovered as pure fresh water.

Oxidation ponds may also be used for the cultivation of fish. In a study carried out by Carpenter *et al.*[7], for example, *Tilapia nilotica*, golden shiners (*Notemigonus crysoleucas*) and channel catfish (*Ictalurus punctatus*) were stocked in four oxidation ponds receiving the effluent from two further ponds. In this system, *T. nilotica* biomass increased from 41 lb to 163 lb in 3 months. *N. crysoleucas* biomass increased from 85 lb to 535 lb and *I. punctatus* biomass increased from 660 lb to 4400 lb, the majority of this gain being over a 6 week period. The presence of fish in the oxidation pond system was also found to significantly improve the quality of effluent. Similarly, in a study carried out by Noble[18] at Rye Mead Sewage Works, Hertfordshire, mirror carp (*Cyprinus carpio*) grown in tertiary treatment lagoons were found to have a maximum production rate of 85.6 g m^{-1} $year^{-1}$. Depending on the stocking density, it was found possible to produce fish of marketable size, and since feeding usually comprises 40 per cent of fish-farm running costs, the production of fish of marketable size in sewage lagoons (where no additional feeding was found necessary) was suggested to be an attractive economic proposition. Unfortunately, concern as to the risk of pathogen and parasite contamination, and over bioaccumulation of potential toxins, largely prevents fish grown in such a manner from being sold for direct human consumption, but a variety of non-consumptive alternatives are available, and include the use of live fish for bait and restocking, and the use of fish for animal feeds.

As already mentioned, sludge treatment and disposal is a costly process, and various ways of utilising sewage sludge have been considered; essentially these fall into three major categories, viz. energy production, agricultural usage and the removal of specific materials of importance.

The digestion of sewage sludge to reduce sludge volume and stabilise the waste has already been considered. Much of the recent interest in sludge digestion, however, has been concerned with biogas production and utilisation. Anaerobic digestion is, unfortunately, rather a slow process and is often accelerated by the burning of a proportion of the gaseous products to elevate the temperature at which the process takes place. The actual amount of gas necessary for this task is dependent on a variety of factors, including the climate, the siting of the digester and the digester's insulation, but in temperate climes about 30 per cent of the total gas production will be required to maintain the digester temperature at the 27–35°C necessary for mesophilic digestion. Biogas may also be used within the treatment works to provide the energy necessary for the rest of the treatment process. For example, two of London's main sewage works, Beckton and Mogden, producing 2.22 and 2.03 million ft^3 gas per day, respectively (1971–72 figures), were able to provide 80–90 per cent of their total energy requirements by utilising their biogas production as available energy. Similarly, the utilisation of biogas at the Yonkers treatment plant in Westchester County, New York, together with waste heat conservation policies, allowed savings in fuel estimated to be $600 000 per annum. Biogas utilisation may, therefore, lead to a reduction in the overall cost of waste-water treatment and may even satisfy local energy demands, but on a national level the value of biogas utilisation looks less impressive.

Pankhurst[20], for example, concluded that if all sewage, animal manure and domestic refuse currently produced in the UK were converted into biogas, it would produce about 7 million m^3/day — i.e. less than 5 per cent of current UK gas demands. Both the production and utilisation of biogas are considered in further detail in Price and Cheremisinoff [21].

Before sewage sludge may be applied to agricultural land as a fertiliser, it is usually necessary for it to be screened, and stabilised by aerobic or anaerobic digestion, or by conditioning with lime and/or dewatering. Unfortunately, the fertiliser potential of sewage sludge is less than may be envisaged: Rothman and Bartlett[23] suggest that in 1968 the fertiliser potential of sludge in the UK represented less than 4.5 per cent of the agricultural nitrogen and phosphate requirements, and less than 1 per cent of the potash requirement. Furthermore, the lack of sludge uniformity, the presence of heavy metals in high concentrations (which may be toxic to certain crop species, and which may be bioaccumulated to produce even higher concentrations in the crops themselves), and the risk of both pathogen and parasite contamination, all reduce the fertiliser value of sludges. Sludge is, therefore, often applied to agricultural land as a soil conditioner rather than as an actual fertiliser.

Agricultural use of sewage sludge also includes its direct or indirect use as an animal feedstuff. Indirect usage involves the use of sludge as a substrate for the growth of bacteria and fungi, which may then be harvested as SCP and used as an animal feed. However, sewage sludges contain a vast number of microbial inhabitants and unless the sludge is sterilised and subsequently inoculated with a single microbial strain, a variety of micro-organisms grow, resulting in an end-product of ill-defined quality. Separation of microbial biomass from waste substrate also poses technical difficulties. SCP production from waste material has usually, therefore, involved microbial culture on 'purer' wastes from specific industrial processes — e.g. spent sulphite liquor, molasses or whey, which are also less likely to harbour pathogens and parasites than are sewage sludges. Reddington and Brown[22], however, were able to grow *Aspergillus niger* on piggery waste liquors, if such wastes were diluted and supplemented with carbohydrate, to produce a final biomass containing 20 per cent wt/wt crude protein. One interesting development in SCP production is that described by Thomas and Evison[27], who deliberately inhibited the methanogenic stage of anaerobic digestion of piggery waste either by overloading the process with raw waste or by the addition of hydrochloric acid so as to reduce pH. The volatile fatty acids produced in the acid-forming stage of the process were then used as a substrate for the growth of the fungus *Candida utilis*, which could then be harvested as SCP.

From the viewpoint of ecological energetics, it is preferable to feed sludge to animals directly rather than indirectly via SCP or traditional crops (this may be considered as analogous to 'cutting out the middleman'), and a number of feeding trials have shown that a variety of animals (including cattle, sheep, pigs and poultry) are able to grow quite satisfactorily on a diet including a sludge supplement. This should, perhaps, not be surprising, since dried activated sludge is known to contain 30–35 per cent crude protein[15], as against 65 per cent for fishmeal and 44 per cent for soya meal. But the health risks associated with the direct feeding of sewage sludge to animals are both complex and controversial, and serve to prevent the widespread use of sludge in such a capacity. Bitton *et al.*[4] provide a thorough review of such health risks associated with the presence of microbes, parasites, toxic metals and toxic organic residues in sludges.

The final pathway of sludge utilisation is the removal of specific materials of importance. It has been calculated, for example, that the sludge produced annually in England and Wales alone contains protein valued at £113 million[23], while Hurwitz[15] estimated that the quantities of amino acids in one ton of activated sludge were worth $12 000–15 000. Methods for the direct extraction of such materials from sludge are essentially physicochemical. Carberry *et al.*[6], for example, describe a process by which sludge proteins were solubilised, separated from impurities using ion-exchange columns and then resolidified by acid precipitation. Using such a technique, it was found that protein could be produced in 5 and 100 million gallon/day plants for 28 and 23 cents/lb, respectively. With respect to materials such as heavy metals, however, the ability of living organisms to accumulate such materials to intracellular concentrations far exceeding those found extracellularly may be used in conjunction with physicochemical processes to effect efficient extraction. Kelly *et al.*[17] further discuss the use of micro-organisms in metal recovery, though not with direct reference to metal in sewage sludges.

THE FUTURE

Pollution is largely considered by the layman to be a necessary consequence of our industrial and urban development. It is clear, however, that economic growth can be reconciled with pollution control so long as public opinion and political will are in favour: from a technological viewpoint aquatic pollution is well able to be controlled by effectively using available technology.

With the rising costs of raw materials, the further development of processes to recycle waste products seems assured. Of the three main pathways of sewage sludge utilisation discussed, it seems that the removal of materials of specific economic importance suffers from the greatest technical difficulties. SCP production, although technically and economically feasible, is unlikely to gain widespread acceptance as an animal and human foodstuff until the health risks associated with such products are better understood, while biogas production is both technically feasible and suffers from few aesthetic objections. At the commercial level, therefore, biogas production is perhaps of greatest interest at the present, and massive plans to utilise biogas are currently being put into effect, especially in China and India (see, for example, Ref. 1).

Undoubtedly, then, there are still various technical difficulties involved in the utilisation of waste-water treatment by-products (sludges) which remain to be solved. But perhaps the greatest problem confronting the 'biologically aware' scientist over the next decade or so is that of educating the layman to understand the fundamental, underlying biological principles and integrating the problems of food production and waste disposal so that waste-derived materials lose their aesthetic objectionability and gain a wider acceptance from the population at large.

REFERENCES

1. Agarwal, A. (1979). Can biogas provide energy for India's rural poor. *Nature, Lond.*, **281**, 9–10
2. Baig, N. and Grenning, E. M. (1976). The use of bacteria to reduce clogging of sewer lines by grease in municipal sewage. In *Biological Control of Water Pollution* (Eds. Tourbier, J. and Pierson, R. W. Jr.), pp. 245–253. University of Philadelphia Press
3. Benemann, J. R., Wiesman, J. C. and Oswald, W. J. (1979). Algal biomass. In *Economic Microbiology*, Vol. 4, *Microbial Biomass* (Ed. Rose, A. H.), pp. 177–206. Academic Press, London
4. Bitton, G., Damron, B. L., Edds, G. T. and Davidson, J. M. (Eds.) (1980). *Sludge — Health Risks of Land Application*. Ann Arbor Science Publishers
5. Boon, A. G. (1978). Oxygen transfer in the activated sludge process. In *New Processes in Waste-water Treatment and Recovery* (Ed. Mattock, G.), pp. 17–33. Ellis Horwood, Chichester
6. Carberry, J. B., McCaffery, T. F. and Clunie, R. M. (1978). Protein recovery from waste activated sludge. In *New Processes in Waste-water Treatment and Recovery* (Ed. Mattock, G.), pp. 379–404. Ellis Horwood, Chichester
7. Carpenter, R. L., Coleman, M. S. and Jarman, R. (1976). Aquaculture as an alternative waste-water treatment system. In *Biological Control of Water Pollution* (Eds. Tourbier, J. and Pierson, R. W. Jr.), pp. 215–225. University of Philadelphia Press
8. Cooper, P. F. and Atkinson, B. (Eds.) (1981). *Biological Fluidised Bed Treatment of Water and Waste-water*. Ellis Horwood, Chichester
9. Cooper, P. F. and Wheeldon, D. H. V. (1980). Fluidized- and expanded-bed reactors for waste-water treatment. *Water Pollution Control,* **79**, 286–306
10. Curds, C. R. (1975). Protozoa. In *Ecological Aspects of Used-water Treatment* (Eds. Curds, C. R. and Hawkes, H. A.), pp. 203–268. Academic Press, London
11. Gasner, L. L. (1979). Microorganisms for waste treatment. In *Microbial*

Technology, Vol. 2, 2nd edn (Eds. Peppler, H. J. and Perlman, D.), pp. 211–222. Academic Press, London

12. Hall, D. O. (1979). Solar energy conversion through biology. *Biologist,* **26**, 67–74
13. Hawkes, H. A. (1963). *The Ecology of Waste Water Treatment*. Pergamon Press, Oxford
14. Hintz, H. F., Heitman, H., Weir, W. C., Torell, D. T. and Mayer, J. H. (1966). Nutritive value of algae grown on sewage. *Journal of Animal Science,* **25**, 675–681
15. Hurwitz, E. (1957). The use of activated sludge as an adjuvant to animal feeds. *Engineering Bulletin Extension Series* No. 94, 395–441
16. Joint Working Party from the Advisory Council for Applied Research and Development, The Advisory Board for Research Councils and The Royal Society (1980). *Biotechnology: Report of the Joint Working Party*. HMSO, London
17. Kelly, D. P., Norris, P. R. and Brierley, C. L. (1979). Microbiological methods for the extraction and recovery of metals. In *Microbial Technology: Current State, Future Prospects* (Eds. Bull, A. T., Ellwood, D. C. and Ratledge, C.), pp. 263–308. Cambridge University Press
18. Noble, R. (1975). Growing fish in sewage. *New Scientist,* **67** (No. 960), 259–261
19. Oswald, W. J. (1977). Determinants of feasibility in bioconversion of solar energy. In: *Research in Photobiology* (Ed. Castellani, A.), pp. 371–386. Plenum Press, London
20. Pankhurst, E. S. (1980). Biogas. *Gas Engineering and Management*, January, 3–14
21. Price, E. C. and Cheremisinoff, P. N. (1981). *Biogas Production and Utilization*. Ann Arbor Science Publishers
22. Reddington, S. and Brown, D. E. (1978). Microbial conversion of piggery waste to a protein feed supplement. In *New Processes in Waste-water Treatment and Recovery* (Ed. Mattock, G.), pp. 366–378. Ellis Horwood, Chichester
23. Rothman, H. and Bartlett, D. (1977). Sewage — pollutant or food? An assessment of alternative ways of disposing of sewage. In *Technology Assessment and the Oceans* (Eds. Wilmott, P. and Stringerland, A.), pp. 161–167. IPC, London
24. Sidwick. J. M., Lewandowski, T. P. and Allum, K. H. (1975). An economic study of the Unox and conventional aeration systems. *Water Pollution Control,* **74**, 645–656
25. Smith, J. E. (1981). *Biotechnology*. Edward Arnold, London
26. Tebbutt, T. H. Y. (1973). *Water Science and Technology*. John Murray, London
27. Thomas, K. and Evison, L. (1978). Process control of anaerobic digestion for SCP recovery. In *New Processes in Waste-water Treatment and Recovery* (Ed. Mattock, G.), pp. 349–365. Ellis Horwood, Chichester
28. Wheeldon, D. H. V. and Bayley, R. W. (1981). Economic studies of biological fluidised beds for waste-water treatment. *Biological Fluidised Bed Treatment of Water and Waste-water* (Eds. Cooper, P. F. and Atkinson, B.), pp. 306–328. Ellis Horwood, Chichester

29. Williams, N. V. and Taylor, H. M. (1968). The effects of *Psychoda alternata* (Say.) (Diptera) and *Lumbricillus rivalis* (Levinson) (Enchytraeidae) on the efficiency of sewage treatment in percolating filters. *Water Research,* **2**, 139–150
30. Wolverton, B. C. and McDonald, R. C. (1979). The water hyacinth; from prolific pest to potential provider. *Ambio,* **8**(1), 2–8
31. Working Party on the Control of Pollution (1972). *Pollution — nuisance or nemesis?* Report of the Working Party. HMSO, London

26
MICROBIOLOGY AND ANAEROBIC DIGESTION

David B. Archer

INTRODUCTION

Anaerobic bacteria convert organic matter to methane and carbon dioxide in the absence of oxygen. This occurs naturally in many environments, which include intestinal tracts and sediments. In biotechnology the degradation of organic wastes by anaerobic bacteria is termed anaerobic digestion, and this process may be used to treat domestic, agricultural and industrial wastes in order to produce methane for use as a fuel and to avoid pollution. The major degradative steps in methanogenesis from wastes are much the same (Figure 1) irrespective of either the waste composition or the type of digester (e.g. high-rate reactor, animal slurry digester or landfill). Because there are a number of steps involved in anaerobic digestion, there are necessarily a number of different bacterial species involved (i.e. the anaerobic digester houses a mixed culture). The microbiology of the anaerobic digestion process is consequently very complicated and yet the prime requirement of the process is that it be reliable. Control of industrial processes which utilise pure species need not be simple, and yet in anaerobic digestion the problems are exacerbated by the number of bacterial species, the lack of knowledge of the interactions between species, variations in the composition and amount of waste, and so on. The need for reliability is paramount and yet the need for other desirable features of anaerobic digesters should not be underestimated. For example, in the treatment of dilute soluble wastes start-up times should be minimised and treatment rates should be high. There is a need for rationally based control procedures in anaerobic digestion, and these can be developed from an improved understanding of the bacteria involved and of their interactions.

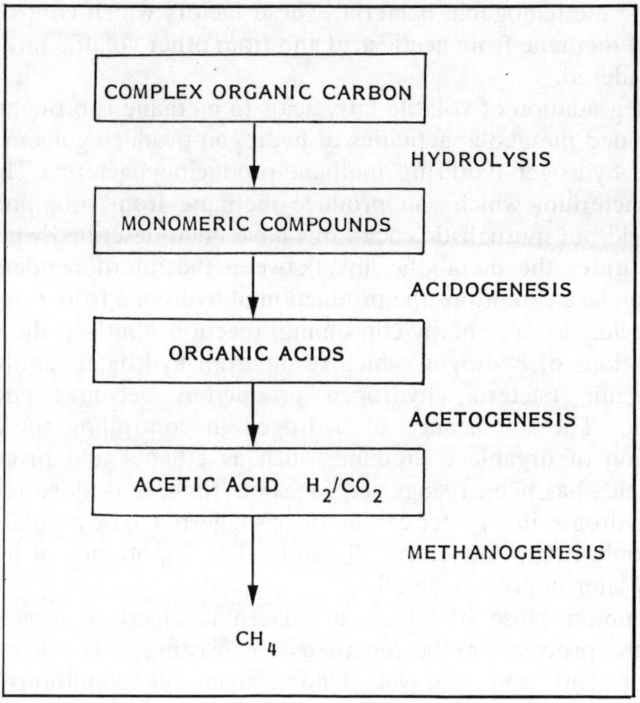

Figure 1 Major metabolic stages in methanogenesis from wastes

THE MICROBIOLOGICAL APPROACH

With such a complicated mixed culture it is necessary to define the most appropriate area for study. In other words, the control points in the sequence of events outlined in Figure 1 should be considered as a priority for study. The control points are those reactions which are slow or which are particularly sensitive to perturbation. Of the four metabolic stages in Figure 1, only the acidogenic stage has not at some time been claimed to be the rate-limiting step. Solubilisation of insoluble organic matter is slow, but even in landfills, where most of the waste is solid, there are still problems with production of acid leachates and reduced methane yields. Thus, the conversion of volatile fatty acids to methane constitutes a major problem in landfills and is also the source of most problems in anaerobic digesters treating soluble wastes. The degradation of volatile fatty acids, such as propionic and butyric, to methane requires several species of bacteria. The acetogenic bacteria degrade volatile acids to acetic acid and hydrogen, but do so only in the presence of the methanogenic bacteria which remove the hydrogen. Acetic acid is degraded to methane and carbon dioxide by the aceticlastic methanogenic bacteria. These factors which control the production of methane from acetic acid and from other volatile fatty acids are now considered.

The degradation of volatile fatty acids to methane is brought about by the combined metabolic activities of hydrogen-producing acetogenic bacteria and hydrogen-oxidising methane-producing bacteria. There is no known bacterium which can produce methane from propionic acid or butyric acid but mutualistic cocultures have been described where hydrogen constitutes the metabolic link between the interdependent species. Under standard conditions the production of hydrogen from propionic and butyric acids is an energy-consuming reaction, but at the very low concentrations of hydrogen which result from hydrogen removal by the methanogenic bacteria, hydrogen production becomes energetically favourable. The importance of hydrogen in controlling the anaerobic degradation of organic compounds such as ethanol and propionic and butyric acids has been recognised for some time, and more recently the level of hydrogen in digester gas has been suggested to be a suitable marker in the monitoring of anaerobic digesters. The importance of hydrogen is discussed later in greater detail.

A common cause of failure in anaerobic digestion is acidification. Indeed, the process may be regarded as consisting of two halves — acid production and acid removal. Under anaerobic conditions it is the methanogenic bacteria which prevent acidification by conversion of acetic acid to methane and carbon dioxide. These methanogenic species grow slowly and have a narrow pH range, usually 6–8, in which they are capable of growth. Slight acidification inhibits the methanogenic bacteria, which results in further acidification, because the acidogenic bacteria are less sensitive to low pH. It is clear, then, that the hydrogen-removing and the acetate-removing methanogenic bacteria are equally important in the anaerobic digestion process.

There are approximately 30 known species of methanogenic bacteria, although not all have been found in anaerobic digesters. There are only two genera known to produce methane from acetic acid: *Methanosarcina* and *Methanothrix*. The most common genera of hydrogen-utilising methanogenic bacteria found in anaerobic digesters are *Methanobacterium, Methanobrevibacter, Methanospirillum* and *Methanococcus. Methanosarcina* utilises hydrogen in addition to acetate. The methanogenic bacteria are a morphologically diverse group of archaebacteria unified by their ability to derive energy from methane production. They are not alone in their anaerobic use of hydrogen and acetic acid as energy-yielding substrates. Alternative hydrogen-removing reactions are given in Table 1. The extent of hydrogen removal by reduction of carbon dioxide to form acetic acid rather than methane is not known, although bacteria which carry out this homoacetogenic reaction have been isolated. Sulphate

Table 1 Fates of hydrogen in anaerobic digesters

	$\Delta G°'$ (kJ per reaction)
$4H_2 + 2HCO_3^- + H^+ \rightarrow$ Acetate $+ 4H_2O$	− 104.6
$4H_2 + HCO_3^- + H^+ \rightarrow CH_4 + 3H_2O$	− 135.6
$4H_2 + SO_4^{2-} + H^+ \rightarrow HS^- + 4H_2O$	− 151.9
$H_2 + NO_3^- \rightarrow NO_2^- + H_2O$	− 163.2

provides a route for the removal of hydrogen and acetic acid by the sulphate-reducing bacteria and is a common problem associated with anaerobic digestion of sulphate-rich wastes.

There can be many microbiological approaches to the study of anaerobic digestion but such research should clearly revolve around the methanogenic bacteria. Much more needs to be known about these microorganisms and their interactions with other species within the anaerobic digester. In that way control procedures can be developed which ensure that a digester runs reliably and at an optimum treatment rate. Two examples of the microbiological research being done in Norwich are described below. The first of these develops the importance of hydrogen in digester monitoring and the second describes a means of controlling the coupling between methanogenesis and cell growth.

HYDROGEN IN ANAEROBIC DIGESTION

Although hydrogen is recognised as a controlling intermediate in the degradation of some organic compounds, a detailed description of the dynamics of mutualistic cocultures, involving interspecies hydrogen transfer, has been lacking. A mathematical model of mutualism has now been developed by G. Powell and this allows some predictions to be made concerning the degradation of compounds such as propionic acid. Two main predictions of the model are that the members of mutualistic cocultures grow with identical specific growth rates and that it is, under most circumstances, the properties of the methanogenic bacterium which determine the specific growth rate of the coculture. A methanogenic bacterium which has either a high affinity (low K_s) for hydrogen or a high maximum specific growth rate will confer a high specific growth rate upon the coculture. These predictions have been confirmed experimentally in model cocultures degrading ethanol to methane.

The consequences for anaerobic digestion of this work are many. For example, many attempts have been made to estimate the maximum specific growth rates and K_s values for bacteria degrading propionate in anaerobic digesters. The values obtained must now be regarded as apparent values of the system rather than being attributable to the propionate-degrading species *per se*. The apparent values are now known to include a number of terms which are described by the growth equations for mutualistic cocultures. Short hydraulic retention times in an anaerobic digester may select for a faster-growing species of methanogenic bacterium, but such a selection, if from a small initial population, would not be rapid. Differing nutritional requirements of methanogenic species may then also become a factor in adapting a mixed population to higher treatment rates. These are questions which arise from this work and which require answering experimentally.

The value of measuring hydrogen in digester gas has a firmer foundation from the models of mutualism. Others (for example, F. Mosey, of the Water Research Centre) have suggested that a rise in the hydrogen levels in digester gas are an early indicator of a rise in volatile fatty acid concentrations in the effluent stream from the digester which can lead to a lowering of pH. As I have already reasoned, it is necessary to prevent even a small lowering of pH, because this inhibits the methanogenic bacteria preferentially. Measuring the pH directly may not be a sensitive guide to the pH in the immediate vicinity of the methanogenic bacteria but measurement of hydrogen in the gas is an appropriate monitoring parameter, because a rise in hydrogen levels is an early warning of a lowering in pH. A rise in the steady-state levels of hydrogen indicates that remedial action must be taken, and this can take the form of either temporarily reducing the load on the digester or adding alkali to the digester. Development of the models for the importance of hydrogen in anaerobic digestion should lead to a prediction of hydrogen tolerance of the system and predict the extent of the action to be taken when the hydrogen levels are excessive.

It was recently shown by the laboratory, in collaboration with Biomechanics Ltd, that the concentration of hydrogen in the gas from a pilot-scale anaerobic digester increased rapidly in response to a volumetric shock loading; for example, from 10 to 100 ppm in 4 h. Research is in progress to develop a deterministic model of the process which predicts the most appropriate course of action (to prevent depletion of buffering capacity and acidification), based on levels of hydrogen in the gas.

COUPLING OF METHANOGENESIS TO CELL GROWTH

Start-up of anaerobic digesters can take several months. This is undesirable, because effluent is not treated effectively and other, expensive, measures must then be taken to treat the effluent. The slow start-up cannot be fully accounted for by the known specific growth rates of the bacteria involved, and it appears that the growth rates of the bacteria are far from maximal. One explanation for this situation is that although methanogenesis occurs, cell growth is not tightly coupled to methanogenesis. Such uncoupling may be an advantage in a fully commissioned digester, where production of excessive biomass is an embarrassment. Control of the coupling would obviously be desirable in the management of an anaerobic digester.

In the eubacteria it has been shown that coupling of the energy derived from metabolism to cell growth can be controlled by levels of phosphate. Recent work from Norwich has shown that phosphate also plays an important role in the coupling of methanogenesis to cell growth in mixed cultures and in a pure culture of *Methanosarcina barkeri*. In the presence of sufficient phosphate methanogenesis is coupled to cell growth, but under phosphate-limiting conditions methanogenesis continues to occur without cell growth. Landfill leachates, some effluents from starch production and some molasses-derived wastes are naturally phosphate-deficient and, for rapid start-up of digesters treating such wastes, it is recommended that the waste be supplemented with phosphate. The phosphate additions can be omitted, to advantage, when the digesters are fully commissioned: then methanogenesis can proceed at high rate without producing excess biomass. Many effluents are, of course, sufficient in phosphate, although landfills may benefit from addition of phosphate. If more was known of the nutritional requirements of mixed cultures, and the methanogenic bacteria in particular, it might be possible to have additional controls on the coupling of methanogensis to cell growth.

SUMMARY

Effective control of the mixed bacterial cultures responsible for the anaerobic digestion of wastes is required to ensure that the process is reliable and operates at high treatment rates. The methanogenic bacteria are the most important group of bacteria in anaerobic digestion. The hydrogen-utilising methanogenic bacteria regulate the degradation of volatile fatty acids such as propionic and butyric. The acetate-utilising methanogenic bacteria regulate the pH in digesters. The degradation of volatile fatty acids to methane is slow and sensitive to the pH in the digester and, as a consequence, conditions must be optimised for the hydrogen- and acetate-utilising methanogenic bacteria. Study of the methanogenic bacteria is required at very many levels: in pure cultures, mixed cultures, metabolic aspects, physical aspects (e.g. formation of flocs, biofilms, granules), and so on. Improvement of the properties of the methanogenic bacteria may be possible by genetic manipulation. Research into these areas is being conducted at Norwich and I have presented above just two examples of microbiological research which hold immediate promise for the control of the anaerobic digestion process.

BIBLIOGRAPHY

1. Archer, D. B. (1983). The microbiological basis of process control in methanogenic fermentation of soluble wastes. *Enzyme and Microbial Technology,* **5**, 162–170
2. Archer, D. B. (1985). Uncoupling of methanogenesis from growth of *Methanosarcina barkeri* by phosphate limitation. *Applied and Environmental Microbiology,* **50**, 1233–1237
3. Archer, D. B., Hilton, M. G., Adams, P. and Wiecko, H. (1986). Hydrogen as a process control index in a pilot scale anaerobic digester. *Biotechnology Letters,* **8**, 197–202
4. Mosey, F. E. (1982). New developments in the anaerobic treatment of industrial wastes. *Water Pollution Control,* **81**, 540–550
5. Powell, G. E. (1985). Stable coexistence of syntrophic associations in continuous culture. *Journal of Chemical Technology and Biotechnology,* **35B**, 46–50
6. Whitmore, T. N. and Lloyd, D. (1986). Mass spectrometric control of the thermophilic anaerobic digestion process based on levels of dissolved hydrogen. *Biotechnology Letters,* **8**, 203–208

27
THE ANAEROBIC DIGESTION OF FOOD PROCESSING WASTES

David A. Stafford

INTRODUCTION

It has long been thought that anaerobic digestion will one day offer a suitable treatment system for a number of waste streams, including many industrial waste-waters — in particular, those produced from the food processing industry. The activated sludge and aerobic trickling filter systems have hitherto taken the leading role in waste treatment because of their supposedly high reputation for reliability and performance. Anaerobic digestion, on the other hand, has had a reputation of being sensitive to toxicity and of less reliable performance in terms of COD reduction. In more recent years, this common misconception about anaerobic digestion, where so-called toxic materials cannot be tolerated, has been shown not to be true. Indeed, in many instances in anaerobic treatment where there is a loss in activity, recovery is fairly rapid. Many of the misconceptions are due to particular problems associated with the treatment of municipal sludge solids, where methane bacteria and other microbes were shown to be inhibited by a number of toxicants. These inhibitions were caused by increases in volatile fatty acid concentrations, which produced what was known as 'sour digester'. There was also a concomitant reduction in pH which exceeded the buffering capacity of the system[12].

With the doubling in the world's population by the year 2000, increases in requirements for energy and food will produce strains to food production and associated processing industries. Any savings, therefore, in energy and waste utilisation by these indusries will be of paramount importance. If this can be coupled with a cheap form of waste treatment and residue reuse, it would further improve the benefits to the food industry.

Waste materials contain polymers such as lipids, carbohydrates and proteins, which are capable of being hydrolysed to long-chain fatty acids, and which in turn are fermented to short-chain fatty acids and eventually into methane in the methanogenic phase. Short-chain fatty acids such as formate and acetate are reduced to methane and carbon dioxide. The metabolic pathways derived from long- and short-chain fatty acids and alcohols, through to acetate, carbon dioxide and methane are demonstrated in Figure 1[15].

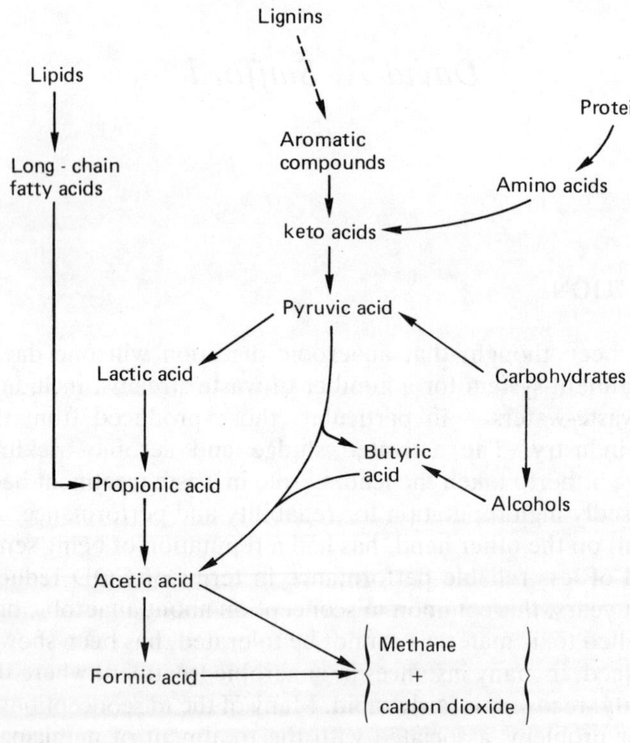

Figure 1 Some of the biochemical pathways during which organic compounds are transformed to methane during anaerobic digestion

The dehydrogenation which occurs after hydrolysis of longer-chain fatty acids that may be present or produced in some digesters is shown in Figure 2. Certain important parameters have been shown to influence the rate of gas production. One of these is the concentration of the fraction

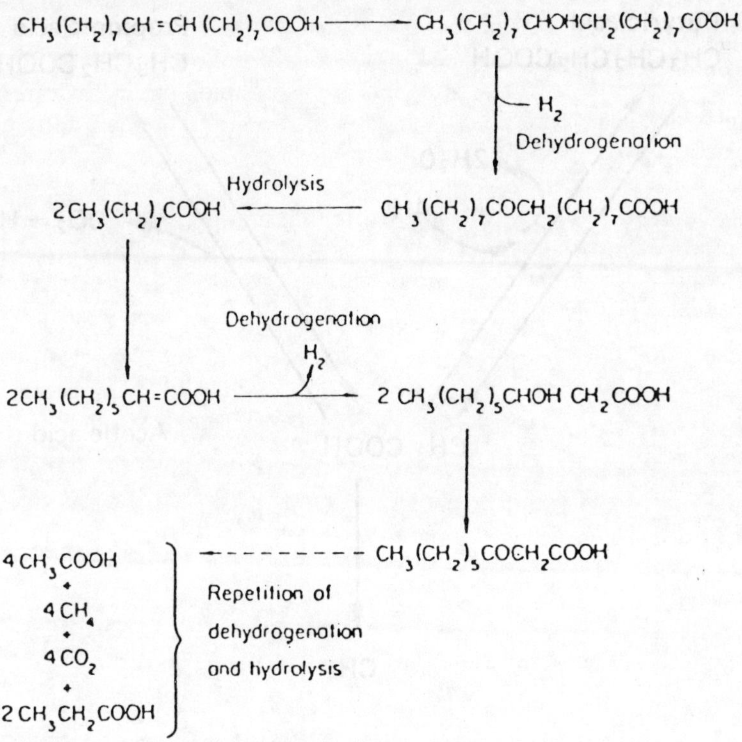

Figure 2 Dehydrogenation and hydrolysis of long-chain fatty acids

known as volatile fatty acids (VFA). In particular, for some thermophilic digesters, the ratio of propionic and butyric acids to acetate has been shown to be important[2]. More recently, this has been shown to be also true for mesophilic digesters where the ratio of propionic acid plus butyric acid over acetate ought to be less than 80:1 to produce good gas yields[13]. Above this ratio inhibitions of gas production occur. Some of the metabolic relationships with the three fatty acids in anaerobic digesters are shown in Figure 3. With respect to the effect of toxicants on the overall anaerobic digestion process, it would appear that this could be complex where the original toxicant plays a minor triggering role, and sets off other metabolic sequences not related to the actual toxicity of the trigger compound. However, the phenomenon of toxic inhibition has given the impression that methanogens are very sensitive to toxicants. This may not be true, and the following cases show the relationship between methanogens in digesters and the overall performance of such a system.

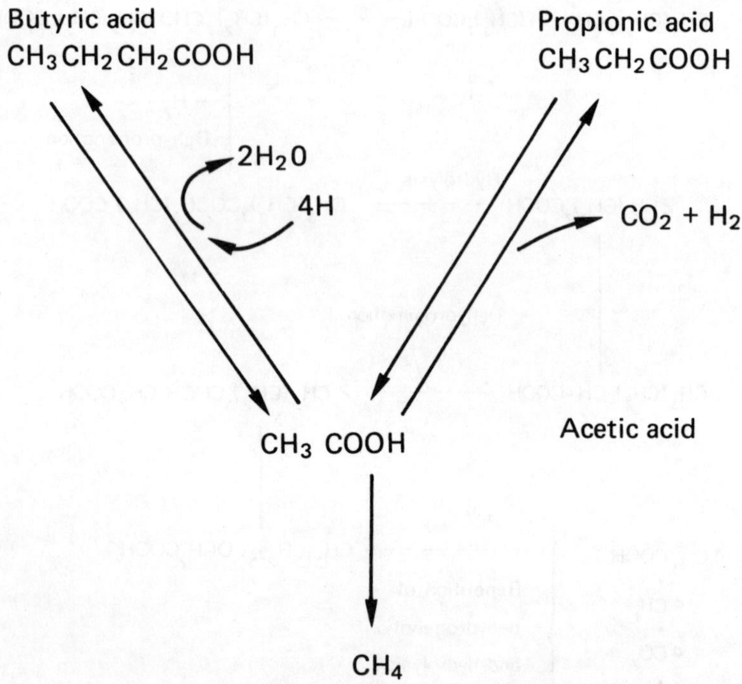

Figure 3 Proposed general metabolism of VFAs in anaerobic digestion of waste material[13]

CONTROL AND INHIBITION OF ANAEROBIC DIGESTERS

Because of the inclusion in some industrial waste-waters of particular toxicants, the anaerobic digestion process is often assumed to be the wrong technology for the treatment of such waste-waters. It is true that methanogens, like many micro-organisms, can be inhibited by toxicants, but the process itself is not necessarily inherently less effective or slower than the rival aerobic process, which is also sensitive to toxicants. Indeed, one of the major advantages of anaerobic digestion is the slow growth of microbes in the system and thus the reduced requirement for disposal of sludge produced. However, should washout occur due to hydraulic overloading, it may take some time for regrowth to occur. Similarly, if the microbes are killed, regrowth may take some time and reseeding may be necessary.

If reversible toxicity occurs, all that is required is acclimation of the sludge in the presence of the toxicant or, on its removal, rapid metabolic recovery. Certainly, in terms of process economics and biochemical engineering control, an understanding of the gross effects of toxicity on the biomass is important to the operational strategy[12].

pH Inhibition

In many cases a 'sour digester' is produced by a fall in pH. One of the triggering mechanisms which can cause changes in pH include a fall in temperature or even a sudden rise in temperature. Similarly, a short exposure to a particular toxicant can cause a sudden increase in fatty acid production because of the transitory inhibition of methane production. This phenomenon is often reversible, and it has been shown[12] that in a digester where the pH fell to 5.5 for 12 h there was a recovery 1 day after the pH was restored. After an inhibition of some 50 per cent in performance, as measured by methane production, methanogenesis was restored to the normal rate. If, however, pH was dropped to 5 for a longer period (15 days), the pH inhibition dropped to 75 per cent of normal within 3 days and held reasonably steady for the period of 15 days. When the pH was restored to normal, the gas production recovered over a period of 3 weeks. It would appear, therefore, that the rate of recovery of gas production was a function of the period of the pH depression and not necessarily of the degree of the pH depression. If the pH was increased, say to 9 or 9.5, and then returned to normal, the gas recovery was far more rapid than during decreases in pH. It must be borne in mind that the type of reactor involved will also influence such changes in performance; for example, anaerobic filters with fixed films would tend to have a pH in the 'micro-niche' nearer neutrality than the pH of the feed solution itself. This is unlikely to be so for the completely mixed contact type reactors, where the microbial cells are intimately mixed with the incoming waste.

Digester Types and Performance

The types of reactors available for anaerobic digestion of industrial and food processing waste-waters are shown in Figure 4. A number of toxic compounds have been shown to be inhibitory to the activity of methanogens in anaerobic digestion. These are shown in Table 1[12]. With these compounds, it was shown that the anaerobic filter proved to be a better reactor type than the completely mixed contact reactors, in that many of the compounds were more metabolised in this type of reactor. This was related in some cases to washout of biomass in the completely mixed reactor. Indeed, only mildly toxic compounds, such as butyric, maleic and fumaric acids, hydroquinone, catechol, resorcinol, ethyl acrylate and phenol, were all degraded well in anaerobic filters. Acclimation was also shown to be good in these filter systems.

PLUG FLOW DIGESTER

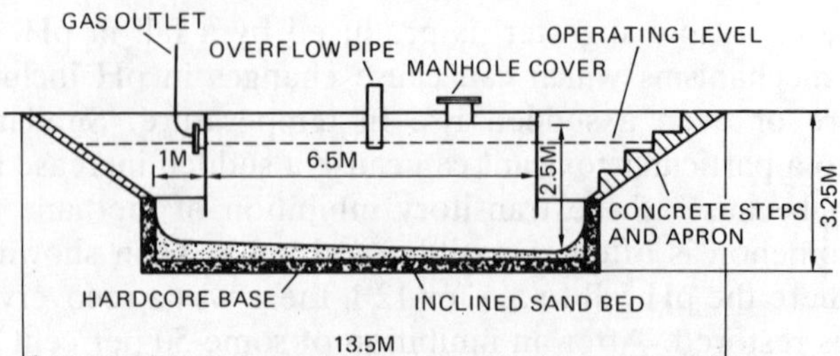

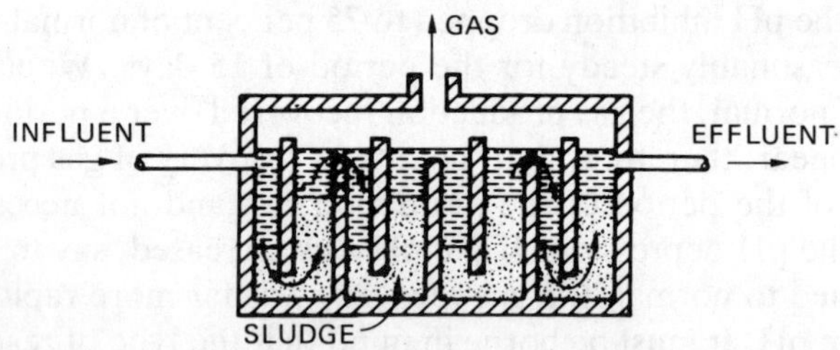

ANAEROBIC BAFFLED REACTOR

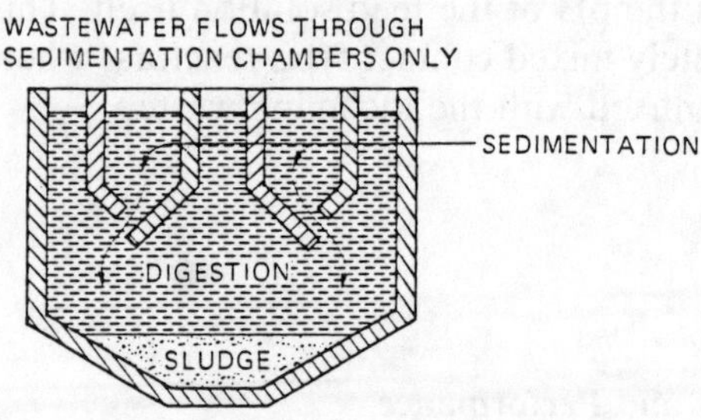

IMHOFF TANK

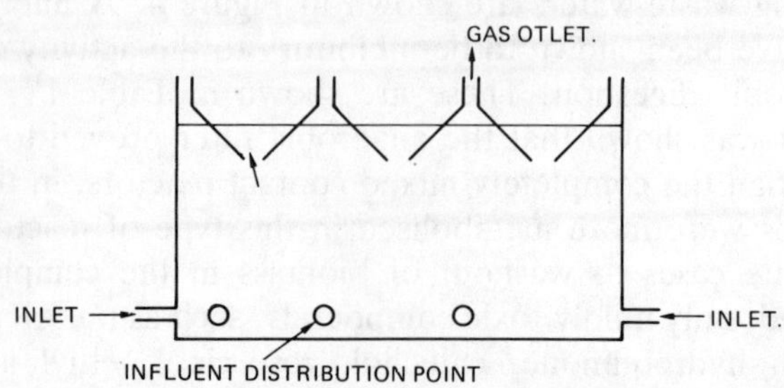

THE UPWARD FLOW SLUDGE BLANKET DIGESTER.

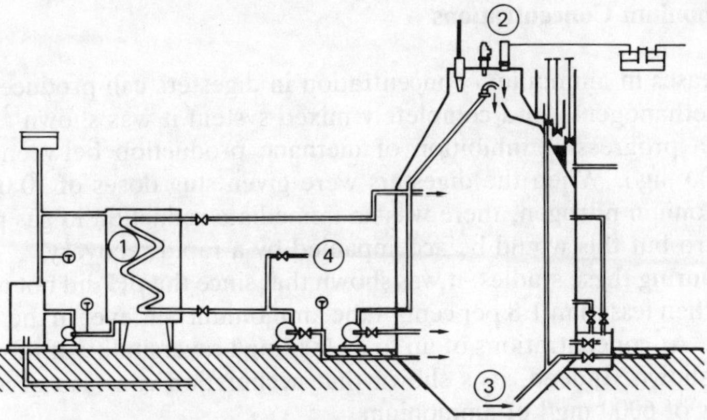

CLASSICAL INDUSTRIAL COMPLETELY MIXED DIGESTER.
1: MIXED LIQUOR HEATING SYSTEM BY LIQUOR RECIRCULATION.
2: GAS OUTLET. 3: MIXED LIQUOR OUTLET. 4: LOAD OUTLET.

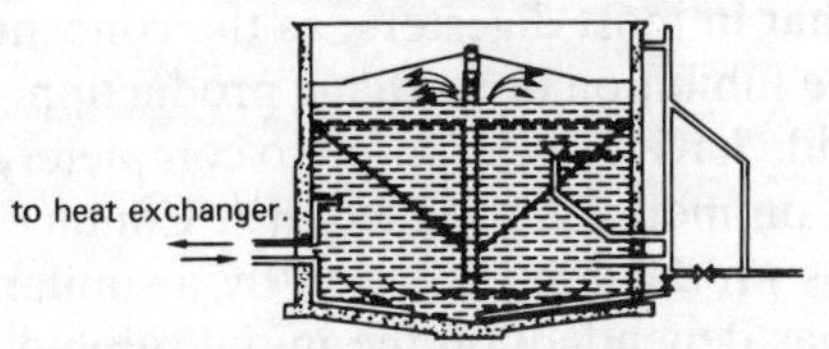

SIMPLIFIED LAYOUT SHOWING MIXING BY SCREW PUMP AND DRAUGHT TUBE. SCUM FORMATION IS ALSO CONTROLLED BY SCATTERING THE DIGESTER CONTENTS OVER THE SCUM SURFACE.

Figure 4 Types of anaerobic digester configurations

Table 1 Relative toxicity of petrochemicals assayed

Compound	*Conc. resulting in 50% activity* (mM)	*Compound*	*Conc. resulting in 50% activity* (mM)
1–Cl–propene	0.1	2–Cl–propionic acid	8
Nitrobenzene	0.1	Vinyl acetate	8
Acrolein	0.2	Acetaldehyde	10
1–Cl–propane	1.9	Ethyl acetate	11
Formaldehyde	2.4	Acrylic acid	12
Lauric acid	2.6	Catechol	24
Ethyl benzene	3.2	Phenol	26
Acrylonitrile	4.0	Aniline	26
3–Cl–1, 2–propandiol	6.5	Resorcinol	29
Crotonaldehyde	6.5	Propanol	90

Ammonium Concentrations

Increases in ammonium concentration in digesters can produce inhibition of methanogens. In a completely mixed system it was shown[12] that there was a progressive inhibition of methane production between 4000 and 14 000 mg/l. When the digesters were given slug doses of 10 000 mg/l of ammonium nitrogen, there was an immediate reduction in gas production to zero but this would be accompanied by a rapid recovery.

During these studies, it was shown that since the pH did not arise above 7.5, then less than 1.8 per cent of the ammonium was ever in the un-ionised form. At concentrations of up to 24 000 mg/l reversibility of this inhibition was shown. Indeed, it is shown that acclimation can also occur even at levels of 6000 mg/l of ammonium.

Sulphide Inhibition

It has been shown that in most digesters, as the concentration of sulphide increases, so does the inhibition of methane production. Concentrations of up to 500 mg/l sulphide have been exposed to completely mixed reactors to determine the effect on methane production[12]. Contact reactors showed a rapid decrease in gas production, followed by a similarly rapid recovery. This recovery rate was dependent on the initial sulphide concentration. If the sulphide could be removed from solution, recovery was accelerated further[12]. For concentrations of sulphide up to 400 mg/l, when continuously added to digesters, no significant reduction in performance was noted. At concentrations above 800 mg/l, there was about a 30 per cent decrease in gas production. At 100 mg/l of sulphide and above, greater reduction in gas yields were noted.

With certain slug dosings of sulphides, increases in gas production have been seen, and more recently it has been shown that sulphur compounds are reduced readily by methane bacteria to yield energy, and this appears to occur preferentially to CO_2 reduction[3].

Many food processing waste-waters contain sulphur compounds, including sulphate, and during their reduction energy is released. It has been shown that the methane bacteria present in anaerobic digesters will favour the reduction of sulphate, since more energy is released than with CO_2 reduction[3]. The following energies are released during the oxido/reductive reactions in anaerobic digesters receiving organic as well as inorganic sulphur compounds:

Methanogenesis

$$CO_2 + 4H_2 \rightarrow CH_4 + 2H_2O \ \Delta G = -135 \text{ kJ}$$
$$CH_3\,COOH \rightarrow CH_4 + CO_2 \ \Delta G = -28.5 \text{ kJ}$$

Sulphate reduction

$$SO_4 + 4H_2 \rightarrow 2H_2S + 2H_2O + 2OH \quad \Delta G = -154 \text{ kJ}$$
$$SO_4 + CH_3\,COOH \rightarrow H_2S + 2H_2O + 2OH \quad \Delta G = -43 \text{ kJ}$$

It is also stated that the sulphate-reducing bacteria are less sensitive to pH and temperature variations. Wherever sulphate is present in an anaerobic reactor or the waste fed to it, the sulphate-reducing bacteria will grow and produce hydrogen sulphide, which is toxic for other bacteria. However, this toxicity depends mainly on the free hydrogen sulphide in solution. It has been shown that a complete breakdown of methane production at sulphate concentrations of 340 mg/l occurs[21]. It would appear, however, that free hydrogen sulphide would inhibit methane production even at lower concentrations than this. It is thus a matter of importance to understand the mechanisms involved in the inhibition of methanogenesis in the presence of hydrogen sulphide, particularly using anaerobic treatment processes for industrial and food waste-waters. It has been shown recently that the ratio between the chemical oxygen demand (COD) and sulphate-sulphur is important with respect to the problem of hydrogen sulphide toxicity[3]. At COD:SO_4 ratios below 100, toxicity occurs and is thus an important parameter to measure. In some cases this problem can be reduced by dilution.

The undissociated H_2S inhibits methanogenesis at low concentrations (below 50 mg/l), which corresponds to H_2S fractions in the gas below 2 per cent. If the temperature is not changed, the partial pressure of H_2S in the gas is proportional to the free H_2S concentration in the liquid. Since the percentage of H_2S in the gas phase can be easily determined by gas chromatography, this seems the best way to check H_2S toxicity during the anaerobic waste-water treatment process. The problem of H_2S toxicity, therefore, can be reduced first by diluting the waste-water to contain lower concentrations of sulphate; second, by raising the pH value to 7.5–8; and third, by raising the COD removal efficiency by optimising the overall digester operation.

It has been well known that the acute toxicity of H_2S can be reduced by adding metals such as iron, but this can be costly. However, it may be possible in future to inhibit the sulphate-reducing process by adding certain other metallic salts, such as molybdate salts[5]. Molybdate is a specific inhibitor of the sulphate-reducing bacteria. Nanninga and Gottschalk[5] also show that formate, ethanol and hydrogen gas can all be used as electron donors for the sulphate-reducing bacteria. The acid-fermenting bacteria are also inhibited when molybdate is added. What is likely to be happening is that the acid-fermenting bacteria are inhibited by the accumulation of hydrogen gas as a result of the sulphate-reducing bacteria being themselves inhibited and unable to utilise the hydrogen released by the fermenting bacteria.

Other Inhibitors

Other inhibitors are also able to reduce gas production, including heavy metals such as copper. At 60 mg/l of copper, methane production is dramatically reduced over a period of time, but recovery is possible. The presence of sulphides will precipitate copper, and thus, when present together, toxicities caused by the presence of each can be mutually reduced. Oxygen has also been shown to inhibit methane production[4], as have other chemical toxicants such as cyanide and chloroform[12].

SPECIFIC INDUSTRIAL WASTE-WATERS AMENABLE TO ANAEROBIC DIGESTION TREATMENT

Citric Acid Fermentation Effluents

During the production of citric acid from molasses using *Aspergillus niger*, from each tonne of citric acid produced during the process, between 18 and 20 tonnes of waste-water are produced with an average COD of 30 000–50 000 mg/l[11]. As with many waste-waters containing a number of compounds such as glucose, fructose, acetate, lactate, citrate, ethanol, propionate and butyrate, it was shown that increases in both yield and the rate of methane production occurred with increases in concentration. However, exceptions to this rule included propionic acid, citric acid and fructose when concentrations exceeded 50 mM. All of these carbon sources are present in citric acid wastes. Whether they are produced as intermediates in the process of anaerobic digestion or whether added externally, the rate of removal of each component was always the same, and this rate depended upon the concentration.

It is, therefore, important to stress that the microbial population has to reach a steady state in order to degrade any intermediate organic carbon compound at a particular rate. Since this is a function of the population numbers, long sludge retention times will tend to increase the rate of removal of these particular aged organisms in anaerobic digesters and mean the possibility of lower hydraulic retention times and faster volume throughput of wastes.

This is an important function, which we shall discuss below.

Proteinaceous Industrial Waste-waters

Other industrial waste-waters are also capable of being treated with this anaerobic process. Such waste-waters contain organic polymers either in solution or in suspension as solids, together with soluble and insoluble inorganic substances. Since a number of metabolic sequence stages are known within the anaerobic process, it may be that there are benefits in separating the metabolic steps in different reactors. For the treatment of solids, the fermentative degradation step or hydrolysis stage is often the rate-limiting step for the microbial generation of methane from organic waste. Similarly, in other treatment systems, the final step of methanogenesis can also be shown to be rate-limiting[18].

As proteins are degraded during hydrolysis, amino acids are released which are deaminated to keto acids and then fatty acids. First, pyruvate is hydrolysed to acetic acid, carbon dioxide and hydrogen. The hydrogen is used to reduce CO_2 to methane.

1. Deamination of alanine to pyruvate

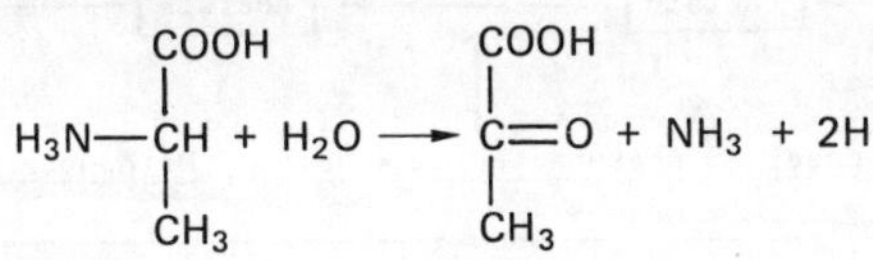

2. Decarboxylation of pyruvate to acetate

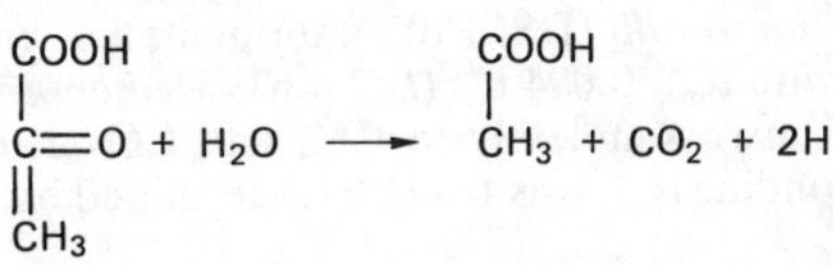

The recommendations for a two-stage process to treat proteinaceous wastes were summarised as follows:

(1) The overall process appears to be more efficient in the two-stage process, in that only 25–30 per cent of the organic compounds of the sludge remained compared with 50 per cent from the single-stage reactor.

(2) The solids loading for the single-stage reactor were 1.3 kg total solids (TS) per m^3 reactor per day, whereas in the two-stage process this was increased to 2.6 kg m^{-3} day^{-1}. The gas yield in the single stage was 0.3 m^3/kg solids added and with the two-stage system this was 0.45 m^3/kg TS added[18].

Dairy Effluents

One of the major sources of waste arising in the food processing industry is, of course, that produced from the making of cheese and other dairy products. Lactose is the main organic compound in the waste, between 4 per cent and 5 per cent (weight/volume of whey permeate), and is highly suitable for the process of anaerobic digestion. Some recent work has shown that it is possible to degrade lactose to methane using three individual species of bacteria along a linear metabolic pathway. This is shown schematically in Figure 5[11]. Thus, it is possible to define the exact microbial population that could degrade certain components of industrial waste-waters. More recently in our own laboratories, we have demonstrated that it is possible to obtain high methane yields at low retention times with lactose present in whey permeate. The data shown at different retention times of biogas yields and COD removal are listed in Table 2.

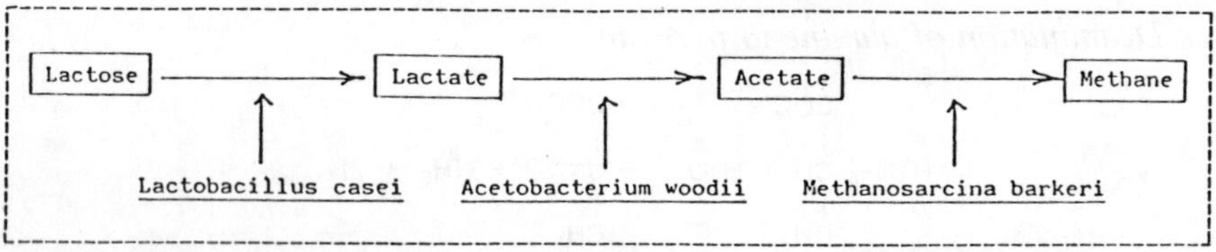

Figure 5 Scheme of food chain for anaerobic methane production from lactose[11]. *Lactobacillus casei* (DSM 20208) converted lactose (50 mM) to lactate (90%) and some acetate and ethanol under pH-static conditions (pH 6.4) with u_{max} 0.196 h^{-1} (t_d:3.5 h). *Acetobacterium woodii* (DSM 1030) fermented D- and L- lactate (55 mM) completely to acetate with u_{max} 0.094 h^{-1} (t_d:7.4 h). *Methanosarcina barkeri* (DSM 804) split acetate (125 mM) completely to CH_4 and CO_2 at constant pH of 6.3. Doubling time corresponding u_{max} was 16–21 h (determined by gas production rate)

The ability of anaerobic digester microbial consortia to degrade this waste-water can be related to a function of the design of the plant. Previously, it was thought that retention times of between 15 and 20 days were necessary to degrade such high-strength soluble organic waste-waters. However, it is now possible to obtain hydraulic retention times of 5 days and less, and still obtain good gas yields and high COD removal. This has been shown to be due to the increase in microbial biomass using a new type of digester known as the hydraulic reactor[13]. This reactor is similar in many ways to the 'Clarigester', developed in South Africa in the 1950s[9]. We have shown with this type of reactor in the laboratory that loading rates of up to 15 kg COD/m^3 reactor are possible per day (Figure 6).

Using the Up-Flow Anaerobic Sludge Blanket (UASB) reactor (Figure 4), Voetberg[19] has shown that, with a two-stage system, the first stage being primary liquefaction and acidification, loading rates of between 25 and 30 kg COD were obtained per m^3 per day. The influent COD in this case was 20 000 mg/l and the COD removal efficiency was 85 per cent. However, at the high loading rates a large washout of the non-granular

Table 2 Studies of dairy waste treatment—anaerobic digester

Hydraulic retention time (days)	*Effluent feed*				*Digester*				*Biogas yields*		
	pH	*TS* (%)	*COD*	*VFA*	pH	*TS* (%)	*COD*	*VFA*	Vol./Vol. per day	*VS* m^3/kg (*COD*) *added*/day	% CH_4
15	3.52	3.31	41 200	7250	6.78	2.82	21 400	6000	0.28	0.165 (0.091) COD	–
12	3.45	3.78	52 000	4750	7.45	2.80	8200	2675	1.30	0.512 (0.300) COD	63.1
10	3.39	3.96	56 000	4500	7.51	2.80	8200	2675	1.77	0.559 (0.316) COD	62.5
9	3.59	3.58	50 000	4000	7.19	3.02	12 600	4500	1.86	0.607 (0.335) COD	67.1
7.5	3.58	3.58	50 000	5750	7.56	2.54	10 400	3500	2.14	0.589 (0.321) COD	61.2
5.6	3.57	3.59	49 000	6000	7.61	2.33	9500	3300	2.13	0.433 (0.242) COD	60.0
5.0	3.72	3.17	47 000	4100	7.61	3.00	7900	4400	2.73	0.512 (0.290) COD	58.9

THE HYDRAULIC DIGESTER.
THE CONFIGURATION ENSURES HIGH HYDRAULIC AND ORGANIC LOADING WITH HIGH BIOGAS YIELD.

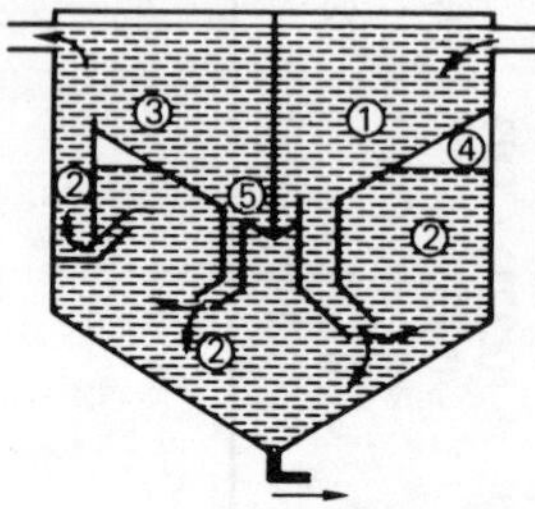

Figure 6

sludge was observed. This shows the possibility of instability using the UASB process at high loading rates with dairy waste effluents.

Other Wastes

Other waste-waters which had been thought to be treatable with only aerobic processes are now capable of being treated anaerobically. Such wastes include that produced from the pulp and paper industry, whose waste-waters have a composition of 11 500 mg/l COD and also contain hydrogen peroxide and sulphate. At a hydraulic retention time of 5 days and a COD loading rate of 2 kg m^{-3} day^{-1} a digester produced a COD reduction of 60 per cent and a methane yield of 1.1 m^3/kg COD added[20]. The gas phase contained 81 per cent methane, and despite the high sulphate concentration the sulphide content at the gas phase did not exceed 0.3 per cent. With the treatment of vegetable canning wastes, it has also been shown that the total productivity of digestion depends on the performance of the liquefaction stage. Again a two-stage system has been used where carrot peeling wastes were able to produce 2.4 m^3 methane per m^3 digester per day at 35°C. The liquefaction stage was carried out at 60°C.

Comparing a one-stage system with mesophilic and thermophilic treatment, the former produced 1.35 m^3 methane per m^3 reactor per day, and the latter 2.27 m^3/day. Unfortunately, no information with respect to hydraulic retention time was given[10]. In our own hands we have shown gas rates of two volumes of biogas per volume of reactor per day at a methane concentration of 56 per cent for mesophilic digestion, and for the thermophilic reactor approximately the same gas yields with methane at 60–65 per cent. These were both operated at retention times of 20 days. It remains to be shown whether the thermophilic reactor is capable of lower hydraulic retention times when compared with the mesophilic reactor.

For the treatment of waste arising from the manufacture of wheat, gluten and starch from flour, waste-waters contain a mixture of amino acids, proteins, carbohydrates, hemicelluloses, pentose gums, suspended starch granules, insoluble protein and bran particles, all of which produce a COD of between 10 000 mg/l and 25 000 mg/l. With anaerobic digestion, a loading rate of 2.6 kg COD was obtained per m^2 per day and the methane production yield of 0.33 m^3 per kg COD. The methane content in the biogas was about 65 per cent and the BOD reduction was 95 per cent. Although this work was carried out in the laboratory, the maximum loading rates obtained in a full-scale system of 1.3 kg COD per 3 m^3 per day were also recorded[8]. The size of the full-scale plant was 1000 m^3.

A number of other waste-water streams are capable of being used as substrates for anaerobic digestion, including waste from coal gasification[17] and other waste-waters containing organic substrates which were not thought possible feeds for anaerobic processes.

Similarly, newer designs are coming on the scene with the Hydraulic[13] and Baffled[1] Reactors as well as the new Anatron (R) and Mars (R) systems[16].

RECOMMENDATIONS AND FUTURE DEVELOPMENT

A number of design types have become available on the market and these are being evaluated at both pilot and full scale around the world. The types of digesters that have been developed include the contact-type systems, which are conventional high-rate reactors; anaerobic filters; fluidised-bed systems; and the hydraulic reactor types developed in Cardiff. All of these designs have particular applications with respect to both the volume of waste applied and the amount of organic material present both in solution and as suspended solids. These parameters indicate the type of reactor that will be used by the food processing manufacturer.

It has been shown to be possible, in some cases, with the value of biogas produced, per unit volume of waste arisings, that a payback time of 6 months to 5 years may be possible, depending on the quality of the waste. This payback time may be further increased if a value for the alternative treatment is included in the calculations. These alternative treatments include aerobic bioprocessing, which also costs money to run in oxygen and energy costs.

It has been shown with the very-high-rate reactors that retention times of 1 day or less can be achieved with organic loading rates of up to 20 kg COD per m^3 reactor per day.

Particular experience is drawn from the dairy and food canning industries, and experience with a pilot-plant dairy waste anaerobic digester will provide data on the operational factors influencing the control of such digesters.

It is important that the fundamental activity of micro-organisms be more fully understood together with the related metabolic and biochemical controls. These data need to be applied to design parameters and kinetics of such reactors, together with an understanding as to whether these can be improved. For example, it is possible to quote loading rates of up to 5 or 10 kg COD per m^3 per day, while other workers quote up to 15 kg COD per m^3 per day. It is important to know the fundamental control parameters that allow loading rates of one dimension in one type of reactor and a different set of loading rates in other reactors[6].

Table 3 gives some idea of the types of waste and the constituents of such wastes which are amenable to anaerobic digestion[7]. It is thus a matter of obtaining basic data on the overall process involved; studying in the laboratory the basic performance and control parameters; and then going through pilot-plant stage to full-scale in order to determine whether in the treatment of new waste-waters the data obtained in the laboratory can also be obtained from pilot and full-scale systems. In this way, we can build up for industry a strong base for the application of anaerobic digestion to treat a wide range of organic waste arisings[14].

ACKNOWLEDGEMENT

This chapter was first published in the proceedings of the symposium *Profitability in Food Processing 1984 —* onwards *— The Chemical Engineer's Contribution*, published by the Institution of Chemical Engineers as Symposium Series No. 84. These proceedings may be purchased from IChemE Book Sales, The Institution of Chemical Engineers, 165–171 Railway Terrace, Rugby CV21 3HQ. Tel.: 0788 78214.

REFERENCES

1. Bachmann, A., Beard, V. L. and McCarty, P. L. (1984). Performance characteristics of the anaerobic baffled reactor. In *Anaerobic Digestion — 1983* (Ed. Wentworth, R.), Pergamon Press, Oxford
2. Cooney, C. L. and Ackerman, R. A. (1975). Thermophilic anaerobic digestion of cellulose waste. *Env. J. Appl. Microbiol.*, **2**, 65–72
3. Kroiss, J. and Plahl-Wabnegg, F. (1983). Sulphide toxicity with anaerobic waste water treatment. *Proceedings of the European Symposium Anaerobic Waste Water Treatment*, 72–85
4. Lloyd, D., Scott, R. R., Williams, T. N. and Whitmore, T. (1983). Direct measurement of methanogenesis in anaerobic digesters by methane inlet mass spectrometry. *Eur. J. Appl. Microbiol. Biotechnol.*, **18**, 236–241

Table 3 Principal characteristics of the waste examined (from Ref. 7)

Type of waste	*Parameters of raw material*				
	pH	*Dry matter* (g/l)	*Solids in suspension* (g/l)	*Volatile solids* (g/l)	*Chemical oxygen demand* (g/l)
Domestic waste-water primary sludge	6.5–7.0	42–165	40–160	18–72	–
Chaff	–	500–900	500–900	490–850	–
Industrial secondary sludge	7.0–8.4	5–10	4–8	5–7	6.5–8.8
Piggery slurry	7.0–8.0	18–55	13–45	16–20	20–29
Waste-waters from					
flax retting	4.6	4.4	0.5	1.9	7.2
enzyme production plant	8.0–8.6	6–10	0.4	2.5–7.5	3.1–8.3
dairy industry 1	8.4–12.5	3–10	1–3	1.6–2.3	0.5–1.9
dairy industry 2	4.2	6.7	2.3	5.5	9.8
brewery	4.2	34.6	0.5	33.3	87.3
yeast production plant	5.1–5.7	100–120	0.5	60–84	115–176
palm oil extraction plant	3.5	31	2.6	28	46
slaughterhouse	6.0–6.5	9–21	4–9	4–10	9.0–12.1
wool industry	6.9–7.3	14–28	5–10	7–17	15–35
margarine industry	1.8–2.2	30–33	1–5	5–8	4–5
vegetable oil production plant	1.3–1.5	50–90	1–3	12–14	10–18
petrochemical industry	2.9–13.9	0–10	0	–	10–70

5. Nanninga, H. J. and Gottschalk, J. C. (1983). Possible interreactions between amino acid fermenting and sulphate reducing bacteria. *Proceedings of the European Symposium Anaerobic Waste Water Treatment*, 34
6. Pipyn, P., Pyle, D. L., Rozzi, A. and Taiganides, E. P. (1981). Waste treatment and sludge disposal. In *Anaerobic Digestion — 1981* (Eds. Hughes, D. E., Stafford, D. A., Wheatley, B. I., Baader, W., Lettinga, B., Hyns, E. J., Verstraete, W. and Wentworth, R. L.), pp. 371–382. Elsevier Biomedical Press, Amsterdam
7. Pipyn, P. and Verstraete, W. (1980). Waste classification for digestibility in anaerobic systems. In *Anaerobic Digestion* (Eds. Stafford, D. A., Wheatley, B. I. and Hughes, D. E.), pp. 151–166. Applied Science Publishers, London
8. Quick, G., Walker, T. S. and Gillespie, G. (1984). Some comparisons between a laboratory scale and a 100 m^3 industrial anaerobic/aerobic waste water treatment system. In *Anaerobic Digestion — 1983*. Pergamon Press
9. Ross, W. R. (1984). The phenomenon of sludge pelletisation in the anaerobic treatment of maize processing waste. In *Anaerobic Digestion — 1983*. Pergamon Press
10. Roy, F., Verrier, D. and Florentz, M. (1983). Importance of the liquefaction step in a two stage methanization. *Proceedings of the European Symposium Anaerobic Waste Water Treatment*, 174
11. Schug, A. and Sahm, J. (1983). Anaerobic degradation of lactose to methane by a defined mixed culture. *Proceedings of the European Symposium Anaerobic Waste Water Treatment*, 33
12. Speece, R. E. and Parkin, G. F. (1984). The response of methane bacteria to toxicity. In *Anaerobic Digestion — 1983*. Pergamon Press
13. Stafford, D. A. (1982). The effects of mixing and volatile fatty acid concentrations on anaerobic digester performance. *J. Biomass,* **2**, 43–55
14. Stafford, D. A., Hawkes, D. C. and Horton, R. (1980). *Methane Production from Waste Organic Matter*. C. R. C. Press, Boca Raton, Florida
15. Stafford, D. A. and Hughes, D. E. (1981). Microbial production of fuels. In *Energy, Present and Future Options* (Eds. Merrick, D. and Mansell, R.), pp. 69–92. Wiley, New York
16. Sutton, R. E. and Evans, R. R. (1984). Anaerobic system designs for efficient treatment of industrial waste waters. In *Anaerobic Digestion — 1983*. Pergamon Press
17. Tomei, F. A. and Bloeman, W. (1983). Methanogens from coal gasification waste. *Proceedings of the European Symposium Anaerobic Waste Water Treatment*, 38
18. Trösch, W., Keller, E., Stephan, E. and Chincel, H. (1983). Microbial generation of methane from a proteinaceous industrial waste water — comparison between single and two stage processing. *Proceedings of the European Symposium Anaerobic Waste Water Treatment*, 117
19. Voetberg, J. W. (1983). Laboratory experiments with a two-phase digestion process for solid wastes. *Proceedings of the European Symposium Anaerobic Waste Water Treatment*, 173
20. Welander, T. and Hansson, G. (1983). Anaerobic treatment of a pulp industry waste water. *Proceedings of the European Symposium Anaerobic Waste Water Treatment*, 174
21. Winifrey, M. R. and Zeikus, J. G. (1977). *Appl. Environ. Microbiol.,* **33**, 275

28
WHEY PROCESSING FOR PROFIT — A WORTHY ALTERNATIVE

D. L. Kemp and J. Quickenden

INTRODUCTION

Of the various options available to process or dispose of whey, the system discussed below has a number of positive advantages:

(1) There is no end product to sell with the attendant marketing problems, and the process removes surplus whey, enabling other uses to be more cost-effective.

(2) The major by-product is a fuel gas which can be used directly in the main plant, providing an in-house return on capital investment.

(3) The process transforms the whey into a final effluent which can readily be discharged direct to sewers or further treated for disposal to local water courses, providing considerable savings in existing disposal costs.

As is well known, whey is a very strong, highly polluting material, 1 m^3 of whey being equivalent to the daily pollution production from approximately 600 people. In waste-water terminology, the composition of a typical whey is:

Total solids (TS)	66 000 mg/l
Volatile solids (VS)	59 400 mg/l
Chemical oxygen demand (COD)	65 000 mg/l
Biological oxygen demand (BOD)	45 000 mg/l
Total nitrogen	1 500 mg/l

A waste of this strength would clearly be uneconomic to treat by conventional aerobic means, owing to the high power requirements for aeration, and for this reason whey has not traditionally been treated in aerobic biological systems. In anaerobic digestion, however, whey is broken down, to produce a gas rich in methane, giving the treatment process a large positive energy balance.

FORMATION OF METHANE FROM WHEY

The individual components of whey — lactose, protein and fat — are broken down in the anaerobic digestion process, with the resultant formation of biogas (a mixture of approximately 60 per cent methane, 40 per cent carbon dioxide). The mechanism of the breakdown is as follows: solubilisation of macromolecules into their component parts, conversion of these component parts largely to volatile acids such as acetic acid, and conversion of these volatile acids into methane and carbon dioxide. Each cubic metre of whey produces approximately 38 m^3 of biogas, the major proportion of which comes from the breakdown of lactose. The anaerobic digestion process is suitable for the treatment of all types of whey, from the production of hard cheese, soft cheese, feta cheese, blue cheese, cottage cheese, casein, etc., as well as whey permeate and waste materials from the production of other products, such as coleslaw, etc.

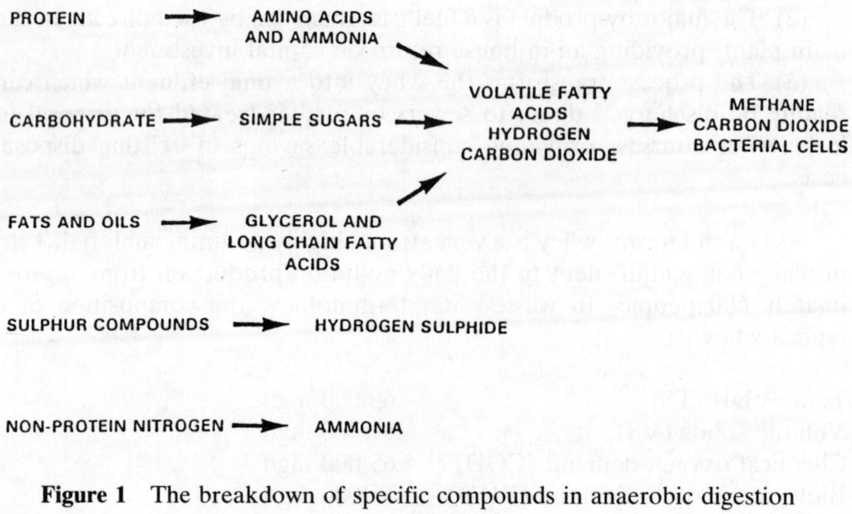

Figure 1 The breakdown of specific compounds in anaerobic digestion

POLLUTION REDUCTION BY ANAEROBIC DIGESTION OF WHEY

Typical digester influent and effluent pollution levels, and the percentage reduction of each parameter, are given in Table 1. As can be seen, considerable reductions in TS, VS, COD and BOD are obtained, and in some instances the effluent may be accepted for direct discharge to a sewer without further treatment. Alternatively, if a Royal Commission Standard effluent is required for discharge to a watercourse, further treatment will be necessary.

Table 1

	Influent	*Effluent*	*% Reduction*
TS (mg/l)	66 000	9400	85
VS (mg/l)	59 400	5900	90
COD (mg/l)	65 000	6500	90
BOD (mg/l)	45 000	2250	95
TSN (mg/l)	1500	1000	33.3

TREATMENT OF THE DIGESTER EFFLUENT

Following the digester, the major pollutant remaining in the waste is nitrogen, the total nitrogen and ammonia nitrogen contents being 1000 mg/l and 750 mg/l, respectively. The nitrogen and BOD remaining in the digester effluent are broken down in an aerobic nitrification/denitrification system, to produce an effluent suitable for river discharge.

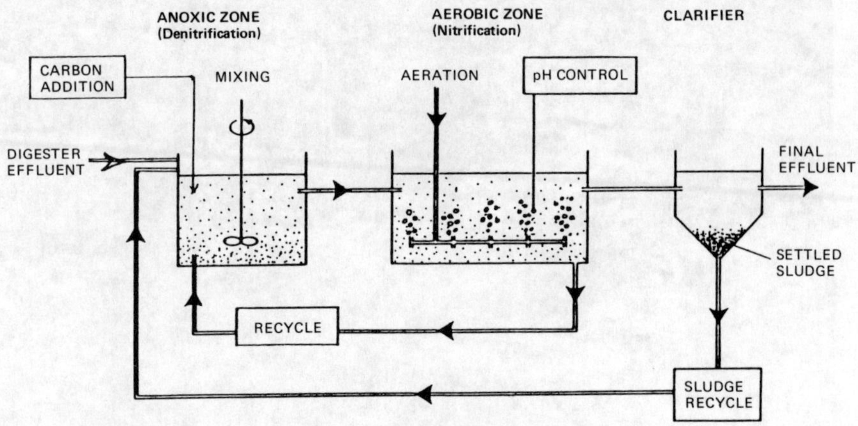

Figure 2

PLANT PERFORMANCE AT SOUTH CAERNARVON CREAMERIES

A diagram of the installation at SCC is shown in Figure 3. Whey production at SCC varies with milk production and utilisation from approximately 50 to 110 m^3/day. The biogas produced varies from 1500 m^3/day at low whey production to approximately 4000 m^3/day at

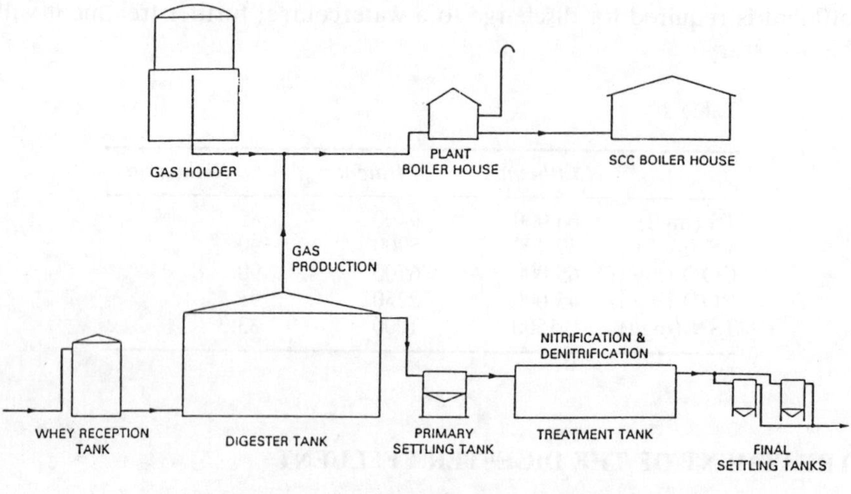

Figure 3 South Caernarvon Creameries digester and after-treatment plant (schematic diagram)

Figure 4 Whey digester plant — South Caernarvon Creameries

maximum production. An overall BOD reduction of 99.95 per cent is achieved.

BIOGAS UTILISATION

Apart from the considerable advantage of removing whey from a grossly oversubscribed market, another major benefit of the anaerobic digestion route is the utilisation of the resulting methane-rich biogas. A number of possible uses can be considered:

(1) As a prime fuel in boiler plant for the generation of process steam and/or hot water.
(2) Electrical power generation with probably waste heat recovery as hot water and the possibility to export excess power to the national grid.
(3) Further treatment and compression for vehicle propulsion.
(4) Sell over-the-wall to other possible users.

Which of the options either singly or in combination are exploited will depend upon the individual requirements and practices of the plant and site being considered. At South Caernarvon Creameries option (4) was not possible, as there was no adjacent large potential user. Option (3) was dropped as being unsuitable, considering the then state of the art and the estimated reduction in payloads. Finally, a combination of (1) and (2) was adopted which gives the flexibility required of a plant with large variations of whey produced, of short-term peak steam demands and a high base load electrical power demand.

Obviously the biogas has a changing value dependent upon where and how it is utilised, and the main feature at South Caernarvon is not only the full-scale digestion of whey, but also the fullest possible utilisation of the energy resulting from the biogas.

POTENTIAL BIOGAS VALUES

With a maximum whey feed of 110 m^3/24 h, at 38 m^3 biogas/m^3 whey, a biogas yield of 4180 m^3 in 24 h can be expected, having a methane content of 62 per cent. This gas has a net calorific value of 21.08 MJ/m^3, which, if used as a boiler fuel, has a heavy fuel oil equivalent of 2560 l/24 h, or alternatively 250 kW of electrical power, with a further potential of 480 kW of hot water at 85–90°C if waste heat is recovered from the engine cooling-water and exhaust.

Heavy fuel oil prices vary greatly, depending upon the supplier and quantity usage, but January 1985 target prices of 25 p/l were common, with average prices of 20 p/l available. Prices as low as 19 p/l and 17 p/l were obtained by some creameries. An equivalent target price for gas oil was quoted at 30 p/l. Thus, using the average price of 20 p/l, the biogas produced by 110 m^3/day of whey had a potential value as a HFO replacement of £512 per day, or £186 880 per year when fully utilised.

The potential value as electrical power is not so readily defined, as it is dependent upon the area electricity board, its scale of charges and the tariff rates that apply in a particular case.

For the purpose of this discussion two alternatives are considered: (1) a reduction in the maximum power demand by 250 kW; (2) the exportation of 250 kW of electrical power using average rates.

Considering the potential saving of reduction in maximum power demand by 250 kW and using tariff number 7N as typical of such an installation:

Reduction in provisional annual charge	£6 437	
Reduction in day time charge/year	£54 914	
Reduction in night time charge/year	£10 156	
Total	£71 507	per year

To this must be added the value of heavy fuel oil equivalent of the potential 480 kW of hot water recovered from engine of 50 l/h, which at 20 p/l is worth £240 each 24 h operation, or £87 600 per year. Therefore, a total potential value of £159 107 per year is available in this mode.

Alternatively, considering the potential saving by the export of 250 kW continuously:

For Dec./Jan./Feb. (average rate 3.52 p/unit)	£19 222
Mar./Nov. (average rate 3.53 p/unit)	£9 235
Apr./May/Jun./Jul./Aug./Sept./Oct. (average rate 2.16 p/unit	£27 605
Total	£56 062

Again the value of heavy fuel oil equivalent of the potential 480 kW of hot water is to be added, giving a total potential value of £148 662.

The 1985 data show that, when possible, direct firing in the plant boilers gives the best possible potential return. Unfortunately, we do not live in a perfect world. The biogas is released at a more or less constant rate, for a given feed, continuously over the 24 h day, while the boiler fuel demand peaks at the start of the process day, which itself is often only of 6–12 h duration. Therefore, to maximise this potential value, it is necessary to carry out a careful energy survey of the intended plant to establish the energy requirements and demand patterns. In most of the studies so far carried out the hourly gas yield is unable to match the peak fuel demand.

Closeness to this match depends upon the individual demand levels and, while methane is easily stored in bell-over-water gas-holders, storage of any useful values is often expensive.

It is possible to reduce the storage volume by compression. For example, 1 m^3 of biogas at 200 mm w.g. (8 in w.g) reduces to 0.23 m^3 at 3.5 bar (50 p.s.i.g.) and 0.005 m^3 at 207 bar (3000 p.s.i.g.). While these are considerable reductions, giving large savings of space for the storage facility, saving in capital cost is marginal, if any, for the gas quantities normally expected with whey processing. In addition, the increased sophistication of high-pressure gas handling has so far made this alternative unattractive.

Fortunately, most creameries have a high base electrical load and a requirement for large quantities of hot water for washing and cleaning, such that a combination of direct firing in boilers and electricity generation with hot water recovery enables optimisation of biogas usage.

BIOGAS UTILISATION AT SOUTH CAERNARVON CREAMERIES

The main feature at SCC is flexibility to maximise the use of the biogas in the areas of highest return for the prevailing conditions of varying whey production. Therefore the operational philosophy is to use as much biogas as possible in the main boilers, diverting to power generation as the steam demand comes off. Hot water at 85°C is recovered from the engine exhaust and cooling water and stored in two tanks to use as either wash water or preheated boiler feed water for the next day.

The new energy centre installed as part of our contract with SCC comprises:

(1) Two Robey Lincoln Two shell boilers, each producing 10 000 lb/h of steam from and at 212°F and 120 p.s.i.g. The boilers are fired with Hamworthy dual/combined biogas/HFO rotary burners type AWD.10, with a three-mode firing pattern (gas only, oil only or simultaneous firing). Flue gases are discharged via a free-standing twin-flue insulated chimney.

(2) An engine generator set incorporating a Dorman 6QG six-cylinder-in-line four-stroke naturally aspirated spark ignition gas engine running at 1500 rev/min. The engine is coupled in line with a brushless, self-regulating, self-exciting alternator rated at 166 kVA. Waste heat recovery from the engine exhaust jacket cooling water and oil cooler produces 240 kW of hot water at 85°C which is stored in two 15 000 m^3 gas tanks.

(3) The existing Atlanta stand-by diesel alternator is also installed in the centre.

Figure 5 Shell boiler fitted with Hamworthy dual fuel (biogas/heavy fuel oil) burners

Figure 6 Engine generator set utilising biogas

The Dorman set has full automatic synchronisation to the mains, with load sharing, so that a minimum input from the main is maintained to meet Manweb's requirements. This is at present set at 30 kW. On mains failure the unit stands-by the back-up emergency generator, to which it will automatically synchronise upon establishment. All the switchgear, controls and electrical wiring were supplied to conform fully with the requirements of the Electricity Supply Board — Manweb. Even with this arrangement, there are times when excess gas has had to be flared. We are therefore studying the practicality of profiling the whey feed rate, to control the times of peak gas production to meet peak requirements.

29
A REASSESSMENT OF THE BRAZILIAN NATIONAL ALCOHOL PROGRAMME (PNA)

Francisco Rosillo-Callé

INTRODUCTION

Towards the end of 1983 Brazil announced that it had built its 'one millionth alcohol car' and celebrated the event by a special show. Over 85 per cent of cars sold last year in Brazil were alcohol-powered cars, and ANFAVEA (The Brazilian National Association of Automobile Manufacturers) estimate that 5 million alcohol cars will be sold by 1990, depending on the world's oil situation.

The achievement of such a widespread development of alcohol cars is a result of the Brazilian National Alcohol Programme (PNA), which set out in 1975 to support a locally produced liquid fuel as an alternative to imported oil. It is well known that the United States government under President Carter in the late 1970s initiated a Gasohol programme which proved something of an economic fiasco. So why should Brazil struggle so hard to substitute alcohol for petrol? Brazil, a non-oil state with enormous foreign debts, was driven to seek ways of reducing oil imports. Dr Camilo Pena, the Brazilian Secretary of State for Trade and Industry, argued that 'our alcohol is competitive with petrol. But if it was not, our option should be for alcohol since Brazil has not the dollars to import oil.' 'The PNA', said Pena, 'has already accomplished its three main strategic objectives: technological development; competitiveness; and economic viability.'

Over 50 billion litres of ethanol have been produced since the creation of the PNA, mostly for fuels (see Table 4); the installed capacity is just over 12 billion litres/year, distributed in approximately 560 distillery projects.

The maturity and viability of the PNA seems to be recognised by the World Bank, which has been a major financial backer. The present economic and financial situation of Brazil should stimulate and strengthen the Programme, since it has forced the government to revise its energy plans further towards oil substitutes as follows:

(1) An increased utilisation of alternative energy resources from 1985 (alcohol, electricity, gas and coal) from the planned 500 000 barrels per day petrol equivalent (b.d.p.e.) per annum to 550 000 b.d.p.e. per annum.

(2) A reduction in imported oil from the previously planned 500 000 b.d.p.e. down to 350 000 b.d.p.e. per annum.

(3) An increase in energy saving from an estimated 200 000 b.d.p.e. to 300 000 b.d.p.e. per annum.

Structural changes in the Brazilian oil market are illustrated in Figure 1. Despite the enormous success of the past few years, total oil imports remain much higher than previously planned.

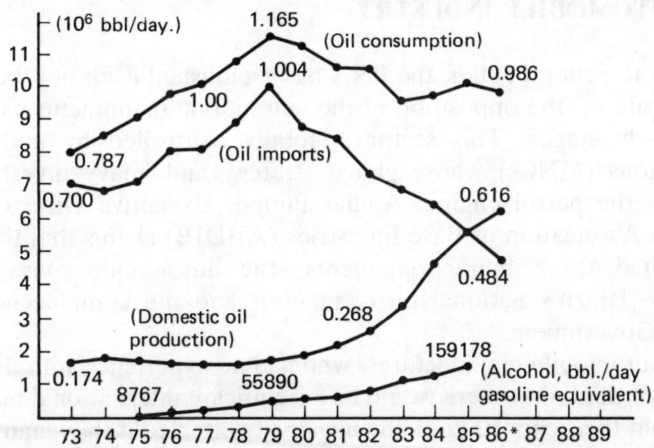

* Estimated;

Sources:

1973–83 *The Financial Times Survey*, November 1984, p. iv.
1984–86 Banco do Brazil: *Brasil Programa Econômico; Ajustamento Interno e Externo*, Vol. 10, February 1986, p. 103, Table 31.
1973–82 (Oil imports): Celestino Rodrigues E., *Solução Energética*, Editoras Unidas Ltda., São Paulo, 1983, p. 37.
1975–85 (Alcohol production), CENAL*, as Table 4, column 3 (*CENAL = Comisão Executiva Nacional do Álcool).

Figure 1 Brazil: Oil consumption, domestic production and imports, 1973–86; alcohol production, 1975–85.

It is often stated that the PNA is inflationary and highly subsidised, but government spokesmen deny this. Dr Bautista Vidal, ex-secretary of the Secretariat of Industrial Technology and a key figure in the establishment of the PNA, believes that because the PNA was able to use the idle capacity of the sugar sectors, over 20 per cent at that time and worth billions of dollars, the Programme is hardly inflationary. L. Navarro Jr., a member of the National Energy Council (CNE) argues that inflation is in fact generated by the 'petroleum dollars' which are subsidised and that the real cost of petrol is at least 46 per cent higher than its market price in Brazil. There are 22 petroleum by-products subsidised in Brazil, which cost 4.5 billion cruzeiros (Cr$) per day (1983), of which petrol fuel receives the highest subsidies of all, although this is questioned by some critics. Furthermore, the benefits of anhydric alcohol sales (alcohol mixed with petrol) gives a profit to the government of at least Cr$ 65 per litre — some critics put it as high as Cr$ 90 per litre because the alcohol is bought at 59 per cent of the cost of petrol — but it is sold in the pumps at the same price as petrol. This differential is more than sufficient to cover any losses by hydrated alcohol sales, according to the supporters of the PNA.

THE AUTOMOBILE INDUSTRY

Contrary to general belief, the PNA has established itself not 'because of' but 'in spite of' the opposition of the automobile manufacturers in Brazil in its early stages. This sector is totally controlled by multinational corporations (MNCs) whose global strategy and heavy investment are based on the petrol engine. Aguiar Puppo, Executive Director of The Brazilian Association of Base Industries (ABDIB) claims that the alcohol engine had three major opponents: the automobile manufacturers; Petrobas (Brazil's national oil company); and the complacency of the Federal Government.

The automobile manufacturers who had no experience with the alcohol car did not think that there would be a significant international market and feared that their exports could be negatively affected if they manufactured alcohol cars. Consequently, the MNCs did not invest in research and development. It was the government, through the Aerospace Technology Centre (CTA), who played the key role. The CTA did all the fundamental research, which was then passed on to the automobile industry and, some argue, was totally free of charge. It was a combination of circumstances involving government pressure, a slump in car sales and government incentives for the alcohol car that finally persuaded the industry to accept the idea. Critics have suggested that despite the fact that there was not any infrastructure for the alcohol car (such as maintenance, experienced personnel and low quality control), most of the problems of the first alcohol cars were caused mainly by the carelessness of manufacturers who were unenthusiastic about the idea and who would have been relieved if the alcohol car proved to be a market failure. However, this attitude changed dramatically after the success of the car with the Brazilian consumers.

The existing ethanol engines are still petrol engines but have been adapted to burn alcohol. An engine based on the thermal properties of ethanol could have a thermal efficiency of 42–45 per cent, against 27 per cent for the petrol engine. The CTA is now developing an ethanol engine for passenger cars and for medium and light trucks.

After a slump in 1982, the PNA is in top gear again and has reached a maturity stage. Alcohol-powered vehicle sales from 1979 to 1985 are shown in Table 1. A further development has been a shift to emphasise the utilisation of alcohol in light- and medium-weight trucks and also tractors. By 1987 all trucks used in the sugar and alcohol sectors should be burning alcohol instead of diesel. This is an important social development, which will spread the PNA's benefits beyond the individual car owner.

However, the need for diesel fuel in Brazil remains a serious problem, since the transport system is heavily based on road transport. The picture is further complicated because the heavy vehicle manufacturers such as Mercedes-Benz do Brazil (MBB), with about 50 per cent of the Brazilian market, demanded that the government should find an alternative liquid fuel suitable for the diesel cycle engine, rather than develop an alternative new engine; this was on the grounds that diesel engines are very efficient for heavy vehicles. Nonetheless, attitudes are changing and resources are being channeled into research and development which which will eventually produce its fruits. Already many experiments with novel fuels have been carried out successfully using vegetable oil and alcohol mixtures. Another approach being actively researched seeks to develop an additive obtained from ethanol itself, which at about 5 per cent level transformed hydrated ethanol into suitable fuel for conventional diesel engines.

Volkswagen do Brazil Caminhoes report that its trucks and lorries fuelled by alcohol are technologically and economically viable; over 1000 trucks are being sold to more than 60 distillers.

ENVIRONMENTAL IMPACT

The PNA distillers produce large quantities of effluent or stillage (12–14 litres of stillage for each litre of alcohol), an estimated 14×10^9 litres in 1985/86 alone. The stillage is still a serious environmental problem in many places. Specific legislation to deal with this problem of stillage, or *vinhoto*, was introduced on 29 September 1978 and on 3 November 1980, prohibiting any discharge into the rivers network system, but the legislation was

Table 1 Total and ethanol-fuelled vehicle sales, 1979–84[a]

	Passenger and mixed use			Light trucks			Heavy trucks			Total		
Year	*Total*	*Alcohol*	*% Alcohol*	*Total*	*Alcohol*	*% Alcohol*	*Total*	*Alcohol*	*% Alcohol*	*Total*	*Alcohol*	*% Alcohol*
1979	828 733	2 271	0.3	95 957	843	0.9	90 235	6	0.0	1 014 925	3 120	0.3
1980	793 028	226 347	28.5	93 768	14 291	15.2	93 465	–	–	980 261	240 638	24.5
1981	447 533	128 679	28.8	68 000	7 563	11.1	65 117	1 065	1.6	580 650	137 307	23.6
1982	556 229	221 761	38.1	85 763	20 814	24.3	49 302	922	1.9	691 294	233 497	33.8
1983	608 499	538 401	88.5	78 085	40 927	52.4	41 148	2 045	5.0	727 732	581 373	79.9
1984	533 040	503 979	94.5	95 615	61 999	64..8	48 880	2 627	5.4	677 535	568 605	83.9
1985[b]	172 453	164 982	95.7	25 864	18 183	70.3	15 687	325	2.1	214 004	183 490	85.7

[a]These figures correspond to the new classification of ANFAVEA (The Brazilian Association of Automobile Industry). From January 1985, vans, ambulances and combi-type vehicles are no longer classified as passenger and of mixed use, but as light commercial vehicles.
[b]January/April.
Source: ANFAVEA/CENAL: *Comissao Executiva Nacional do Alcool, Relatorio,* January/April 1985, Appendix, 50.

seldom put into practice. The *vinhoto* problem, however, has been exaggerated and outside critics often forget to take into account not only the Brazilian conditions as a whole but those of other industrial sectors such as the petrochemical, which have very high levels of pollution. In our opinion, this problem is being tackled. We have been assured quite often by many Brazilians that there is not a *vinhoto* problem. While this may be the case in the State of São Paulo, where the industry itself has developed and implemented modern techniques for utilisation and handling of stillage, it might not be the case in many other parts of the country.

Stillage is increasingly being used successfully as a fertiliser for the sugar-cane itself, with significant savings in the imported chemical fertilisers, particularly nitrogen and potassium. More importantly, new techniques are being developed which would greatly reduce the volume of stillage. In 1982 the average stillage concentration was 3.5–4°Brix, but a 20–30°Brix concentration is not too far away and would reduce effluent volume by about 88 per cent, particularly in the fermentation process. Recently Codistil's Biostill process and Zanini's hybrid processes have achieved a concentration of 60°Brix (sugar content), producing 1 litre of alcohol for 1 litre of stillage. Of course, these processes are still in an experimental phase and it remains to be seen whether or not the PNA can incorporate them with their modulation system. These technological developments and the increasing awareness of the economic value of stillage as a source of fertilisers could resolve the stillage problem in the not too distant future.

SOCIAL IMPLICATIONS

On the whole, the PNA has led to an increase in the social benefits for those working in the sugar and alcohol industry. There is a mandatory contribution of 2 per cent on the sales of ethanol and 1 per cent on the sales value of sugar-cane, for improving housing, education and health for its workforce, and this has led to higher living standards. Salaries are much higher than in other agricultural sectors, although they remain quite low; 46 per cent of the price of a ton of sugar is reported to be labour cost, as compared with 26 per cent in other agricultural sectors. The *bioa-fria*, the Brazilian sugar-cane cutter, has been offered the professional registration card, with the additional benefits it brings.

The creation of jobs in this sector has reversed the previous tendency towards population loss in agriculture. There has been an expansion in the commercial sector, and the resulting job opportunities in the secondary and tertiary sectors have possibly beneficially affected the quality of life as a whole in many communities. Many distilleries have adopted policies designed to stabilise employment and train the workers, and there is a real effort to reduce the seasonability of employment and increase social benefits. Unfortunately, this is not always the case, and the national press has reported abuses by employers avoiding their responsibilities.

Other negative impacts have also been reported — for example, those caused by the speed of implementation of large-scale projects with inadequate technology for local conditions located in remote areas not possessing a proper social infrastructure.

Employment creation appears to be by and large one of the PNA's major positive impacts, and more than once it was put to us that in the present circumstances the PNA has been a major stabilising force in the rural areas. Biago Filho, Director of one of the largest sugar-cane and alcohol production groups in Brazil, says that 1.5 million people depend directly on this sector, which had a GDP of US $15 billion in 1983. Luis Zillo, President of Copersucar, Brazil's largest sugar-cane and alcohol producer, puts the number employed in this sector at 700 000 people directly and about 2 million if their relatives are included. Job creation for the period 1975–85 is illustrated in Table 2. Although these figures may

Table 2 Number of direct jobs created by ethanol production, 1975–86 (in cumulative figures)

Year	*Number of direct jobs*
1975	28 200
1976	32 900
1977	70 500
1978	117 500
1979	159 800
1980	173 900
1981	197 400
1982	272 600
1983	371 300
1984	423 000
1985	428 000
1986	510 000[a]

[a]Estimates: Foster, R. quotes 800 000 jobs for that year. *Financial Times*, 4 May 1986, p. 40.

Sources: 1975–84: The Brazilian Ethanol Producers' Special Committee. *Ethanol: Brazil Export Potential* (undated), p.12, Table 8.

1985–6: *CENAL Relatorio Anual 1985*, CENAL 1986, p. 40.

Note: according to the Brazilian Ethanol Producers, the number of indirect jobs created by the ProAlcool total over 1.8 million.

appear somewhat high, employment creation has been high compared with other rural sectors, though many of these jobs are not necessarily new ones.

The possible conflict between land use for fuel and that for food has also been a major criticism. While one cannot deny the possible implications, this issue has been largely dramatised. In the Brazilian case it does not correspond to reality: evidence shows that sugar-cane expansion has taken place almost exclusively in pasture land. Perhaps land concentration in some areas has been a more serious problem but, ironically, Brazilian-grown food is exported to pay for oil imports. Figure 2 shows total land area and that of main crops. Most of these crops are exported precisely to pay for oil imports and mostly as animal food. Agriculture production has actually increased since the creation of the PNA in absolute and relative terms, according to highly placed sources. Contrary to general belief, sugar-cane does not necessarily need top-quality soil, as has been demonstrated by autonomous distilleries too frequently set up in poor soil. Government officials appear to be generally aware of the possible negative implications. They created the Projeto cana-de-acucar e Producao de

DISTRIBUTION OF
THE CULTIVATED AREA
BY CROPS, 1984

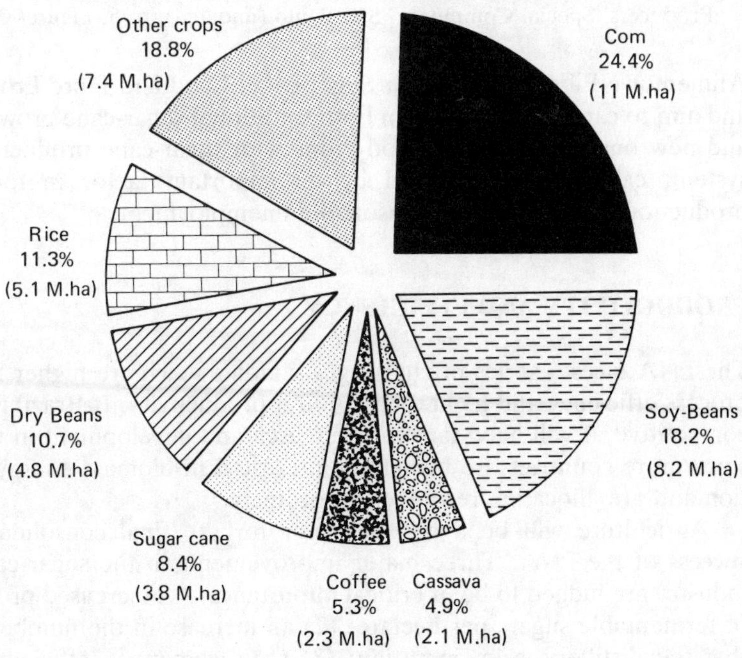

Figure 2 *(continued on next page)*

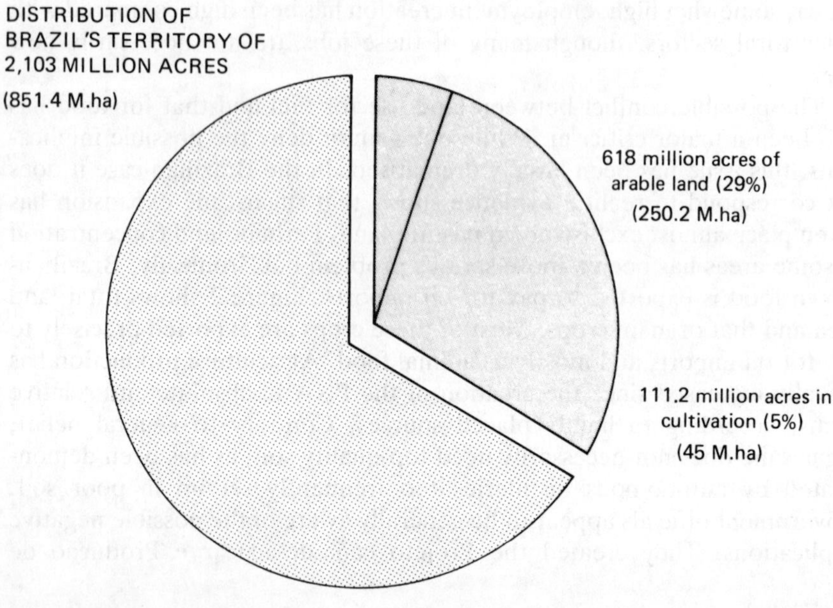

Figure 2 Source: *Ethanol: Brazil Export Potential*, The Brazilian Ethanol Producers' Special Committee, São Paulo (undated), p. 5, Figures 2 and 3.

Alimentos e Fibras (Project for Sugar-cane, Food and Fibre Production), and aim to carry out research in both traditional sugar-cane growing areas and new ones, overlapping food crops with sugar-cane production. This system, called *Rodizio*, could be an important factor in food crops production and in reducing seasonable unemployment.

PRODUCTIVITY AND EFFICIENCY

The PNA faces a series of challenges, among which are higher industrial process efficiency and higher agricultural productivity. If Brazil is to keep competitive, it will need far more research and development in this area, since more countries are becoming interested in biomass energy production and are allocating resources accordingly.

Agriculture will be a crucial factor for the final consolidation and success of the PNA. Three major improvements in the sugar-cane agro-industry are judged to be of critical importance: (1) increased productivity of fermentable sugars per hectare; (2) an increase in the number of days that the distillery is in operation; (3) improvement in efficiency of the industrial processes.

Improved productivity of fermentable sugars per hectare could come from an increase in the sugar content of harvested cane; currently this is 13–14 per cent by weight. It is planned to increase sugar content to 17 per cent over the coming 5–10 years, principally through the development and use of improved varieties and better management practices.

The increase in productivity judged feasible in the medium term would raise the 1984 average productivity from 65 tons per hectare (three harvests in 4 years) to 80 tons per hectare per year. New improvements would permit an additional ratoon crop, four harvests in 5 years, with a consequent increase in productivity from 72.7 to 101 tons per hectare; this is an increase of 39 per cent. A second factor affecting productivity is the number of days a distillery is in operation; currently in Brazil this is 150–180 days per year. It is possible to increase the harvest period to 200 days per year through the use of new varieties and better management practices. A distillery can be operated for 300 days per year if production and milling (the 200 day season) are increased by 50 per cent, and if cane juice concentration can be increased to about 60°Brix and stored for later dilution.

A third major area of improvement concerns increases in efficiency of various steps in the industrial processes of transforming sugar-cane into alcohol. Today's efficiency is about 80 per cent. It is estimated that this efficiency could be increased to approximately 85 per cent, a gain of over 21 per cent (in the State of São Paulo it is much higher), through the development and adoption of improved industrial processes which also reduce the total investment necessary per litre of annual capacity.

These three improvements, when considered together, represent a remarkable increase in overall productivity, with substantial savings in investment cost production. They also represent a key factor in reducing the total land area required for alcohol production. In the long term there is a need to avoid an overdependence on sugar-cane by developing alternative energy crops such as, particularly, cassava.

ALCOHOL CHEMISTRY

The chemical sector is particularly regarded as important in the long-term objectives of the PNA. Their original estimates of providing 1.5 billion litres of ethanol a year to this industry by 1985 was obviously over-ambitious, given the modern plant and overcapacity of the existing petrochemical sector in Brazil. The alcohol consumption in the alcohol chemistry industry for 1978–85 is shown in Table 3.

Brazil has the technological capability to develop this industry but the major problem is the financial situation facing many companies. It was reported to the *2nd Brazilian Congress of Alcohol Chemistry* in September

Table 3 Ethanol use by the chemical industry in 1978–85 for domestic and export products

Year	*Total use* (10^3 m^3)	*Domestic use* (10^3 m^3)	*Percentage domestic*	*Exports* (10^3 m^3)	*Value FOB* ($1000)	*Percentage exports*
1978	101.5	99.5	98.0	2.0	550	2.0
1979	125.7	113.0	90.6	11.8	4 993	9.4
1980	146.3	138.2	94.5	8.1	5 180	5.5
1981	118.5	103.6	87.4	14.9	11 030	12.6
1982	235.8	192.8	81.8	43.0	15 814	18.2
1983	380.8	252.6	66.5	127.4	47 356	33.5
1984	494.5	249.2	50.4	245.3	82 691	49.6
1985[a]	304.7	212.9	69.9	91.8	31 000	30.1

[a]Estimates, January/September.

Source: *Abiquim Noticias*, Vol 7(36), September/October 1985 (ABIQUIM = Associação Brasileira da Indústria Quimica e de Produtos Derivados).

1983 that important technological breakthroughs could easily occur in the development of direct routes to alcohol and in the area of sucrochemistry, due to the research now under way.

Currently ethylene, acetaldehyde, acetic acid, butanol and octanol can compete with similar products in the petrochemical sector, although with current oil prices this may increasingly be called into question. In the short term the development of this sector is seen as complementary to the petrochemical sector. Brazil has begun to make important gas discoveries which are now cheaper than alcohol and can produce ammonia and methanol — important petrochemical building blocks — which may affect the longer-term development of this industry.

There are many other opportunities created by the PNA, which are just beginning to be considered. Sugar-cane bagasse can replace diesel in the distilleries as fuel for generating electricity. Over 15 per cent of all imported oil could be replaced if the 7 million tons of surplus bagasse were burned to generate electricity. Research carried out in the distillery Sta. Lydia S.A. with sugar-cane bagasse with 20 per cent water content suggests that the 200 million tons of sugar-cane produced in 1985 could generate a minimum of 1.18 million tons petrol equivalent. If a production of 40 billion litres of alcohol is achieved in the year 2000, the resulting sugar-cane bagasse could achieve a 15 million tons of petrol equivalent.

According to CENAL (National Executive Commission of Alcohol), from 1975 to 1985 the programme saved the country $8.97 billion (see Table 4). However, the foreign exchange savings due to the ProAlcool are disputed by some analysts, who argue that the opportunity costs of

Table 4 Estimated foreign exchange savings due to the ProAlcool, 1975–85

Year	*Alcohol production* (10^6 litres)	*Gasoline equivalent* (10^6 bbl)	*Equivalent value* (\$ 10^3)
1975	580.1	3.2	42.0
1976	642.2	3.5	50.1
1977	1 387.7	8.2	125.4
1978	2 359.1	13.9	217.2
1979	3 448.3	20.4	431.9
1980	3 676.1	20.6	764.5
1981	4 206.7	22.3	956.7
1982	5 617.9	31.8	1 274.0
1983	7 950.3	42.1	1 525.7
1984	9 201.0	47.7	1 729.5
1985	11 200.0	58.1	1 860.6
Total	50 269.4	271.8	8 977.6

Source: *CENAL Relatorio Anual 1985*, CENAL 1986 (Brasilia), p. 39, Table 42.

producing alcohol are exceeded by the value of alcohol as a petroleum substitute. For example, Barzelay estimates that for the 1978/79 harvest Brazil experienced a net social loss of $0.08 per litre of ethanol. Since some 2.5 billion litres were produced that year, the total social cost was about $175 million, and the social losses for the period 1980–1983 are put at about $214 million.

In 1983 the Brazilian press reported that talks were taking place on the possibility of setting up an 'alcohol production centre' with an estimated investment of US $12 billion to produce 500 000 b.d.p.e. for export to the US market. However, since the US Congress has put an import tax of US $0.5 per gallon on alcohol, there are now great doubts about this market, further complicated by the fall in oil prices.

The second potential market is for hydrated alcohol for chemical products, drinks and pharmaceuticals.

The third and most promising is for a 'synthetic ethanol', produced currently in other countries from naphtha, but at much higher cost than fermentation ethanol. Fermentation alcohol happens to be cheaper and more efficient than naphtha ethanol.

CONCLUSION

With all its faults, the PNA remains the world's largest and most successful biomass energy programme and its importance extends beyond Brazil's national boundaries. It creates much-needed employment opportunities; and it can be an important factor in promoting social equality, improving living standards in the rural areas, modernising agriculture, and increasing energy and technological independence. The PNA is forcing the pace of innovation in the related agricultural and industrial sectors. Brazil can establish itself as a world-wide supplier of process and equipment technology for production of fuel and industrial alcohol. Already Brazil is providing 'know-how' to a number of countries.

Much of the conventional wisdom on energy requirements needs reconsideration. The dismissal of new alternatives, particularly by the conventional energy agencies, tends to be done on debatable grounds. The PNA is sufficiently important to warrant serious reconsiderations with economic development policy in many LDCs. It can serve as an instrument for rethinking economic development strategy which will make greatest use of bioenergy resources, despite the glut in oil prices. The present oil situation presents new difficulties for the development of alternative energy in general and the ProAlcool in particular. This has prompted a reassessment of the programme. Although the PNA is accepted as 'irreversible' and will be kept as a 'strategic asset', some important changes will be introduced. No new projects are to be set up in the next 2 years, and the price differential between alcohol and gasoline could be reduced further, together with other tax incentives to the alcohol car. To reduce surpluses, an increase in alcohol blending, from the present 22 per cent to 25 per cent, may be tried; and 5 per cent blending alcohol-diesel is also under consideration. Another alternative would be to expand alcohol exports, when possible, particularly to the USA and the EEC. The PNA remains vulnerable, but it seems safe to say that any decision regarding the programme would not be taken solely on the basis of unpredictable oil prices. Given the great ramifications of the ProAlcool, wider socio-economic and political considerations will undoubtedly be a deciding factor; these range from job creation and greater energy independence to international sugar prices.

BIBLIOGRAPHY

1. Anon. (1983). *Alcohol Fuels: Options for Developing Countries*. Advisory Committee Report on Technology Innovation, Board of Science and Technology for International Development. National Academy Press, Washington DC
2. Barzelay, M. (1986). *The Political Market Economy; Alcohol in Brazil's Energy Strategy*. University of California Press, London
3. Geller, H. (1985). Ethanol from sugarcane in Brazil. *Annual Review of Energy,* **10**, 135–164
4. Greenshields, R. N. (1984). Gasohol: The United States power alcohol scene. *Industrial Biotechnology Wales,* **3** (5), May
5. Rosillo-Callé, F. and Rothman, H. (1984). The Brazilian National Biotechnology Programme. *Biotechnology,* **2** (5), May
6. Rosillo-Callé, F. and Rothman, H. (1986). The Brazilian National Alcohol Programme: some technological and socio-economic aspects. *Outlook on Agriculture,* **5** (4) (in press)
7. Rothman, H., Greenshields, R. N. and Rosillo-Callé, F. (1983). *The Alcohol Economy: Fuel Ethanol and the Brazilian Experience*. Frances Pinter, London; Kentucky University Press, Lexington

30
ETHANOL IN EUROPE: THE LONG ROAD TO PROGRESS

Jayne Heaford

EC ETHANOL BECOMES NEWS

The topic of bioethanol (ethyl alcohol of agricultural origin) is no newcomer to the European political arena. Proposals to introduce the use of bioethanol have been discussed since 1972, with a number of academic studies and debates taking place, resulting in no positive action.

However, in the past 4 years the bioethanol cause has suffered mixed fortunes, with a massive upsurge in interest, followed by disappointment at the lack of assistance by governments. Yet in some countries there are recent tangible signs that biofuels will find their role in the energy scenario of the future.

The belief that bioethanol use in Europe would lag behind that of other regions was based on the assertions that European nations had either indigenous fossil fuels or economies strong enough to sustain imports of these. The fact that ethanol from substrates such as sugar-beet used oil in the production processes thus reduced the net energy ratio obtained and therefore the economic viability. Moreover, Europe's climate and smaller land areas are less favourable for producing large quantities of biomass or energy crops needed for the ethanol feedstocks because farmland is required for food crops.

AGRICULTURE OUTPACES FOOD DEMAND

Clearly, agriculture in the 1980s has outdated the 'food versus fuel' argument: structural oversupply in farm products is now a global problem. While European manufacturing industries have contracted since 1980, agriculture has expanded concurrently with the economic recession, changed diets and static populations, which have decreased demand.

In 'A Future For Community Farming' the EC Commission identified the priority of developing agro-industrial products that involve agriculture in technological change. The key to success is that markets must exist for the ensuing products. Bioethanol presents the opportunity of converting something in excess (i.e. surplus grains, sugar, potatoes, etc.) into items we will increasingly need to import (e.g. liquid transport fuels plus a range of products arrived at through its major process routes, which include synthetic fibres, textiles, paints, tyres and plastics of great varieties).

In principle, bioethanol's versatility is not disputed; opinions divide when there are attempts to develop its potential now.

This attitude brought about a situation whereby European chemical companies, wishing to expand their fledgling biotechnology industries, were forced to transfer their production plants outside the EC in order to buy substrate sugar at world prices, which are vastly lower than those prevailing within the Community as a result of the Common Agricultural Policy's pricing system. After studies confirmed the damage being done to the European companies, a concession was made for the use of sugar in the biotechnology companies, but while the chemical sector is favourably disposed to taking agricultural products as substrates, it is a small market compared with that of the energy sector. It is in trying to take ethanol use into the energy sector that the advocates of biofuels get caught in the crossfire between agriculturalists, industrialists and politicians.

AN ABUNDANCE OF TRANSPORT FUELS?

Oil analysts world-wide assure us that oil prices will remain low for the foreseeable future, a prediction perhaps made less assuring by the fact that none contemplated the 1986 oil price would crash to levels that would devastate the oil industry and destabilise producers' economies. The fact is that only countries which have developed broad-based and integrated energy policies can be confident of stability.

The EC Statistical Office states that the Community's oil reserves will last only to the end of this century, and natural gas about 5 years longer, if both continue to be produced at 1984 levels. Twelve years from now is not so distant that alternative transport fuels should be dismissed as unnecessary. Although lack of political foresight hampers balanced energy policies, one cannot underestimate the extent of the problems to be addressed.

In assessing bioethanol for fuel or chemical use, politicians rely on information from diverse sources: agriculturalists, the oil and petroleum sector, biotechnologists, motor industry chemical engineers (combustion processes) and mechanical engineers (engine design). All have spokesmen who are experts and highly protective of their own field.

Also, the EC must serve the interests of many nations with differing situations as regards energy supplies and agricultural organisation; the amount of Community budget spent on the latter has been so highly publicised that public opinion is influencing political decision-making.

It was the increasing influence of the European Green movement that pushed through the Schleider Resolution 1349/84 and heralded the start of an ethanol lobby capable of affecting policies. The resolution advocated using bioethanol in motor vehicles for environmental reasons and won wide support from Christian Democrats, Conservatives and other groups in the European Parliament.

1984 saw the proposal that the EC wine lake could be used as fuel oil in boilers or be added to gasoline, and this had a mixed reaction from politicians and public.

The resistance of the petrochemicals sector was seen in BP Chemical's argument as meaning that to use wine-alcohol as feedstock for chemicals would seriously disrupt their business. The ethyl alcohol market was said to be static and it was claimed that to release such an amount without disrupting the market would take about 8 years. The real significance of this event was that if wine-alcohol was even being considered, it had to open the door for sugar and grains, since wine as a substrate was 3–5 times more costly than other substrates.

But it was in March 1985 that the pro-ethanol lobby saw what it hoped would be a new and immediate market for ethanol when EC Directive 85/210/CEE called for unleaded petrol to be made available to consumers in the Community from 1989 and recommended a reduction in the current permitted level of lead.

BIOETHANOL IN THE OXYGENATES MARKET

Campaigners for the removal of lead from petrol say that the decision was stalled for 20 years as a result of lobbying from the oil and motor industries, which is an indication of the political strength of these sectors.

Although detrimental to the environment and human health, lead in petrol is a magic ingredient: small amounts of lead compounds added to petrol give the high-octane fuel needed for modern high-compression engines to run without 'knock'. Since 1920 lead has been used, but vast sums have been invested in the search for alternative 'anti-knock' agents: one research programme tested 980 compounds. Some materials with a degree of anti-knock potency were discarded as damaging to cars with catalytic exhaust reactors or because they were at least as damaging environmentally as lead.

Octanes can be improved by reforming processes within the oil refineries but this implies increased investment and consumes much more crude oil. A BP spokesman in 1982 estimated it would require between £1000 million and £2000 million spread over 10 years and take a further 1.5 million tonnes of crude oil per annum. However, these figures were given when the industry was still resisting the idea of lead's removal. At that time, the UK Petroleum Industry Association argues, surplus refining facilities existed only in basic crude oil distillation costs, and no meaningful estimates could be made of the investment needed to super-refine.

That left the route of incorporating high-octane chemical compounds as a substitute for 'anti-knock agents'. These 'oxygenates' must be used in concentrations of more than 5 per cent to be effective, and are the alcohols (ethanol, methanol, tertiary butanol (TBA)) and the ethers (methyl tertiary butyl ether (MTBE)).

In trying to stake a claim to the oxygenates market, the pro-ethanol lobbies faced many difficulties. The technical assessments of the various oxygenates, which are raised at seminars and put forward to governments, are based on research undertaken by the oil industry. It is known that the level of secrecy in this industry is such that even government committees find it virtually impossible to obtain the information they need and it is understandable in the sense that oil majors are formulating their own additive packages which, if successful, are highly valuable in the market-place.

Thus, the agriculturalists were less able to argue the case for ethanol in fuel, as this is a new subject for them, whereas the oil sector holds the technical data on the various options. Without going into the merits and disadvantages of the competing oxygenates, suffice it to say that the main obstacles to using ethanol were stated by the oil industry as being availability and price.

Oil spokesmen stated they would wish to buy in ethanol at 20 ECUs per hectolitre, whereas the cost of ethanol was calculated at 49–50 ECU/hl from sugar-beet, 55 ECU/hl from wheat and 389 ECU/hl from wine (figures given at February 1986). At that time farmers acknowledged that they could not close the gap without the help of subsidies and decided to address the problem of increasing competitiveness through yields and process technology. As well as increasing the value of by-products they tried alternative raw materials such as Jerusalem artichoke, sorghum, chicory, wood and straw.

Obviously, the oil industry would not increase its own costs to alleviate agricultural problems, so hopes rested on some form of government assistance to give ethanol a push.

The report 'Perspectives for the Common Agricultural Policy' of July 1985 stated that ethanol could be envisaged as an additive (up to 5 per cent without manufacturers having to indicate the actual quantity used) or as an auxiliary solvent (2–3 per cent with methanol) in petrol, adding that incorporating large quantities of bioethanol into motor fuel would present a number of advantages for agriculture. But it sounded a pessimistic note about the large-scale subsidies which would be needed to finance the gap between bioethanol and its competitors. The budgetary options were: defraying the full difference between offer and selling price — implying high costs to the EC budget which would be difficult to estimate because of possible sudden changes in market fuel prices; or defraying part of the difference, with the balance financed by farmers.

Other points were the need to avoid distortion in competition between the various basic products (sugar, wheat, etc.) and between bioethanol and other oxygenates. A network of processing plants would be needed, large enough to have economies of scale and given guarantees regarding supply of materials.

The UK Agriculture Minister wrote to a fellow MP: 'The Commission are rightly cautious about the cost of setting up the processing facilities to produce bioethanol, acknowledging that it would require legislation and incentives to encourage the use of bioethanol to stimulate such investment.'

Impatient with political 'caution', strong lobbying groups were heard in Europe throughout 1985, the most publicity going to the Italian agribusiness conglomerate Ferruzzi when Managing Director Raul Gardini announced plans to invest in building a network of up to 12 ethanol plants in France and Italy, if he could secure the political backing and incentives needed. He put together a highly professional study document to back his claims that the project was viable and won influential support.

Meanwhile, in France, West Germany and Austria, groups from the sugar and wheat producers and also environmental pressure groups were pressing for government incentives for ethanol, presenting figures to show there was adequate supply of substrates already.

European Member of Parliament R. M. Seligman quoted figures showing that the total agricultural surpluses produced under the CAP regime occupies 8–9 million hectares out of a total 152 million hectares. From 1 hectare producing 4 tonnes ethanol and 1 tonne biomass residue, the 8 million hectares could give 32 million tonnes ethanol per annum. As 90 million tonnes of gasoline were consumed annually in the EEC, theoretically, he reasoned, 35 per cent could come from the land currently producing food surpluses. He noted the price oil majors wished to pay — 250 ECUs per tonne seemed unfair, since petrol costs 350 ECUs per tonne to produce.

Availability is now not considered a problem but the Committee of Agricultural Organizations in the European Community (COPA) do not wish agricultural products for industrial use to be seen merely as a way of dumping surpluses, saying that the wider economic advantages should be considered, such as reducing energy imports, creating new skills and maintaining stability in rural areas. This last point is overlooked in the current proposals to reduce surpluses by taking land out of use, retiring farmers or offering direct grants. From the farming communities' viewpoint, a healthy agricultural sector has a stabilising effect on a country.

There has been a growing recognition that the way to convince governments of bioethanol's usefulness was by taking the broader view through macro-economic studies which were hitherto lacking. For example, in Sweden it is admitted that alcohol produced at the Skaraborg plant, which works on the Biostill process of continuous fermentation, costs more to produce than if it were imported. But the Swedes consider that the value of by-products which save on imports makes the project worth while. Employment is created and they have concluded that the addition of 4 per cent alcohol to petrol gives rise to no technical problems whatsoever.

More studies are needed by by-product specialists, who argue that if governments were willing to forego some of their treasury revenue from petrol tax and promote the use of ethanol, this would be offset by: saving in EC intervention costs; preserving irreplaceable oil for high-value uses; protein-rich semi-wet by-products could be used as animal feed, saving on the import of soya, etc; the molasses currently imported which supply the market for carbon dioxide would no longer be necessary; straw could be used as boiler fuel; and employment would be created in building distilleries, eventually exporting the technology to less developed nations. The cynics suggest that all this takes more energy than merely collecting petrol tax revenue, and so the decisions are repeatedly deferred.

Certainly, it was a severe blow to the pro-alcohol lobbyists when, on 6 August 1986, the proposal to promote the use of ethanol from agricultural sources was rejected following the year's fall in petrol prices. According to the Energy Committee, the production of 3.4 million tonnes of bioethanol would cost the EC 1.3 billion US dollars more than imported petrol.

THE THIRD REVOLUTION

If politicians and farmers tend to think 5 years ahead, scientists do not. It may be politically untimely to voice plans to assist agriculture while there are more votes to be caught by implying that the answer is to cut production by taking land out of use. In reality, that barely addresses the problem. On present estimates EC cereal production will produce a surplus of around 80 million tonnes by the early 1990s; and European farmers recently began planting a new winter wheat variety which produces crops 20 per cent larger than previously. Also, Europe trades with the world. Countries once on the verge of starvation now no longer need to import grains; new varieties are pushing the world's corn belts to once inhospitable terrains.

Agriculture's 'Third Revolution', that of plant genetics, is estimated to be more powerful than the change from animal to mechanical production and the second production revolution of using pesticides, fertilisers and chemicals. Land taken out of use would soon be offset by increased yields and production elsewhere. New outlets must be created.

The EC's most wide-ranging practical study on the production, costs and use of bioethanol as a petrol additive has been researched by independent contractors. It examines the Community as a whole, reviews the technical issues and considers the outlook from 1986 to the year 2000.

Meanwhile, those discouraged by the lack of political encouragement should remember that, contrary to popular belief, the ProAlcool project programme in Brazil was started by farmers and entrepreneurs, in the face of official scepticism.

It may be wiser to pay less heed to the headlines and watch what is happening on the ground — such as the commissioning of the new bioethanol plant in West Germany to produce some 250 000 hectolitres of ethanol for fuel from 1988. Then there is the biofuel-fed power station being built at Abruzzo in Italy. This plant, funded by the Community's directorate-general for energy research and development at a cost of 227 million US dollars, will process in its first stage about 320 000 tonnes of organic waste and energy crops, some of which will produce ethanol as an octane-enhancer in petrol. The station will be a strategic testing-ground to test the viability of basing new industries on processing a wide range of agricultural products. If successful, 10 more could follow in Italy, with up to 100 in Europe as a whole.

The casebook on bioethanol is far from closed.

BIBLIOGRAPHY

1. Alternative outlets for renewable carbohydrates. Peter Shearer to European League for Economic Cooperation, Paris, 1 January 1986
2. Biomass Fuels in a European Context. R. M. Seligman, MEP, to EEC Conference, Venice, 24 March 1985
3. Royal Commission on Environmental Pollution, Ninth Report, *Lead in the Environment,* April 1983

4. *Changing Agriculture in the European Community*. EEC Background Report ISEC/B6/86, 27 February 1986
5. House of Lords Session 84–85, 12 Report Select Committee on the European Communities — Common Market in Wine
6. Consumers Association: Explanatory Notes on Lead in Petrol and Vehicle Emissions. Consumers Association, April 1985 (B-48)
7. F. O. Licht. International Molasses and Alcohol Report, May 1986
8. *Financial Times* 19 February 1985, 10 October 1986; *Sunday Times*, 29 June 1986; *International Herald Tribune*, 11 September 1986

31
FOSSIL FUELS — NOT THE END OF AN ERA

D. F. S. Natusch and D. W. Richardson

INTRODUCTION

The notion that biomass will be widely used as a feedstock for producing non-petroleum fuels has been frequently postulated. This notion has a world-wide following and is based on the widespread occurrence and essential renewability of biomass resources.

People in some parts of the world rely entirely on biomass for cooking and space-heating. It is, however, quite rare for biomass to be used as a feedstock for transport fuels production. This occurs only when there are reasons other than normal market forces for that production, *and* when sales of the resulting fuels are encouraged by government subsidies, sales incentives and/or the relaxation of taxes. With these premises in mind, a discussion of the factors which determine this situation is relevant.

BIOMASS CONVERSION PATHWAYS

Carbohydrate Conversion to Ethanol

Only the sugars and starches of agricultural crops can be easily converted to ethanol. The process is accomplished by the anaerobic fermentation of sugar solutions with specific ethanol-producing yeasts.

Crops which contain sugar are usually bulky. Sugar-cane and sugar-beet, for example, contain only 20 per cent sugar. The balance of the crop is water (about 75 per cent), cellulose (about 5 per cent) and inorganic salts (about 1 per cent).

Crops which contain starch, such as potatoes or Jerusalem artichokes, are also bulky and contain up to 20 per cent starch. Grains contain up to 70 per cent starch when green and 80–85 per cent starch when dry. The process of hydrolysing the starch to sugar, which is accomplished by using enzymes, is carried out in a mash with water and produces a sugar solution of 20–25 per cent.

A direct consequence of utilising these bulky crops which yield sugar solutions of only 20–25 per cent strength is that the processing plant must be sized to handle the large crop residue, which is mainly water. This adds considerably to production costs. In addition, only half of the carbon present in sugar is converted to ethanol (the remainder is lost as carbon dioxide) and the resulting ethanol (about 10 per cent) must be distilled from the dilute solution:

$$C_6H_{12}O_6 \rightarrow 2C_2H_5OH + 2CO_2$$

Cellulose Conversion to Ethanol

Cellulose, together with hemicellulose and lignin, are the major constituents of wood. The cellulose and hemicellulose are polysaccharides, which are complex polymers of sugars and can be used as feedstocks for ethanol production, provided that they can be hydrolysed to simple sugars. The lignin contains no polysaccharides and passes through the various processes essentially unaltered.

Mild acid hydrolysis will hydrolyse the predominantly pentose (5-carbon) sugar polymers of hemicellulose. More severe conditions are needed to hydrolyse the predominantly hexose (6-carbon) sugar polymers of cellulose. Significant quantities of degradation products and extremely dilute sugar solutions result. Furthermore, acid hydrolysis is not yet considered to be a commercially available process.

Enzymes produced by certain fungi have the ability to cleave the polysaccharides of cellulose into sugars which can be fermented to ethanol. Unfortunately, the process is slow and is complicated by the need to produce the enzymes; it generally produces sugars (and, hence, ethanol) in dilute solution. Moreover, enzyme hydrolysis of cellulose is not yet available as a commercial process.

In its most elegant form, direct microbial conversion involves the use of a single cellulosic microbe to yield a single product. In practice this concept is not achieved and a combination of microbes is used to produce a mixed product — usually ethanol plus acetic acid, glycols, etc. Direct microbial conversion is currently only at the research stage.

THERMAL PROCESSES FOR WOOD CONVERSION

Pyrolysis

The pyrolysis of wood, which is carried out by rapidly heating the wood in the absence of air, produces pyrolysis liquids, together with char and carbon dioxide. These liquids must be refined to remove compounds containing oxygen and nitrogen before being further refined into petroleum substitutes. Pyrolysis liquid yields of up to 60 per cent have been claimed in research pilot plants[2].

Gasification

The gasification of wood involves controlled combustion to produce the maximum possible amounts of carbon monoxide and hydrogen. In the same process much of the carbon in the wood will be converted to carbon dioxide, which has to be stripped prior to further processing. This processing involves catalytic conversion either to methanol or to Fischer Tropsch liquids. In the latter case the Fischer Tropsch liquids need subsequent refining into conventional products.

Yields from gasification are estimated as 62 per cent of methanol or 33 per cent of Fischer Tropsch liquids based on wood input[1]. These thermal processes contrast with the chemical/biological processes of hydrolysis/fermentation, because they do not take place in dilute solution and, as a consequence, the plant size does not have to be designed to cope with large quantities of water. The unit production costs of such a plant are therefore smaller.

SPECIAL PROCESSES — OILS/FATS AND ESTERS

Triglycerides (Oils and Fats)

The feedstocks used in the production of oils and fats are oil seeds and animals, respectively.

Oils are produced when dry seeds are crushed, heated and subjected to high pressure, which squeezes the oil from the ruptured cell walls of the seed. The yield of oil from seeds is usually between 20 and 50 per cent of dry seed weight and the residue (meal) can be used as a stock food. The processing plant for oil seeds produces high-yield marketable products in oil and meal. The plant does not have to be sized to cope with large flows of water.

Fats are produced from the trimmings of animal carcasses and the rendering of waste animal tissue. The rendering process involves boiling the waste tissue, followed by the collection of the melted fat from the top of the resultant mixture. Plant for this process must be sized to handle large volumes of water, but the marketable co-products of meat meal, bone meal and/or blood meal are also produced.

Esters

Both methyl and ethyl esters can readily be produced from fats and oils. The esterification process is a simple chemical reaction which is carried out by using a surplus of the esterifying alcohol. The surplus alcohol is recovered by distillation. Ester yield is usually more than 95 per cent based on oil or fat input. The plant must be sized to cope with the surplus alcohol, which is recycled.

INHERENT DRAWBACKS OF BIOMASS

Density of Occurrence

The growth of plants is controlled by soil fertility, temperature, moisture, incident energy and the ability of the plant to utilise these inputs to produce dry matter (DM). In general, these criteria limit the production of biomass to 10–20 tonnes DM per hectare per year, with a typical yield in New Zealand of 10–12 tonnes DM per hectare per year. For a large processing plant, this low density results in crops having to be collected from a wide area. The cost of delivery to the processing plant therefore rises in proportion to the distance travelled for crop collection.

Dry Matter (DM) Content

The DM content of biomass, which in the context of this chapter means that portion of the feedstock which is usable, varies from about 10 per cent (green crops, fruit) to 50 per cent (trees) and from 75 per cent (ripe grain) to 100 per cent (fats and oils). In general, transport and processing costs are high in relation to the proportion of the product which is available for use.

Some of the processes described earlier require high DM content of their feedstock for efficient processing. If the biomass is not available at the required DM content, additional processing plant must be provided for drying it prior to use. The result is an increase in energy costs.

Biomass Market Values

The traditional markets for processed biomass are seasonal, and include food, clothing, shelter, heat and quality of life products (soaps, cosmetics, etc.). The values that can be assigned to processed biomass depend upon: its suitability for the market in question (e.g. food or fibre); the quality of biomass for that market; and the supply of biomass to the market.

Since the grower of biomass will aim to market his product to get the maximum return, he may choose different markets for his product each time he sells. The price he gets will be affected by seasonal variation, by product quality and by market demand.

In contrast, the owner of a plant which is processing transport fuels from biomass needs a guaranteed supply of feedstock over a period of perhaps 15–20 years. Market fluctuations could result in the fuel manufacturer having difficulty in buying feedstock at an economic price.

Zero-cost (Waste) Markets

Large quantities of biomass end up as waste products, often with a disposal cost attached. It is frequently assumed that these wastes have a zero or even negative value. If, however, we can find ways of utilising materials which are currently seen as waste, that waste becomes a marketable product — with a positive value. The cost of producing fuel from biomass may then be reduced.

Multiple Ownership

A processing plant for producing transport fuel from biomass will need to be sized so as to take advantage of the economies of scale. This must be balanced against the increased collection costs associated with a feedstock which may be spread over a wide area. This could make it uneconomic to use biomass on a large scale for transport fuels production.

In addition, there is a need to ensure an adequate supply of biomass for the life of the processing plant, before the plant can be financed. A large number of growers must agree to supply their product at an economic cost, over the life of the plant. They have to do this in the face of multiple market values for their product.

This factor also contributes to the argument which suggests that it may be uneconomic to use biomass for transport fuels production.

COMPARISON WITH FOSSIL RESOURCES

Processing

Fossil resources such as natural gas, petroleum, coal and lignite are generally processed to the desired products by means of thermal processes. Because most thermal processes can be operated on a large scale, the unit cost of a product from fossil resources is minimised. By way of contrast, biomass conversion processes currently available generally use feedstocks with a low content of usable input. As a consequence of this, a large plant is required for a relatively small product output.

Fossil feedstocks such as some natural gases, petroleum and some coals are essentially totally usable, and consequently require no surplus materials handling capacity to cope with unusable inputs.

Occurrence

The collection costs for a fossil feedstock will be substantially smaller than for biomass, because of the density of occurrence of fossil resources. A 1-m-thick coal seam, for example, will produce about 10 000 tonnes DM/ha, which is a thousandfold increase over that of typical biomass.

Energy Market Values

Fossil resources which can be used for the production of transport fuels will have two market values — one as an existing source of thermal heat and one as a feedstock for a transport fuels processing plant. Furthermore, unlike biomass, these values do not fluctuate greatly with seasonal changes, or with the volume of the resource available.

Ownership

The costs of finding and producing fossil resources are generally large, and the necessary funds are often provided by a single owner — usually a company or a government. Consequently, the single ownership of fossil resources assists both the provision of feedstock and the arrangements for funding a processing plant. The reverse situation is, however, usually the case for biomass resources.

CRISIS IMPLEMENTATION

Currently biomass feedstocks suitable for transport fuels production are readily available in the form of surpluses and wastes. The volume of this biomass is small in relation to total biomass production and insignificant in terms of transport fuel replacement. It is, nonetheless, available.

In a crisis scenario our values would change. Ethanol is one of the fuels which is most easily derived from biomass and which could be used to extend petrol supplies. Similarly, fuels to replace or extend diesel could be most easily produced from vegetable oils and/or esters. Both of these fuel types can be produced from simple processing plants which can be set up rapidly and start producing fuel after one growing season.

This is in marked contrast to the sequence of exploration and discovery through to the production of fossil fuels. For example, in the case of the Kapuni and Maui gas fields the time lag was 10 years. In a crisis situation the speed with which a fuel can be produced is all-important. Biomass would then have a time production advantage over other feedstocks.

THE MARKET REALITIES

As has been stated earlier, it is clear that there are a number of fundamental disadvantages to the large-scale use of biomass for the production of alternative transport fuels. However, there are at least two countries producing and selling fuel derived from biomass, and a third country is planning a similar programme.

Brazil, United States and Canada

Ethanol production in Brazil is based on the need to conserve overseas exchange. At the same time it provides a use and therefore gives value to the surplus sugar crop. The ethanol is alleged to be heavily subsidised and is sold at the energy equivalent price for petrol.

The *Gasohol* programme in the United States is encouraged by a strong farm lobby which pushes for continuing agricultural production to provide adequate farm incomes. Much of this production is surplus to food market requirements. Gasohol is petrol blended with 10 per cent ethanol. The economics of the ethanol production depend on the production of a high-value by-product (distillers' dried grains) as well as the relaxation of Federal and State taxes. Without these factors the price of ethanol would not be competitive.

The possible use of wood to produce alcohol fuels in Canada relies on the need to replace large tracts of very poor (degraded) forest with good-quality forest. It is planned to use the wood cleared from this poor forest as biomass feedstock. This is an example of a low-valued resource being utilised to pay for its own replacement by a more valuable resource[3].

It should be noted that in all three countries special circumstances or concessions exist which allow, or will assist, the fuel produced from biomass to be sold at a price which is competitive with that of petrol. These special circumstances and/or concessions appear to be needed to mitigate the previously described inherent drawbacks of biomass as a transport fuel.

Tallow Esters

The mooted use of tallow methyl esters as a component of the New Zealand diesel pool is the only case where the normal disadvantages of biomass do not apply. Tallow is automatically collected at meat processing works when animals are slaughtered. The primary reason for the existence of meat works is to process meat. The collection of tallow is therefore of secondary importance and its collection cost is covered by that of the meat.

Unlike other biomass sources, tallow has limited markets — food, chemicals, soap or fuel. It also has few owners — the meat processing companies. In addition, the by-product, glycerol, could meet all the costs of converting the tallow to tallow esters[4].

All these factors combine to suggest that in New Zealand, at least, the production of tallow esters to be used as diesel fuel could be both feasible and economic. It is possible that if a biomass resource elsewhere in the world had the same advantages as tallow, the production of biomass specifically for fuel could be economic.

CONCLUSIONS

The idea of using renewable biomass as a feedstock for alternative fuels production processes is unlikely to become a reality in a normal market scenario without either government subsidy or some form of tax incentive. This situation arises because biomass feedstocks are inherently disadvantaged when compared with fossil feedstocks, owing to their occurrence, handling costs, multiple markets and multiple ownership.

The proposed New Zealand tallow ester extension of diesel may be feasible, because the discussed disadvantages of biomass do not, in general, apply to the tallow resource.

The era of fossil fuels is not over and fossil resources will be paramount in the medium to long term.

Biomass-derived transport fuels could have an important role in a fuels crisis situation, because they can be implemented rapidly.

REFERENCES

1. Bechtel Pacific Corporation (1985). *Liquid Fuels by Wood Conversion* (Report to the Liquid Fuels Trust Board)
2. Gorton, C. W., Georgia Institute of Technology (1983). Personal communication
3. Gravel, G., Biosyn, Montreal (1984). Personal communication
4. Liquid Fuels Trust Board (1985). *World Glycerol Markets: Impact of the Development of Large-scale Ester-diesel Fuel Business* (Report No. 2041, Wellington)

32

APPLICATIONS OF MICROBIAL ELECTROCHEMISTRY

H. Peter Bennetto, Gerard M. Delaney, Jeremy R. Mason, Sibel D. Roller, John L. Stirling and Christopher F. Thurston

INTRODUCTION

Although electrical interactions of living cells have been known since Galvani's time, the advantages of interfacing electrotechnology with biology have not yet been fully appreciated. The workings of the cell involve many types of charge-transfer process (e.g. biochemical 'redox' reactions, ion-transport and diffusion), and in some senses the micro-organism can be regarded as an electrochemical machine[2] which, if interfaced with an electrochemical system, can be manipulated by 'taking' electrons from the biosystem or 'pushing' them back in. Since electrochemistry is itself becoming increasingly versatile and is ideally suited to 'fine tuning' by computer control, it is to be expected that such application will broaden the scope of process-control, products and devices available to biotechnology. Some likely areas for exploitation of electron transduction from micro-organisms are summarised below.

Of the various applications listed in Table 1, 'biosensors' has received the lion's share of attention, mainly because of the suitability of sensing devices for rapid development and commercial exploitation. In this chapter we describe recent studies in which micro-organisms were used for sensing glucose, but an illustrative account of the bioelectrochemical background is also given to draw attention to other fruitful areas of development.

Table 1 Electron and energy transduction in microbial systems; applications

Electricity generation:	Bio-fuel cells, Bio-batteries, Biosensors
Biomass conversion:	Treatment of wastes and effluents
Bio-electrosynthesis:	Electrochemically assisted microbial products

WHOLE CELLS AS CATALYSTS FOR BIOSENSORS; DIRECT AND INDIRECT WHOLE-CELL SENSORS

Microbial biosensors follow in the wake of enzyme sensors[1,21], but have many advantages for a very wide range of substrates[14,21]. Most of the whole-cell sensors so far proposed work in an 'indirect' mode — i.e. in conjunction with a conventional sensing element such as a pH or an oxygen probe. The alternative method described here involves a 'direct' transduction of the signal which is coupled to the substrate-dependent production of electrons by the organism. Transfer of electrons is usually effected with the aid of a 'redox' mediator, which is first reduced by the biocatalyst and is then reoxidised at the working electrode. The merits of these two approaches in relation to enzyme electrodes are discussed more fully elsewhere[3]. One important difference results from the nature of substrate oxidation in the cells, which provides reducing intermediates from several degradative stages, and in principle can provide a bigger electrochemical signal than a single-enzyme reaction. For example, in one type of 'direct' glucose sensor[11] the oxidation of glucose to gluconolactone using glucose oxidase yields two electrons per molecule of substrate:

$$\beta\text{-D-glucose} = \text{D-gluconolactone} + 2H^+ + 2e^-$$

In contrast, the complete oxidation of glucose in a whole cell may be represented by the equation

$$C_6H_{12}O_6 + 6H_2O = 6CO_2 + 24H^+ + 24e^-$$

Recent evidence shows that a substantial proportion of this large electron yield may be realised in practice[15,22].

ELECTRON TRANSDUCTION FROM ENZYMES AND MICRO-ORGANISMS

It is interesting to examine the intrinsic differences between the mechanisms of redox-mediated generation of signals from enzymes and whole cells (Figure 1; substrate not shown). Electrons may be taken from reduced

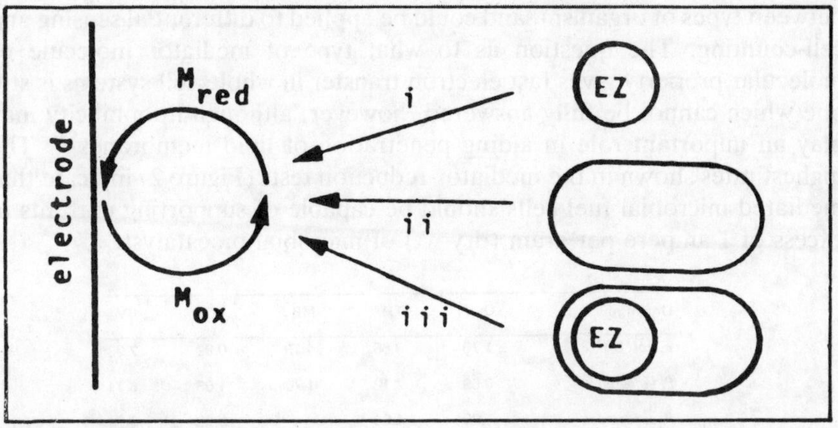

Figure 1 Electron transduction from enzymes and whole cells. Process i represents mediated electron transfer from an enzyme; ii, transfer from a micro-organism; iii, transfer from a particular intracellular enzyme or reduced intermediate

enzymes at a donor site, or a limited number of sites, provided that there is relatively free access for the mediator (process i). Electron transduction from a whole micro-organism, represented by process ii, presents different problems, however, because access of a mediator to sources of reducing power within an organism is restricted by the cell walls and membrane, and these sources may be varied in number and location (enzymes, pyridine nucleotides, quinone intermediates, cytochromes). Since complex, interactive redox states are present, it is useful for many purposes to view the electrons as residing in a 'reducing pool' within the cytoplasm. But in principle it is also possible for a mediator to interact with a particular intracellular electron donor, as represented by process iii. Such factors also relate to applications of protein electrochemistry and electron transduction to (and from) cytochrome centres[10], and are relevant to the design and construction of electrochemical biosensors.

GENERATION OF ELECTRICITY BY MICRO-ORGANISMS; MICROBIAL FUEL AND CELL STUDIES

The crucial role of redox mediators in promoting transfer of charge from inside the organism has been elucidated by studies of mediator reduction, a process which will affect the efficiency and rapidity of signal transduction in direct biosensors. The considerable variation for organisms and mediators is illustrated by a selection of data in Figure 2 taken from recent work[19]. Such differences in reactivity might be used to differentiate between types of organism, and could be applied to differential sensing and cell-counting. The question as to what type of mediator molecule or molecular property gives fast electron transfer in whole-cell systems is still one which cannot be fully answered, however, although lipophilicity may play an important role in aiding penetration of lipid membranes[5,6]. The highest rates shown in the mediator-reduction tests (Figure 2) indicate that mediated microbial fuel cells should be capable of supporting currents in excess of 1 ampere per gram (dry wt) of microbial biocatalyst.

Organism	Q_{O_2}	TH	MB	BCB	BV
E. coli B/r	1.76	3.65	4.16	0.99	2.23
P. vulgaris	0.69	7.10	1.70	1.03	6.11
P. aeruginosa	1.03	1.54	–	0.63	0.31

substrate: glucose
TH = thionine; MB = methylene blue; BCB = brilliant cresyl blue; BY = benzyl viologen;
Q_{O_2} is the respiratory quotient, or rate of reduction of dioxygen. (It should be noted that the reductions involve 4, 2, 2, 2, and 1 electrons for O_2, TH, MB, BCB and BV).

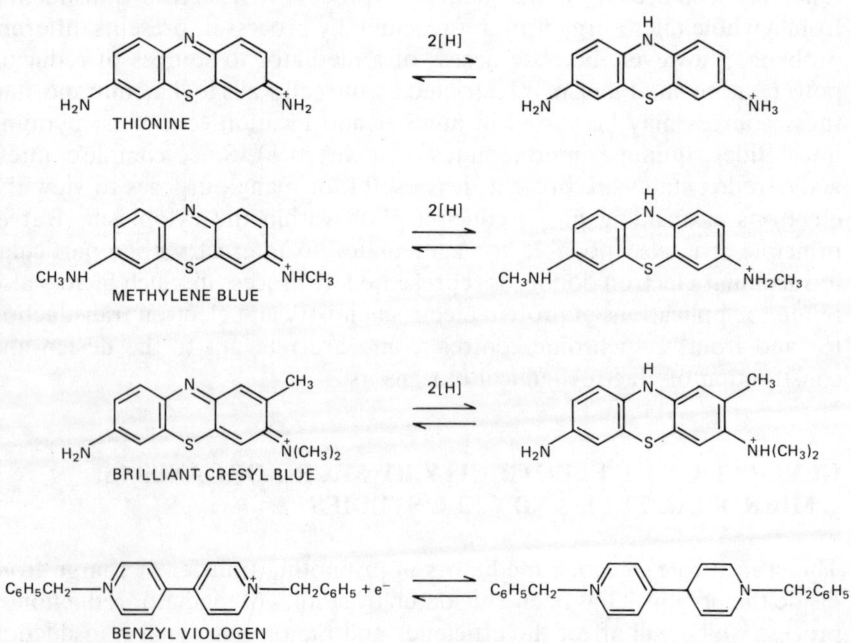

Figure 2 Rates of respiration and mediator reduction by micro-organisms (μmol g^{-1} s^{-1}, 30°C)

Microbial fuel-cell studies have demonstrated that electrons can be abstracted from micro-organisms with remarkable efficiency when a suitable mediator is included in the anolyte[2,7]. The electrochemical cell is formally represented as shown below:

The mode of operation of the microbial fuel cell[2,19] is shown in Figure 3, and the typical discharge behaviour observed and the effect produced by further addition of substrate are shown in Figure 4 for the thionine-mediated glucose/*Proteus vulgaris* cell[15]. The amount of electricity drawn from the cell across a fixed load resistance, given by the area under the graph, is proportional to the quantity of substrate added. A recent detailed analysis of this system[22] has established the fate of glucose in the anode compartment, and confirms that the coulombic yield from oxidation of glucose is about 50 per cent, while the efficiency of energy conversion is somewhat less, on account of polarisation effects which reduce the output voltage. (As expected, the yields and other cell characteristics vary for different mediators.) These results were obtained with 30 mg freely suspended bacteria, which generate a signal of milliamp or hundreds of mV proportions. Figure 4 also illustrates the rapid regeneration of power on a subsequent addition of substrate (0.5–5 min), which shows the potential for a fast-sensing device, in contrast to the rather slow response for many indirect whole-cell sensing methods (generally about 20 min)[16].

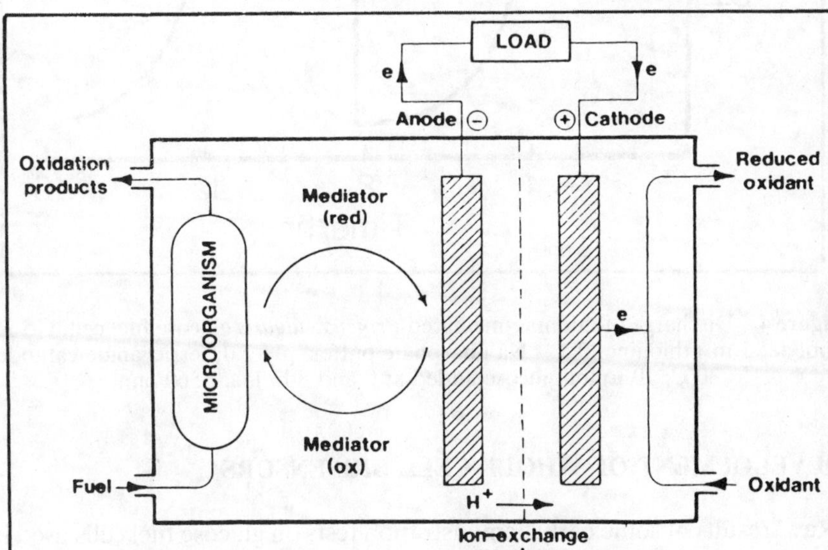

Figure 3 The microbial fuel cell: schematic representation

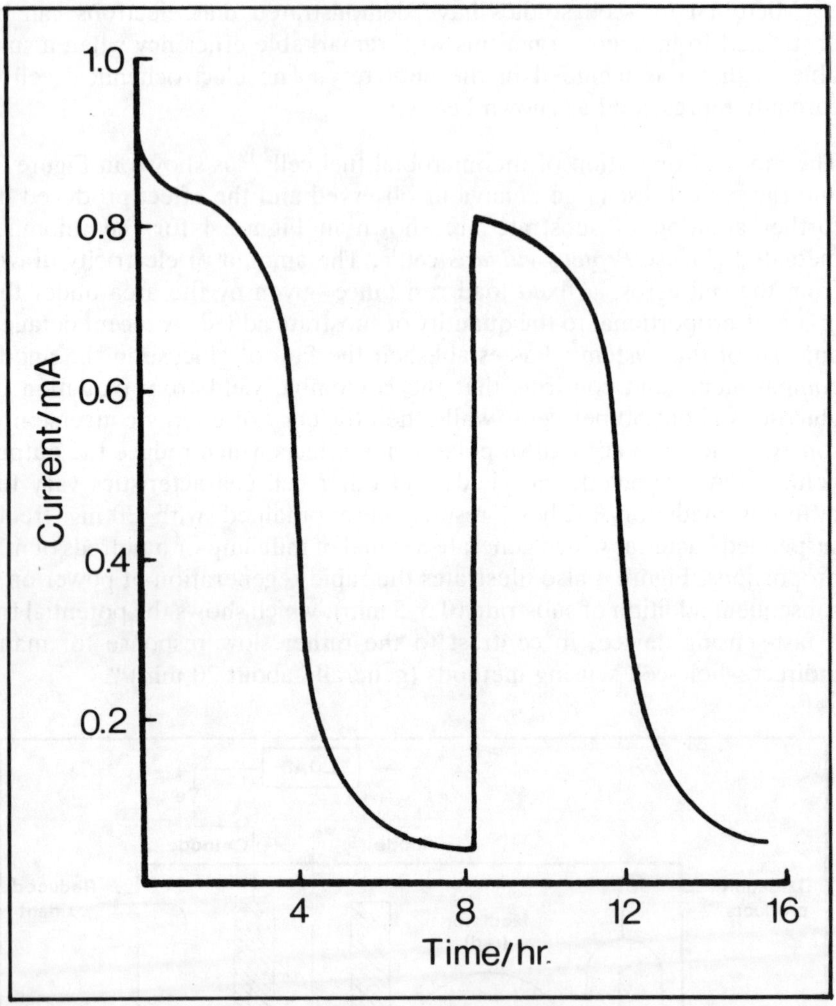

Figure 4 Discharge of thionine-mediated *Proteus vulgaris* glucose fuel cell. (15 ml anolyte; 1 mM thionine; 0.1 M Na phosphate buffer, pH 7.0; ferricyanide cathode; 30°C; 10 μmol glucose added at 0 and 8 h; load, 560 ohm

DEVELOPMENT OF WHOLE-CELL BIOSENSORS

'Raw' results of some early demonstration tests on glucose fuel cells used in a sensing mode are presented in Figure 5, which shows VDU displays of current–time plots for eight cells 'fed' with different amounts of glucose. The glucose dependence of coulombic outputs, maximum currents and rates of response were examined, and the repeatability of measurements

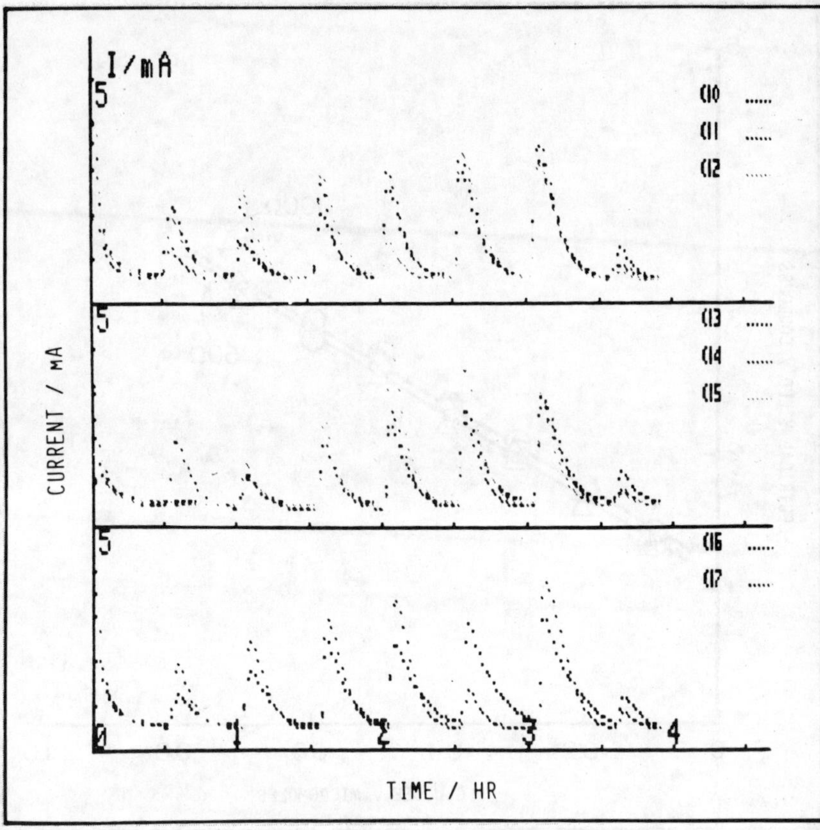

Figure 5 Computer plot of outputs from eight glucose fuel cells. Different amounts of substrate were added to the cells C10–C17. (Experimental conditions as for Figure 4 but 100 or 200 ohm load)

was tested by making further varied glucose additions (see regenerated 'bumps' in the figure). The good correlation of coulombic output with quantity of glucose added, taken from some of the plots in Figure 5, is shown in Figure 6, where an intercept results from the endogenous output of the organisms. Figure 7 shows a test of the correlation between rates of emf, increase and glucose content, taking all the results of Figure 5. Despite the scatter of points (some of which result from variations in volume, etc., in the different cells), even these relatively unrefined experiments show the potential of the method. In subsequent experiments with *P. vulgaris* (and with other organisms/substrates) such plots showed good linearity over an extended range.

Development of these systems as direct amperometric sensor probes is under way in a number of laboratories. The main design modifications

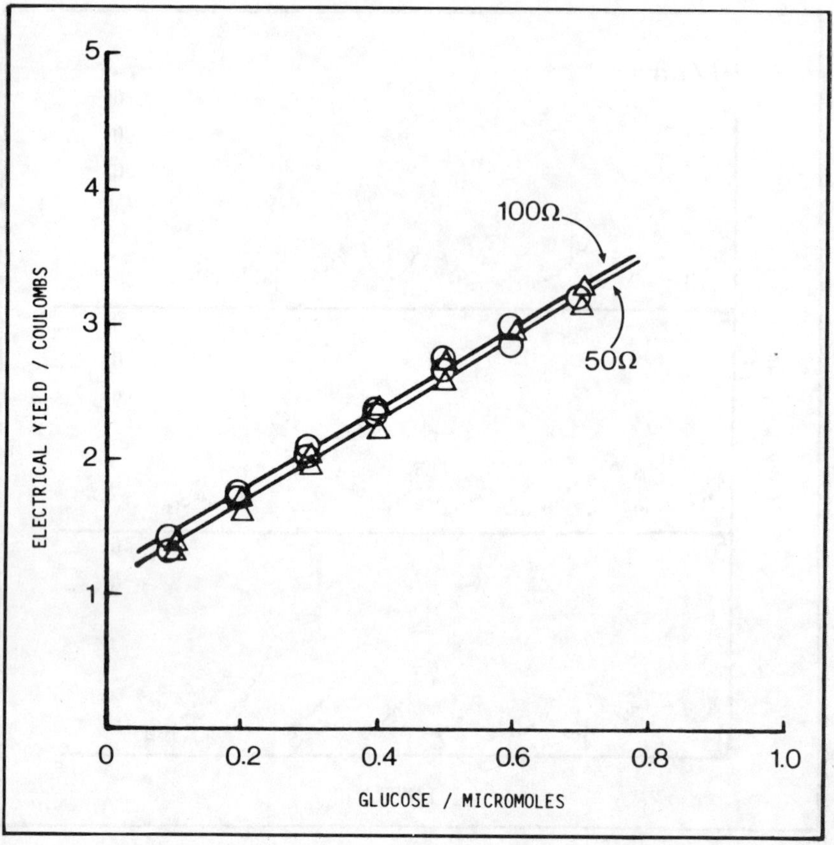

Figure 6 Dependence of fuel cell output on glucose added (coulombic yields were calculated from areas under the curves shown in Figure 5)

required are miniaturisation, and immobilisation of organisms and other components by absorption, physical entrapment or covalent attachment[12,23] to give a convenient format and improved stability. There are numerous ways of 'tailoring' the electrode surface to make it compatible with the organisms while at the same time allowing for facile electron transfer (e.g. with the use of redox polymers[18]). Problems of selectivity and interference can be tackled by changing or modifying the mediator and other components, by adjusting the experimental conditions (e.g. load, potential of the working anode), by inclusion of filtering layers[20], and through the use of mutants or genetically modified organisms.

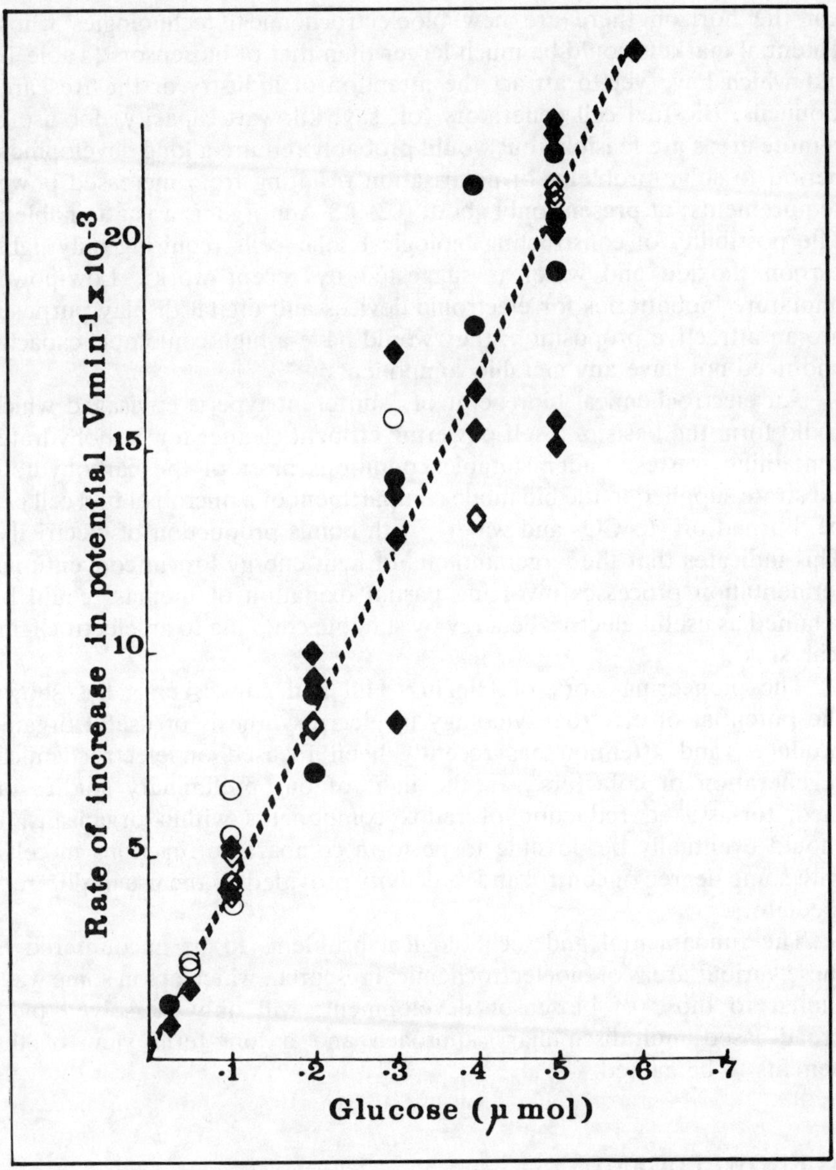

Figure 7 Rate of voltage change on adding glucose as a function of glucose added to cell. (Initial rates taken over a 1 min interval. Sets of points from 4 series of glucose additions to different cells)

OTHER APPLICATIONS OF BIOELECTRODE TECHNOLOGY

On the horizon there are new bioelectrochemical technologies whose potential markets could be much larger than that of biosensors (Table 1), but which have yet to attract the attention of industry or the Research Councils. Bio-fuel cell generators (of, say, kilowatt capacity) for use in remote areas are feasible, but would probably require a long development period to solve problems of polarisation resulting from increased power requirements; at present only about 0.2–0.3 A/g organism is attainable[15]. The possibility of constructing biological solar cells requiring only light, carbon dioxide and water is suggested by recent work[8]. Low-power miniature biobatteries for electronic devices and digital display purposes are an attractive proposition; they would have a high coulombic capacity and need not have any metallic components[2].

An electrochemical bioreactor of a different type is envisaged which could form the basis of a self-powering effluent cleaner for carbohydrate-containing wastes: under suitable conditions most of the carbohydrate substrate supplied to the bioanode compartment of a microbial fuel cell can be 'burned off' to CO_2 and water[4], with bonus production of electricity. This indicates that the large amounts of heat energy lost in conventional fermentation processes involving partial oxidation of biomass could be retained as useful electrical energy by suitable coupling to an electrochemical 'sink'.

The pioneering work of Higgins, Hill and co-workers[17] has shown the potential of electroenzymology in electrosynthesis of useful organic products, and attention has recently been focused on electrochemical regeneration of cofactors[9]. In the light of our preliminary studies on mediator-assisted reduction of redox components within organisms, it should eventually be possible to perform comparable reactions in cells, with some degree of control and selectivity provided by the use of different mediators.

The fundamental and technological problems to be encountered in these various areas of bioelectrochemical research, which are in some ways similar to those of biosensor development, will only be solved by a broad-based multidisciplinary approach and a long-term view of the benefits to be gained.

ACKNOWLEDGEMENTS

We thank the SERC (Biotechnology Directorate), who supported the microbial fuel cells project, and Cambridge Life Sciences for supporting (in part) work on whole-cell biosensors. We thank Johnson Matthey Research for loan of electrode materials.

REFERENCES

1. Aston, W. J. and Turner, A. P. F. (1984). In *Biotechnology and Genetic Engineering Reviews* (Ed. G. Russell), Vol. 1, pp.89–120. Intercept, Newcastle-upon-Tyne
2. Bennetto, H. P. (1984). In *Life Chemistry Reports* (Eds. A. M. Michelson and J. V. Bannister), Vol.2, No.4, pp.363–453. Harwood Academic, London
3. Bennetto, H. P., Box, J., Delaney, G. M., Mason, J. R., Roller, S. D., Stirling, J. L. and Thurston, C. F. (1986). In *Biosensors: Fundamentals and Applications* (Eds. A. P. F. Turner, I. Karube, and G. S. Wilson) Oxford University Press (in press)
4. Bennetto, H. P., Delaney, G. M., Mason, J. R., Roller, S. D., Stirling, J. L. and Thurston, C. F. (1986). *Biotechnol. Letters,* **7**, 699–705. (1985). UK Patent Application No. 2165087A; US Patent Application Serial No. 755779
5. Bennetto, H. P., Dew, M. E., Stirling, J. L. and Tanaka, K. (1981). *Chem. Ind.,* 776–778
6. Bennetto, H. P., Stirling, J. L. and Tanaka, K. (1985). *Chem. Ind.,* 695–697
7. Bennetto, H. P., Stirling, J. L., Tanaka, K. and Vega, C. A. (1980). *Soc. Gen. Microbiol. Quarterly,* **8**, 37. (1983). *Biotechnol. Bioeng.,* **25**, 559–568
8. Bennetto, H. P., Tanaka, K. and Matsuda, K. (1984). In *Charge and Field Effects in Biosystems* (Eds. M. J. Allen and P. N. R. Usherwood). Abacus Press, Tunbridge Wells, pp.515–522
9. Bowen, R. and Pugh, S. (1985). *Chem. Ind.,* 323–326; Allen, P. M. and Bowen, W. R. (1985). *Trends Biotechnol.,* **3**, 145–149
10. Cass, A. E. G. (1984). In *Life Chemistry Reports* (Eds. A. M. Michelson and J. V. Bannister), Vol.2, No.4, pp.321–362. Harwood Academic, London
11. Cass, A. E. G., Davis, G., Francis, G. D., Hill, H. A. O., Aston, W. J., Higgins, I. J., Plotkin, E. V., Scott, L. D. L. and Turner, A. P. F. (1984). *Analyt. Chem.,* **56**, 667–673
12. Chibata, I. and Wingard, L. B. Jr. (1983). *Applied Biochemistry and Bioengineering.* Vol.4: *Immobilised Cells.* Academic Press, London
13. Clark, D. J., Calder, M. R., Carr, R. J. G., Blake-Coleman, B. C., Moody, S. C. and Collinge, T. A. (1985). *Biosensors,* **1**, 213–320
14. Corcoran, C. A. and Rechnitz, G. A. (1985). *Trends Biotechnol.,* **3**, 92–96
15. Delaney, G. M., Bennetto, H. P., Mason, J. R., Roller, S. D., Stirling, J. L. and Thurston, C. F. (1984). *J. Chem. Tech. Biotechnol,* **34B**, 13–27
16. Guilbault, G. G. (1984). *Analytical Uses of Immobilised Enzymes.* Marcel Dekker, New York, pp.211–326
17. Higgins, I. J. and Hill, H. A. O. (1979). In *Microbial Technology* (Eds. A. T. Bull, D. C. Ellwood and C. Ratledge), Gen. Microb. Symp. No. 29, pp.359–377. Cambridge University Press
18. Murray, R. W. (1980). *Accounts Chem. Res.,* **13**, 135–141. (1984). In *Electroanalytical Chemistry* (Ed. A. J. Bard). Vol.13, pp.191–368. Marcel Dekker, New York
19. Roller, S. D., Bennetto, H. P., Delaney, G. M., Mason, J. R., Stirling, J. L. and Thurston, C. F. (1984). *J. Chem. Tech. Biotechnol.,* **34B**, 3–12
20. Scheller, F. W., Schubert, F., Renneberg, R. and Miller, H. G. (1985). *Biosensors,* **1**, 135–160

21. Suzuki, S. and Karube, I. (1983). In *Applied Biochemistry and Bioengineering* (Eds. I. Chibata and L. B. Wingard), Vol.4, Academic Press, London, pp.281–310
22. Thurston, C. F., Bennetto, H. P., Delaney, G. M., Mason, J. R., Roller, S. D. and Stirling, J. L. (1985). *J. Gen. Microbiol.*, **131**, 1393–1401
23. Wiseman, A. (1985). *Handbook of Enzyme Biotechnology* (2nd edn). Ellis Horwood, Chichester

33
AN INTRODUCTION TO THE INDUSTRIAL APPLICATIONS OF BIOSENSORS

Charles F. M. Kingdon, David J. Anderton, M. Gronow and M. J. Goldfinch

INTRODUCTION

What is a biosensor? Many authors have attempted to define this potentially ubiquitous device but occasionally one is still faced with the problem of explaining, to those unfamiliar with the term, how a biosensor is different from the numerous biochemical laboratory techniques currently used to define biological activity. 'Is not a spectrophotometer a biosensor?' is often the first question asked. Essentially biosensors differ from existing techniques in three useful and fundamental ways.

(1) In that there is an intimate contact of the biological material (whether it be whole cells, organelles, antibodies or enzymes) with a transducer which converts the biological signal into a quantifiable electrical signal.

(2) In its functional size. The 'sensing' portion of the biosensor is generally small and this allows small test sample size, minimum interference with existing processes following implantation, and access to difficult and possibly dangerous environments, without interrupting process flow.

(3) In that biological material can be selected to suit the industrial needs and operate at various levels of specificity. It may be highly specific;

may be specific to a narrow band of compounds; or may exhibit spectrum specificity. A theoretical illustration of this gradation in specificity would be a sensor sensitive to a single antibiotic (for example, gentamycin), or all aminoglycosides or all antibiotics. This flexibility in choice of biological material allows the user to 'tailor' the biosensor to the industrial process.

The basic design of a biosensor is illustrated in Figure 1 and highlights the arrangement of signal generator to signal transducer, and the simple intervening membranes required to protect the transducer, the biological material and the process stream from cross-contamination. In this short

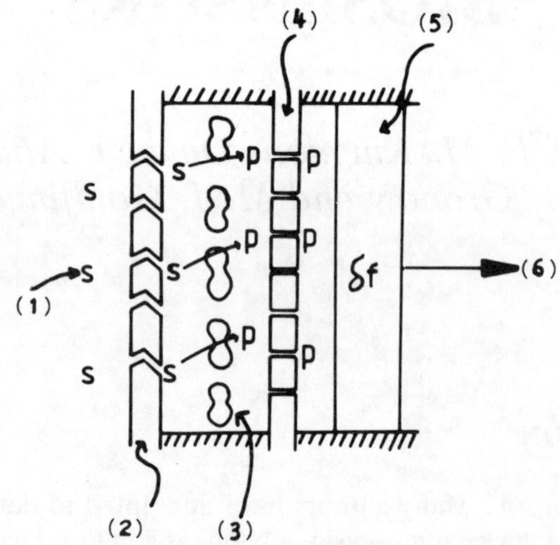

Figure 1 Schematic biosensor (s = target substance, p = product). (1) Discrete sample or process flow. (2) Outer membrane, selectivity to (s). (3) Immobilised biological material. (4) Inner membrane, selective diffusion of (p) and protects transducer. (5) Transducer. (6) Electronic signal

chapter it is not possible to discuss existing transducer technology in depth and the reader is recommended to two excellent review articles for this information[2,11]. A summary of the most common transducer systems is provided in Table 1.

A major impetus behind the development of biosensor technology has been the desire to improve existing health care. Indeed it sometimes appears that all biosensor development is aimed at the diabetic market. This effort has concentrated on the construction of simple-to-use, cheap and reliable biosensors for intended use by the patient, to assay target analyses in complex biological media such as whole blood, serum and

Table 1 A summary of the most common transducer technology used in biosensors

Transducer type	*Biological-initiated change in*
1. Fibre optic and optoelectronic	The nature of transmitted light
2. Conductance	The concentration of a free charged ionic species
3. Amperometric and potentiometric	The current or potential difference. Includes field-effect transistors (FETs), ion-selective electrodes (OISEs) and gas-sensing electrodes
4. Thermistors	The temperature of a system
5. Piezoelectric	The pressure of a system

urine. Thus, there is a wealth of practical knowledge in the construction of accurate yet simple-to-use biosensors which can be readily adapted to the industrial arena.

INDUSTRIAL BIOSENSORS

Biosensors are expected to be widely used in many industries outside of the medical field and some of these applications are listed in Table 2. To review the biosensor potential for each industry would be repetitive and, therefore, it will be divided according to the nature of the immobilised biological material.

Table 2 Major industrial markets for biosensors

1. Fermentation analysis and process control
2. Environmental control, pollution monitoring, etc.
3. Agriculture, horticulture and hydroponics
4. Industrial, military and mining detection of toxic compounds and gases

The biological material used in the construction of biosensors is usually one of the following.

(1) An enzyme or collection of enzymes.
(2) A whole-cell population. This may be a single species or of mixed species.
(3) An organelle population.
(4) Immunoglobulins.

The majority of biosensors use either whole cells or an enzyme or both as the biological signal generator. For convenience these sensors will be considered separately.

Enzyme Biosensors

There are many references to enzyme biosensors in the literature, and many of these refer to devices which could detect and quantify substances of industrial importance (see Table 3). The enzyme sensors listed in the table would also be useful in fermentation control, in the monitoring of biodeterioration, in food processing, and, in the case of an alcohol biosensor, in the detection of drink driving offences. All of the listed biosensors offer a rapid response time and relatively long stability: useful assets in maintaining strict process control and in the reduction of skilled manpower. Using the same types of approach, it is possible to envisage additional enzyme biosensors for the detection of pyruvate, carbon monoxide and formaldehyde (HCHO). The CO and HCHO biosensors will be useful in combustion control, fire detection and pollution monitoring, whereas a pyruvate biosensor would assist in fermentation control.

Table 3 Enzyme biosensors with industrial applications

Target substance (imm. enzyme)	*Transducer*	*Response time*(s)	*Stability* (days)	*Range*
Alcohol[20] (alcohol oxidase)	O_2	30 s	120	5–10^3 mg/dm^3
Sucrose[7] (invertase or mutarotase)	O_2	360 s	14	$2 \times 10^{-3} - 10^{-2}$M
Glucose[1] (glucose oxidase)	O_2	—	21	0.5–4% w/v
Cholesterol[8]	Pt	120 s	~30	$10^{-2} - 3 \times 10^{-5}$ M
Lactate[9]	Pt[$Fe(CN_6)^{4-}$]	200–600 s	~ 7	$2 \times 10^{-3} - 10^{-4}$M

Microbial Biosensors

Microbial biosensors are primarily intended for industrial process monitoring and environmental control. The potential for on-line measurement and control of fermentation processes has been realised; biosensors will monitor both substrates and products on-line and reduce the number of time-consuming and labour-intensive off-line analyses. A selection of

Table 4 Potential microbial biosensors for fermentation and process control

Sensor for	*Microbe*	*Response time* (min)	*Trans-ducer*	*Stability* (days)	*Range*
Assimilable sugars[8]	*Brevibacterium lacto-fermentum*	1–10	O_2	10	Linear above 1 mM
Acetate$_3$	*Trichosporon brassicae*	6–10	O_2	21	Linear up to 22.5 mg/1
Cephalosporin	*Citrobacter freundii*	10	pH	28	Below 50 μg/cm^3
Ethanol[4]	*Trichosporon brassicae*	10	o_2	21	Below 22.5 mg/l
Formate[15]	*Clostridium butyricum*	20	Fuel cell	20	Upt to 1 g/l
Glucose	*Pseudomonas fluorescens*	15–20	O_2	14	Minimum 2 mg/l

Table 5 Potential microbial sensors for environmental control and gas monitoring

Sensor for	*Microbe*	*Respons time* (min)	*Trans-ducer*	*Stability* (days)	*Range*
BOD[6]	*Trichosporon cutaneum*	30	O_2	17	1–41 mg/1
Ammonia + NO^{2-9}	Mixed culture nitrifying bacteria	4	O_2	10	0.1–42 mg/1
Nitrate[12] Nitrite hydroxylamine	*Azobacter vinelandii*	5–10	NH_3	14	10^{-5}–10^{-4} 8 × 10 M
Methane[16]	*Methylomonas flagellata*	1	O_2	20	Up to 6.6 mM
Mutagen screening[7]	*Bacillus subtilis*	90–120	O_2	—	1.6 µg/cm

microbial biosensors which are being developed for fermentation constituents is given in Table 4. A similar list of microbial biosensors for the detection of environmentally important and toxic gases and compounds is given in Table 5.

ADVANTAGES OF INDUSTRIAL BIOSENSORS

The use of biosensors in industrial processes will benefit the manufacturer in many ways:

(1) A biosensor may be compatible with both on-line assay and discrete sampling.
(2) It provides the opportunity to give a rapid response and, therefore, improved feedback control.
(3) It will not interfere with the process stream.
(4) A biosensor has a potential lifetime of days, sometimes weeks, which releases technical staff for other duties.
(5) It facilitates rapid sampling and rejection of below-standard raw materials on delivery.
(6) It will provide low-cost monitoring of stored products and raw materials.
(7) It will give access to remote environments.
(8) Biosensors can be made relatively inexpensive.

THE FUTURE

This chapter mentions many of the biosensors which have been conceived to meet some of the more predominant industrial processes. However, we may find biosensors being used in: food hygiene; detection of CO, CO_2 and CH_4; detection and control of Na,P, K, Ca, Fe and Mg for hydroponic farming; and milk quality control.

As an example, work carried out in collaboration with the Biochemistry Department, University of Cambridge, has resulted in a simple amperometric biosensor for the detection of antibiotics in complex solutions[9]. The adaptation and development of this idea with the appropriate industrial interest will culminate in a biosensor for the rapid detection of antibiotics in milk.

Cambridge Life Sciences plc are involved in the support of research and development of biosensor devices for both industrial and health care applications in their own and external laboratories. It is a policy of the company not only to market selected products which result from these activities, but also to offer facilities to carry out contract work to develop biosensor products for clients with specific problems or who wish to extend their product range.

REFERENCES

1. Aston, W. J. and Turner, A. P. F. (1984). *Biotech. Gen. Eng. Rev,* 1 February, 89
2. Gronow, M. (1984). *TIBS,* August, 336
3. Hikuma, M., Kubo, T., Yasuda, T., Karube, I. and Suzuki, S. (1979a). *Analyt. Chim. Acta,* **109**, 33
4. Hikuma, M., Kubo, T., Yasuda, T., Karube, I. and Suzuki, S. (1979b). *Biotechnol. Bioeng.,* **21**, 1845
5. Hikuma, M., Suzuki, H., Yasuda, T., Karube, I. and Suzuki, S. (1979). *Eur. J. Appl. Microb. Biotechnol.,* **8**, 289
6. Karube, I., Matsunaga, T., Nakahara, T., Suzuki, S. and Kada, T. (1981). *Analyt. Chem.,* **53**, 1024
7. Karube, I., Mitsuda, S. and Suzuki, S. (1979). *Biotechnology,* **7**, 343
8. Karube, I., Okada, T. and Suzuki, S. (1981). *Analyt. Chem.,* **53**, 1852
9. Kingdon, C. F. M. (1984). Poster presentation, Churchill College, Cambridge
10. Kobos, R. K., Rice, D. J. and Flourney, D. S. (1979). *Analyt. Chem.,* **51**, 1122
11. Lowe, C. R. (1985). *Biosensors,* **1**, 3
12. Matsumoto, K., Seijo, H. and Watanabe, T. (1979). *Analyt. Chim. Acta,* **105,** 429
13. Matsunaga, T., Karube, I. and Suzuki, S. (1980). *Biotechnology,* **10**, 235
14. Okada, T., Karube, I. and Suzuki, S. (1981). *Eur. J. Appl. Microb. Biotechnol.* **12**, 102
15. Satah, I., Karube, I. and Suzuki, S. (1976). *Biotechnol. Bioeng.,* **18**, 269
16. Satah, I., Karube, I. and Suzuki, S. (1977). *Biotechnol. Bioeng.,* **19**, 1095
17. Shinho, T., Sugiura, M. and Kamo, N. (1970). *Analyt. Chem.,* **42**, 118
18. Yao, T. and Musha, S. (1979). *Analyt. Chim. Acta,* **110,** 203

34
IMMUNOASSAY KITS FOR FOOD ANALYSIS

J. C. Allen

INTRODUCTION

Although immunoassays (IA) have been in use in clinical diagnostics for some 20 years, they are only now making an impact on food analysis and in public analysts' laboratories. The reasons for this are several. The science of immunology grew up within a clinical environment, and clinicians were the first to appreciate its therapeutic and diagnostic potential. Immunologists were able to meet the need of the diagnostician to selectively detect the presence of minute quantities of an agent in a patient's circulation. And food analysts, being mostly chemists by training, were not generally aware of the development of IA and of its potential in their area.

There is a further reason. Clinical samples, such as blood, urine or cerebrospinal fluid, all have a known history and are universal in character, whereas food matrices are an order of magnitude more complex, ranging from liquids and pastes to solids and powders. The components of a food may have been processed by a variety of heat treatments or extractive procedures. The food may have been stored frozen or it may have been stored sterilised in a can at 120 °C. All this suggests that it might be exceedingly difficult, if not impossible, to apply IA to routine food analysis.

The answer is that it is by no means impossible to use IA for the analysis of foods, and certain assays, particularly those based on enzyme-linked immunoabsorbent systems (ELISA), may well have considerable applications in the food industry.

Our own group at The North East Wales Institute is heavily engaged on a joint SERC Biotechnology Directorate/industrially supported programme in this area. As a result of this research, a new Grand Metropolitan Biotechnology company, Biokits, has now been established on the Institute-linked Newtech Science Park in North Wales. This successful example of academic/industrial co-operation in biotechnology has led to the market launch of a range of food IA kits, including assays for soya protein in meat products[3], meat species detection and ocluratoxin. A range of other lists is in the development stage.

In order to assess the future of applying IA to the food industry, let us first take a critical look at the needs of the food analyst and public analyst, in terms of their requirements for *routine* as opposed to non-routine or research techniques.

REQUIREMENTS FOR THE FOOD ANALYST

Ideally, routine analyses should have the following qualities.

Cheap

Academics or researchers often overlook the sheer cost of quality control (QC), which can become a real burden to the food industry. Such costs will depend on: the level of skill needed, i.e. the cost of staff; the time taken for the analysis; the cost of equipment; the throughput (for instance, it is only possible to put one sample on an expensive gas chromatograph at any one time); and the cost of consumables.

IA can be relatively inexpensive compared with conventional chemical analyses. Recent developments in ELISA systems obviate the need for the expensive scintillation counter and radioisotopes required for radioimmunoassays. (Radioactive materials are unwelcome in food factories, anyway.) Although the cost of production and testing of a good antiserum —which is the key to a good assay system—is high, so little is used in each assay that the cost per test can be quite low.

Easy

This also has a bearing on cost. The easier the analysis, the more reliable, but usually the less specific it becomes. However, ELISAs are both easy to perform *and* specific. The procedures involved are so routine and straightforward that relatively unskilled personnel can perform them reliably.

Rapid

The quicker the test, the greater the throughput. It is worth remembering that although instrument manuals will often state that GLC or HPLC analyses require only 20–30 min to perform, there is often a prior 'work-up' procedure of 24 h or more, which may be multistage and thus require attention from a skilled technician as well as being lengthy. It is tempting to compare this with the rapidity of IA in clinical diagnosis, where a whole ELISA plate of 96 tests can be completed in an hour or so, starting with the initial serum sample. However, it would be wrong to assume that all foodstuffs can be so readily analysed. It would appear that the problem is not nearly so intractable as it has been made out to be, but it seems likely that the performance of food IA will greatly depend on careful consideration and development of the correct 'sample work-up' procedures.

Accurate and Reproducible

Accuracy is often overemphasised: it is not always necessary to measure to 1 part in 10 000. Provided that the accuracy is sufficient for the purpose, sensitivity and selectivity can be far more important. Companies supplying clinical diagnostic kits have their own inbuilt QC schemes with standards being supplied in each kit. In this way, regular checks on both intra- and inter-laboratory values can be made.

At a meeting on IA in Food Analysis at the Food Research Institute, Norwich, it was suggested by some that food IA might well have inherent intra-laboratory variations of 10–20 per cent and inter-laboratory variations of up to 50 per cent. This would be regarded as quite intolerable by clinical diagnostic firms, who would regard a 5–10 per cent variation as normal, and it should be deemed equally unacceptable in food analysis. Our own experience suggests that the 5 per cent limit is not going to be impossible to achieve, but an enormous amount of development work is necessary to achieve this, in terms of both refining sample 'work-up' and development of a robust, reliable kit.

Sensitive and Selective

The sensitivity required depends on the purpose of the analysis. Selectivity is equally important. Is the required analysis for total protein or just whey protein; total aflatoxin or just aflatoxin M? ELISA can be superbly sensitive and selective.

Meaningful

Food analysis is beset by a vast array of 'factors'. These are used, for example, to convert Kjeldahl N to protein, lean meat from total meat, etc. To accept such a variety of such empirically determined factors, when results of litigation may depend on them, seems inappropriate. IAs, since they are so specific, are inherently less ambiguous.

THE KEY FEATURE OF IMMUNOASSAY KITS

The feature which distinguishes IA kits from other modern techniques of analysis is that in IA the high technology is built into the molecules and not into the apparatus. This has an important corollary for the practising analyst, which is that the person who actually conducts the test — or, indeed, even the supervisor — need know nothing about immunology, IA principles or antibody production. This feature is well understood by senior pathologists, but is not generally appreciated by food analysts or public analysts.

CURRENT APPLICATIONS

Unless one includes the precipitating antisera which have been available for meat species identification for some years (and which use a relatively crude procedure), routine IAs for food analyses have only been available for the past year or so. However, several prototype assays had been described, including the soya bean protein assay of Professor Hitchcock and his team at Unilever, which has now undergone 'ring tests', with encouraging results, both in the UK and in Europe[1].

The group at the Food Research Institute are also active and are principally concentrating on food toxin analysis. Their ochratoxin assay, which we have now developed in collaboration as a commercial kit, is well-researched and has great potential[2]. Certainly, it has enormous advantages over the cumbersome and lengthy classical procedure.

In North Wales we are developing several lines. We have already reported our casein assay at a recent meeting at The National Institute for Research in Dairying. It is accurate and rapid and can be used simply on a diluted milk sample. It may also be used to determine the presence of undeclared casein added to meat, as an extender — i.e. to 'boost' the protein N value.

However, the potential applications of IA to foods stretch far wider, and include meat species, vitamins, emulsifiers, antioxidants, antibiotics, sweeteners, etc., although many man-years will be necessary to develop robust kits for all these.

It seems likely that the principal initial applications for IA kits will be:

(1) High-throughput QC — e.g. in creameries, or checking grain, etc.

(2) Public analysis, port health authorities and trading standards officers. Here it might be appropriate to develop simple qualitative, 'litmus paper' tests.

Whether such kits will find the same acceptance as in medicine will, of course, depend on their reliability, robustness — and price.

FUTURE DEVELOPMENTS

Conventionally produced antibodies are likely to be used at first, and they may be inherently more suitable than monoclonals for some purposes. This may be particularly true for assay of heat-treated foods, where the presence of a variety of antigens may be needed to adequately detect those portions of partially denatured proteins which retain antigenic determinants.

It is possible that monoclonals, or even defined mixtures of monoclonals, may be preferable in some instances. Monoclonals have the advantage of being very selective indeed, and can be produced in large amounts.

Nevertheless, it seems more likely that initial future developments will concentrate not so much on the antibody as on improved methods of sample 'work-up' and on means of detection. Most food IAs have been ELISA up to now, but it may well be that other techniques will be more appropriate for certain purposes. For example, the aggregation of antibody-coated latex spheres has been used to good effect for semi-quantitative clinical diagnostic assays, and may prove useful in food analysis. Immunofluorescent techniques may also be used. For specialised applications dealing with exceedingly low analyte concentrations (such as detection of substances related to anabolic steroids in animals), it may be necessary to develop cascade-amplified systems.

CONCLUSIONS

IAs are not a panacea. However, their advantages are not yet fully appreciated by public analysts or in the food industry, and, once these become more generally accepted and, more important, once robust kits become commercially available, it seems likely that they will assume considerable importance.

REFERENCES

1. Hitchcock, C. H. S., Bailey, F. J., Crimes, A. A., Dean, D. A. G. and Davies, P. J. (1981). *J. Sci. Food. Agric.*, **32**, 157–165
2. Morgan, M. R. A., McNerney, R. and Chan, J. W-S. (1983). *J. Assoc. Offic. Anal. Chem.*, **66**, 1481–1484
3. Rittenburg, J. H., Adams, A., Palmer, J. and Allen, J. C. (1986). *J. Am. Offic. Analysts Soc.* (in press)

Part III
Wider Aspects of Biotechnology

35
GETTING RESULTS FROM R & D

Charles M. Moseley

INTRODUCTION

Biotechnology is not merely a new scientific discipline, nor is it merely a novel application of science in technology; it is much more than either. It is the first visible manifestation of a new synthesis of knowledge, cutting across the boundaries of traditional disciplines and sometimes circumventing them altogether.

There are profound consequences for the management of commercial innovation. The key requirement for success in industrial biotechnology projects, succinctly expressed in the Common Market *FAST Report*, is skill in the creation and management of flexible and adaptable '. . . interdisciplinary teams, focused on the determined pursuit of a clear and significant joint objective' (Ref.3, p.31). Unfortunately, while this is easy to advocate, it is extraordinarily difficult to achieve in practice in any area of innovation, and for reasons associated with the fundamentally revolutionary nature of biotechnology it is particularly difficult in this field.

One of these reasons emerges when we attempt to define what is meant by interdisciplinary (or multidisciplinary) in the context of biotechnology. As mentioned above, biotechnology is part of a new synthesis of knowledge and this itself has made traditional disciplines obsolete. The new ones which are emerging are internally more cohesive than the old, and in this sense biotechnology is not at all inter-(multi-)disciplinary.

What it is instead is *at odds with* the traditional disciplines and this is what creates the problem for the commercial application of biotechnology. The specialists who are employed in such work have been trained and are experienced in disciplines whose boundaries and internal logic are paid scant respect by, or may be completely at odds with, the real needs of the project.

This deep-seated problem is rarely recognised within industry. Its implications are difficult to see and grasp, and it is frequently obscured by other more obvious difficulties. One of these is that biologists as a group tend to be commercially naïve and often find it difficult to work within the constraints of commercial reality. Managers tend to find this to be a sufficient explanation for lack of progress and fail to probe further and recognise those barriers that lie below the surface.

Obviously this problem is not confined to the United Kingdom, but it is particularly acute here because of the highly structured professionalism and disciplinary specialisation fostered by the British educational system and by British industry. Our education institutions are traditionally hostile to curricula change, and the professions are equally resistant and jealous of professional demarcation lines. Smaller economies such as Canada and Denmark tend to produce generalists and so may in some ways find it easier to adapt to biotechnology than larger economies, which tend to produce and employ specialists. (The adaptive value of generalism in a changing environment will be readily apparent to biologist readers of this chapter.)

While there is considerable literature on many aspects of technological innovation, there is not, so far as I am aware, any significant body of published work relating to the management and organisation of biotechnology R & D. Also, much of the research on technological innovation has not led to results which are of prescriptive value. Thus, managers are largely left to their own resources in trying to overcome the kind of problems described above in creating and maintaining effective research groups.

This chapter attempts to make a small contribution to this problem by briefly describing some aspects of the management of one bioscience R & D project, based on the author's experience as Project Manager over a five-year period.

PUBLISHED RESEARCH ON THE MANAGEMENT OF TECHNOLOGICAL INNOVATION

There has been considerable academic effort expended on elucidating the factors related to success and failure in technological innovation, and there is now a substantial body of published research on this topic. A number of references are given in the papers by Johne[4–6] which may be used as a point of introduction to the literature.

The findings of most investigators are very similar. Those factors related to understanding the needs of users together with good internal co-operation and co-ordination between the various specialist functions (engineering, marketing, etc.) are the most important criteria for success. This is not surprising, and it is difficult to disagree with such findings. However, they are difficult to translate into operationally useful prescriptions for management action. Recognition of the importance of good internal co-ordination, for instance, begs the question of how this can actually be achieved.

There is a separate body of literature which offers advice on how to manage R & D in practice. This work — mostly social-science-oriented — is easy to disagree with, as most of it is at odds with the above findings. Far from taking into account the crucial importance of control and co-ordination, a 'tranquil greenhouse environment free from inhibitory management policies' is typically recommended[7]. Lee effectively criticised this naïve view by showing that there is little evidence to support it. His opinion[7], based on '. . . personal observation of several R and D labs and the university scene plus scientist recruitment experience' is that '. . . many of the recommendations for the greenhouse environment for creative people are based on the self-referenced values and personality needs of the social scientists . . . [and] . . . an absence of quality standards or high expectations, financial responsibilities, and time constraints is not likely to be shown to improve organisation research and development laboratory performance any more than such conditions can be expected to improve the performance of manufacturing or marketing departments'.

Recently some investigators have started looking at the actual management practices of technological innovator companies. For example, Johne[4–6] has studied the organisational structures used by experienced product innovators in the British electrical and electronic test and measuring instruments industry. He found very tight, controlled management structures; specialisation, formalisation, standardisation, centralisation and stratification were all high in the development and launch phases of new product development. Certainly, looser approaches were found to be associated with the generation of new product ideas but this is in accordance with common-sense. The 'greenhouse' environment recommended by the theorists is completely opposite to the actual management structures Johne found in successful technological innovator companies.

It is the present author's opinion that the special problems of biotechnology R & D require an even greater degree of management intervention in the development process. In the case to be described, an extremely tight management structure was instituted in a major bioscience R & D project with a considerable degree of success.

THE CASE

'Irish Moss' (*Chondrus crispus*) is a Red Seaweed used as one raw material for the extraction of carrageenan, a hydrocolloid which has applications in the food and other industries. Irish Moss is harvested from the wild in Europe, Atlantic Canada and Maine. It has been apparent for many years that the annual harvest approaches the sustainable yield of the natural resource, so industry forecasts of continuing growth in market demand led the two major Irish Moss processors to initiate programmes aimed at ensuring adequate long-term production levels. Along with other parallel projects, both companies began large-scale investigations into the technical and economic feasibility of commercial on-shore (tank) cultivation. The availability of high-quality water and other environmental considerations led both companies to establish their R & D facilities in Nova Scotia, Canada.

The author joined Genu Products Canada Limited, one of the two companies involved, in 1974, at which time the cultivation R & D establishment was under construction, and remained until 1979, by which time there was a well-developed technology with a large-scale pilot unit operating and under evaluation.

Prior to the Genu Products programme there had been virtually no useful research on Irish Moss cultivation. The sole exception was the important discovery by the Canadian National Research Council that vegetative plants will spontaneously fragment when grown detached in agitated sea-water. Spontaneous fragmentation permits continuous cultivation, which is essential for economic reasons. However, such basic questions as the practicality of year-round operation under Canadian climatic conditions had not been addressed and the systems being promoted by NRC at the time would have produced weed at some 50 or more times its then current market cost! The decision by the parent company (A/S Kobenhaven Pektinfabrik) to support and finance the project was courageous indeed.

By 1979, the Genu group had developed and scaled up to pilot size a reliable and innovative production technology which operated year-round under Nova Scotia conditions without greenhousing and was capable of producing a premium quality *Chondrus* at a unit cost approaching that of the lower-quality natural product. Detailed planning for a 10 000 m^2 commercial production facility was well under way. Even today this remains the most advanced seaweeds cultivation technology development anywhere, and since details were recently published[1] it is being rapidly adopted elsewhere. It is anticipated that the Genu FB Marine Plants Cultivation System will be the basis for all future on-shore seaweeds cultivation technologies.

MANAGING THE PROJECT

Issues and Problems

The Genu Cultivation Programme was an atypical R & D project in one important way: there were clear goals. It is more usual for there to be considerable uncertainty. The innovative process itself is usually one of refining concepts and clarifying the ultimate aims. Our goals, however, were easily visible and could be defined precisely in terms of the maximum acceptable cost per unit production and minimum quality characteristics acceptable for the cultivated seaweed.

Project management is made much easier by unambiguous targets and they also permit — even demand — a highly directed goal-oriented approach.

All other characteristics of the project, however, presented a series of hurdles which at times seemed unsurmountable.

(1) It required research into many different fields: the biology and genetics of *Chondrus*; pests and diseases; sea-water chemistry; carbon nutrition; tank, agitation and support systems design and engineering; and a variety of other subjects. New topics emerged as the programme proceeded. Unfortunately, almost nothing of direct use had been done previously on any of these topics, either for *Chondrus* cultivation or for the artificial cultivation of any other marine plant. At the time even elementary facts about the biology of seaweeds were often not known. While the Company believed at first that they were embarking upon a commercial scale-up, it became apparent very quickly that extensive fundamental research into seaweeds biology was also needed[8].

(2) The growth and production of seaweeds growing in on-shore tanks is dependent upon a very large number of variables. Even if only biological factors are considered, it will be obvious that a highly complex system is involved. The complexity is greater by several orders of magnitude when engineering design variables such as tank configuration and agitation systems are also considered. All of these variables are interdependent, with complex and mostly unpredictable interrelationships.

(3) On-shore tank production systems are operating under seminatural conditions. Input variables such as illumination and sea-water temperature are therefore only partly controllable and can only be predicted within broad limits. As a result, optimum operating conditions are impossible to specify exactly for any given time.

(4) In addition to the complexity and the numerous uncontrolled variables, there were also various unknown factors which affected the system in a variety of ways.

Such problems are not unique. They apply to a greater or lesser extent to all biological process systems, especially those which are operated under seminatural conditions. For example, one of the major problems encountered in applying biotechnology to wastes treatment lies not in the biotechnology involved but in 'the extreme physical and microbiological complexity of the wastes. This often necessitates trials on treatment sites and makes it hard to obtain adequate predictions from anything less than substantially sized test systems.'[2]

Overall progress under these conditions depends upon simultaneous progress in all aspects of the programme because of the interdependence between them. Thus, the crucial factor for success, underlying everything else, is *integration*. Associated with this is the need to avoid dominance of the programme by any one specialism, because this leads to neglect of other equally important aspects. Programmes related to algal cultivation have typically been conceived and run by research biologists and have tended to neglect factors such as, for example, agitation systems design, with a consequent lack of progress.

Another consequence of the complexity of such systems is that traditional academic research approaches are inappropriate. Experiments conducted in outdoor systems are never accurately reproducible, and the overriding importance of the interrelationships between variables demands a *holistic, systems, approach* instead of the classical reductionist approach used in scientific research and with which we are all more familiar and comfortable.

Weaving together the strands from many different lines of investigation requires a very flexible approach and a rapid response to new findings in order to maintain momentum. Lines of enquiry frequently have to be abandoned or changed before completion, because progress elsewhere makes them irrelevant or leads to a change of emphasis. Continuous — daily — review and *co-ordination* of the programme is desirable.

Finally, an additional consequence of the complexity and lack of total control is that it is impossible to fully model large-scale cultivation systems either with small-scale test systems or mathematically. Therefore, pilot- and commercial-scale units cannot be fully designed without full-scale trials. Each stage of scale-up is thus subject to independent development. In practice this process proves to be iterative. The new problems which inevitably arise at each stage have to be reinvestigated at smaller scale, and often require redesign and complete rebuilding of intermediate-scale facilities. An expensive process! The textbook picture of commercial process scale-up:

bench-scale ⇢ pilot unit ⇢ commercial-scale

bears little resemblance to reality. A more accurate picture is:

bench-scale ⇌ pilot unit ⇌ commercial-scale
↓ ↓
bench-scale ⇌ pilot unit ⇌ commercial-scale
↓ ↓

This should, again, emphasise the importance of *integration*, not only between disciplines and commercial functions, but also between scales of development. These requirements (i.e. integration, holistic approach, flexibility and co-ordination), together with the need for strict commercial *secrecy*, determined the approach used at the Project management level and following from the approach the structure of the project organisation.

Approach and Structure

It was obvious that the goal-orientated nature of the project together with the overriding importance of integration meant that tight control was essential. This was instituted from the start, with extensive use of financial constraints and quantitative measures of progress. Although the latter evolved and became more sophisticated as experience was accumulated, this fundamental strategy was found to be effective and was retained.

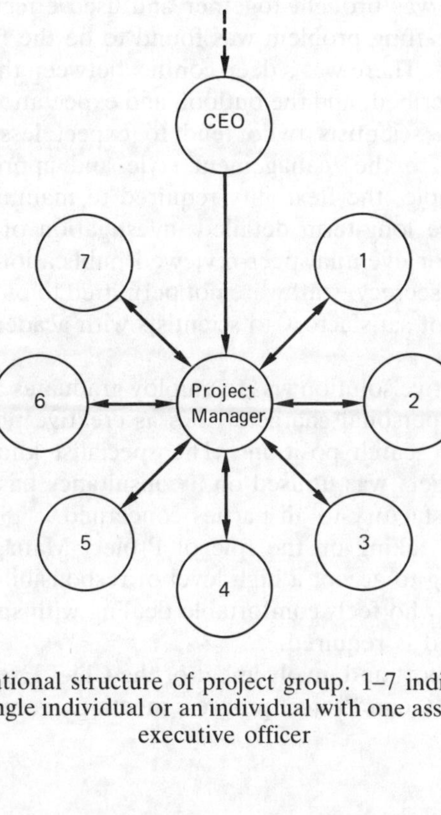

Figure 1 Organisational structure of project group. 1–7 indicate specialist functions: normally a single individual or an individual with one assistant. CEO = chief executive officer

In terms of organisation there was only one structure possible which permitted integration to the extent necessary, coped with the 'multi-disciplinary' nature of the project and also allowed maximum flexibility in responding to new data and insights. This is shown in Figure 1.

Close daily supervision and control were maintained over all facets of the project. In addition, as a matter of deliberate strategy, the total size of the project group was kept small enough to permit this degree of control by the Project Manager. Formal direct interaction between specialist functions was kept to a minimum (there were of course informal links), in order to allow maximum control and co-ordination. As a bonus, this permitted the Project Manager to maintain commercial confidentiality by restricting some research information (for example, details of the specific herbicide formulations used to control algal contaminants) on a need-to-know basis.

Staffing

Staffing is consequent upon structure. The effective operation of the structure described depends on staff at all levels who can live with it. An unusual degree of adaptability, talent and commitment is required, and although finding and identifying suitable individuals was certainly a major problem, a group was brought together and used effectively.

The greatest staffing problem was found to be the finding of suitable research specialists. There was a deep conflict between the requirements of the project, as described, and the outlook and expectations of traditionally educated research scientists, who tend to expect less control and are frequently hostile to the management style and approach used in the project. For example, the flexibility required to maintain rapid progress did not permit the long-term detailed investigation of individual topics which is needed for eventual peer-reviewed publication. Because of this and for reasons of secrecy, staff were not permitted to publish their results. Naturally this is not satisfactory to scientists with academic career aspirations.

The most effective solution was to employ graduates and postgraduates selected for their personal characteristics as creative problem solvers, in permanent staff research positions. The specialist knowledge of senior university researchers was utilised on a consultancy basis. This approach was generally satisfactory to all parties concerned.

The individual taking on the role of Project Manager must be both capable and willing to accept a high level of responsibility. His is the key role. A generalist who feels comfortable dealing with specialists in all the disciplines involved is required.

The active support and involvement of the Chief Executive is essential for success. In the Genu Project progress meetings were held between the

CEO and the Project Manager at least weekly and project activity was discussed in detail. It is also vital that the CEO give full support to the Project Manager, because the latter is in a very exposed position.

CONCLUSION

The approach which has been described was successfully used to cope with the requirements of a major bioscience R & D project. It evolved within one specific project and thus may not be appropriate elsewhere and should not blindly be adopted without careful thought. Nevertheless it is believed that this model has features which make it useful for solving the kind of problems that are found in many biotechnology projects.

In Canada, a Federally financed research institute — the Atlantic Institute of Biotechnology — was established in 1985 with the aim of adopting similar approaches to those developed at Genu Products in assisting local industry to apply and exploit biotechnology.

REFERENCES

1. Bidwell, R. G. S. *et al.* (1985). Tank cultivation of Irish Moss, *Chondrus crispus. Botanica Marina,* **XXVIII**, 87–97
2. Dunnil, P. and Rudd, M. (1984). *Biotechnology and British Industry.* Biotechnology Directorate, SERC
3. EC (1982). *The FAST Programme.* Vol.1. *Results and Recommendations.* Brussels, December 1982
4. Johne, F. A. (1982). The organisation of product innovation in high technology manufacturing firms: Some preliminary results. Working Paper No.35, City University Business School (ms)
5. Johne, F. A. (1983). The organisation of product innovation in high technology manufacturing firms. Working Paper No.50, City University Business School
6. Johne, F. A. (1984). Organisation structures of experienced product innovators. Working Paper No.60, City University Business School
7. Lee, J. A. (1980). *The Gold and the Garbage in Management Theories and Prescriptions.* Ohio University Press, Athens, Ohio
8. Moseley, C. M. and Norve, H. (1977). Some restraints in onshore seaweeds cultivation. Read at *IX International Seaweeds Symposium,* Santa Barbara, California

36
ENABLING MANPOWER IN BIOTECHNOLOGY: AN INDUSTRIAL PERSPECTIVE

D. J. Parsons

INTRODUCTION

Debate on the economic potential of biotechnology as a UK growth industry has largely focused on the strategic role to be played by both private and public sector investment. Yet the extent to which the UK economy in the longer term is likely to benefit from the full potential of novel biotechnological processes and products will also depend on the availability of suitable, highly qualified specialist manpower. Rather less attention has been paid to this issue. A research programme conducted by the Institute of Manpower Studies on behalf of the Biotechnology Directorate of the Science and Engineering Research Council casts some light on this manpower dimension. This chapter looks at some of the principal findings of the study in the context of current industrial demand for highly qualified manpower in biotechnology and likely future needs.

ISSUES OF DEFINITION

The term 'biotechnology' invites any number of interpretations from an industrial perspective. To understand the scope of the research findings, it is important to establish the definition followed in the IMS study. There are a number of competing scientific definitions of biotechnology, but the one adopted by the research team was that of the Spinks report[4]: 'the application of biological organisms, systems and processes to manufacturing and service industries'.

Within this wide framework the research focused on commercial organisations active in 'novel' biotechnology, and especially on areas such as applied fundamental research, R&D, plant and process engineering related to biotechnology and industrial service organisations. Those involved predominantly in equipment development and specialist biotechnology investment companies were excluded. 'Traditional' areas of biotechnology such as in the brewing and food industry were also excluded. Some 60 commercial organisations in the UK active in novel biotechnology were identified during the course of the study, of whom 42 fell within the scope of this definition.

The research also embraced relevant activities within higher education and research centres. This was of some relevance to 'industrial' biotechnology, since the dividing line between commercial organisations and institutions of higher education and research centres are increasingly seen as extending into quasi-commercial activities involving contract research, extensive academic consultancy in the UK and overseas, joint ventures and other entrepreneurial developments. In consequence, these sectors increasingly draw on a similar pool of skills to industrial needs.

The focus of the IMS research, reflecting SERC's major interests, was on postgraduate scientists, technologists or personnel similarly qualified by virtue of experience. The term 'biotechnologist' is not used here. Not only does it not have a common meaning, but also it is rarely used in employing organisations. Biotechnology by its nature is an interdisciplinary activity embracing a range of bioscience and bioprocess skills. Relevant highly qualified manpower (HQM) in industry was more commonly defined by work experience and application than by discipline or speciality.

EMPLOYMENT DISTRIBUTION

The IMS research estimated a total of 1500–2000 highly qualified biotechnologists working in Britain in the new biotechnology industries. Of these, only just under half were employed by commercial organisations. One in five were employed in institutions of higher education on contract research teams and joint ventures, and in related activities. The remainder were spread across a range of research centres in the UK, although only one of these was exclusively involved in biotechnological research.

In industry, the distribution of HQM employed wholly or substantially in biotechnology is characterised by firms with only relatively small numbers of technical staff at this level. In the IMS research two-thirds of the firms had less than 25 such personnel. These did not include technicians and ancillary staff wholly or partly related to biotechnology activities, marketing and administrative staff, and managerial posts not directly involved in R&D or production.

More specific information for 18 case study companies is shown in Tables 1 and 2. This suggests that although these are characterised by small establishments and divisions, the overall employment was dominated by

Table 1 Highly qualified biotechnology manpower in case study companies: Spring 1983

	Number of case study firms			
Numbers wholly or substantially employed in biotechnology	*Health care and pharm.*	*Industrial services, etc.*	*Other sectors*	*All firms*
Under 10	—	3	2	5
10–24	2	3	3	8
25–49	—	—	1	1
50–74	2	1	1	4
75–99	—	—	—	—
100 and over	—	—	—	—
Total companies	4	7	7	18
Total employees	132	130	118	380

Source: IMS Study, 1983

Table 2 Manpower distribution in the case study companies

Posts	*Pharm.*	*Industrial*	*Other industry*	*Total*
Scientific posts at:				
PhD postdoctoral	61	70	66	197
level	9	5	6	20
Other postgraduate				
Graduate	60	55	50	163
Technician posts	50	44	92	186
Technical posts not classified above	—	—	14	14
No. of firms represented	3	6	6	15

the six largest, which together employed nearly three-quarters (71 per cent) of the HQM. It is difficult to be precise about the industrial characteristics within this overall distribution, but over a third (35 per cent) were concentrated in health care and pharmaceutical activities, and nearly a quarter (23 per cent) in industrial service companies. Other evidence suggests that this is broadly representative of the industrial sector as a whole.

Reported employment trends mainly emphasised moderate growth, but showed substantial variation within firms. For example, the research identified an intake of 40 HQM recruits per annum to permanent posts across the 16 case study firms able to supply data, with nearly 90 per cent of these going to health care/pharmaceutical or industrial service activities (Table 3). Recruitment needs overall may be dominated by a handful of expanding companies in the health care/pharmaceutical and industrial service sectors. In the IMS research, aggregate recruitment between 1980 and 1983 was dominated by just four companies — two new start-ups and two expanding pharmaceutical divisions. Together, these accounted for three-quarters of all recruits. Elsewhere, HQM manpower demand was low, typically in the range of just 1 to 2 vacancies at a point of time often related to replacement demand, rather than to additional or new recruitment. In two of the case study organisations, HQM levels in novel biotechnology had fallen since 1980 as a result of corporate rationalisation programmes.

Table 3 Approximate annual level of external recruits to highly qualified biotechnology posts in the case study companies (since 1980)

No. of HQM recruits p.a.[a]	*Health care and pharm.*	*Industrial services*	*Other*	*All firms*
Nil	—	3	1	4
1–5	1	2	3	6
6–10	2	2	1	5
10 or more	—	1	—	1
No. of firms represented	3[b]	8	5	16
Total recruits	20	40	9	69
No. of HQM recruits per firm	6.6	5.0	1.8	4.3

[a] Excluding temporary and short-term contract posts.
[b] Excluding those organisations unable to provide data.

RECRUITMENT CHANNELS AND CONSTRAINTS

Entry channels and qualification levels for bioscientists and bioprocess engineers were rather different. Across the case study firms first-degree bioscientists were mostly relevant to higher technician and project assistant posts. Employers' need for HQM often emphasised PhD or post-doctoral scientists. Recruitment of teaching course postgraduates on conventional MSc or conversion courses was not generally significant to companies.

Recruits to junior bioscience posts were mainly newly qualified post-doctoral and more occasionally PhD students. In contrast, relatively few senior recruits came direct from higher education, and most team leaders in R&D were attracted from parallel industrial posts or occasionally from research centres. The importance of 'internal' recruitment varied substantially between firms, according to their size, corporate manpower and career policies, but in some cases this was constrained by the limited transferability of bioscience specialists across discrete project areas.

In contrast, for most firms the main initial entry for bioprocess engineers was of first-degree chemical engineers. Only in the industrial service companies were these commonly recruited direct to specialist bioprocess engineering functions. In most other areas of industrial biotechnology — and especially in process engineering contractors — a high degree of flexibility was required among recruits across conventional engineering and bioprocess engineering applications. Individuals with MScs in bioprocess engineering were often regarded as overspecialised for such recruitment needs.

Although recruitment focused on manpower in the UK, about a quarter of the case study companies had recruited UK nationals from abroad. This reflects the international labour market in which experienced specialists operated and was mostly characteristic of senior bioscience posts in specialised industrial service companies. Few overseas nationals were recruited to industrial biotechnology in the UK.

Widespread recruitment constraints were not a feature of industrial biotechnology. Nonetheless, two-thirds of the case study organisations reported at least one form of recruitment difficulty for HQ biotechnological manpower — generally a perceived shortage of suitable experienced specialists. The main difficulties related to a shortfall of small numbers of experienced personnel in aspects of fermentation technology, with particular difficulties for suitably experienced microbial physiologists, bioscientists with relevant experience in relevant scale-up and downstream applications, and senior bioscientists with suitable R&D experience in special aspects of plant and tissue cell culture.

These shortages were, however, highly selective and related to senior technical staff, where the premium was on experience in an applied environment rather than academic knowledge or training, however specialised. Inevitably, in a rapidly developing technology with many novel techniques and where lead times in 'training' may exceed 10 years, there will be such difficulties. Even in areas where several employers reported a better density of applications, such as genetic engineering, considerable time and effort were often needed to find an applicant of suitable calibre and with appropriate technical and project management experience.

Such manpower shortfalls, however limited in scale, are not a new problem in industrial biotechnology. Earlier reports[3,4] identified the need to train HQM in certain specific disciplines relevant to biotechnology. Both reports made it clear that in addition to meeting manpower demand in the newer areas of molecular biology, there was an urgent need for specialists in microbial physiology, fermentation and some applied areas such as downstream processing. Subsequent developments suggest that while industrial demand for bioscientists in areas such as recombinant DNA, gene sequencing, DNA synthesis and the use of monoclonal antibodies has largely been met, the problems for fermentation and related specialists seem rather more persistent.

There is evidence that these problems were compounded in some areas by the relative unpopularity, with PhD students, of some 'dirty hands' specialities within bioprocess engineering, and the unfashionable image of others, such as microbial physiology. While the scale of this problem in terms of the annual level of recruitment demand for suitable specialists is limited, its extent and significance are potentially much wider. Many of the organisations who had not experienced specialist shortfalls — perhaps through lower recruitment needs — acknowledged that it would require an extensive effort if it was necessary to replace certain specialist personnel or fill new senior posts.

Few industrial employers reported recruitment difficulties at the technician or scientific assistant levels. First-degree constraints were not an issue other than in support areas of electronics and instrumentation, which reflected national shortfalls more than problems specific to biotechnology.

It is notable that in some of the companies internal constraints compounded difficulties experienced through a shortfall of suitable applicants. In particular, in a highly competitive market, organisations with rigid salary structures often found it difficult to attract personnel, and for many this added to the problems in attracting back UK nationals from posts in the USA.

FUTURE MANPOWER NEEDS IN BIOTECHNOLOGY

Evaluating future levels of HQ manpower demand involves many difficulties. Most industrial employers have relatively short time horizons when assessing future manpower demand. Consequently, any attempt to project industry's recruitment needs for suitable HQ manpower more than a year or two ahead encounters special problems of obtaining information, however speculative, from firms. These difficulties are compounded by the economic and technological uncertainties which surround even the short-term development of biotechnological products and processes in the UK and elsewhere.

Nonetheless, if precise quantitative forecasts are impractical, and misleading, some assessment is needed to evaluate any future constraints to the recruitment of appropriately skilled manpower. Furthermore, it is important that any such evaluation should take account of not only commercial organisations but also related demand in research centres and relevant posts in higher education. The Institute of Manpower Studies has made some broad projections drawing on available information from this study. Given the range of uncertainty involved, these are presented not as a single estimate, but according to three scenarios of sectoral development, one based on assumptions of 'relative stability', others on 'moderate growth' and 'significant expansion'. The assumptions underlying these scenarios are described in the full report of the study[1].

The three scenarios (Table 4) have very different implications for the scale and mix of job growth, but whatever the scenario, biotechnology is not going to provide many new jobs at professional levels. More specific observation suggests:

(1) Even in a situation characterised by stability rather than employment growth, some recruitment will be needed for new skills, given the limited technological transferability of scientific personnel within project teams. Losses to posts overseas or to non-biotechnology functions in the UK would also need to be accommodated.

(2) Wastage rates are likely to rise in line with any general expansion in the biotechnology labour market. In a situation where certain highly specialised skills and abilities are in shortage, this could serve to exacerbate

Table 4 Projections of annual HQM demand for the UK, 1984–90

	Range of new posts per annum		
Economic scenario[a]	*Commercial organisations*	*Research centres*	*Higher education*
Relative stability	15–20	10–20	0–5
Moderate growth	40–100	30–50	5–10
Significant expansion	120–130	60–100	40–50

[a]See full report for specification.
Source: IMS Survey (and estimates), 1983.

any recruitment difficulties experienced, in particular, by expanding employers.

(3) The major contributor to increased HQM demand is likely to be industry, whose recruitment needs are likely to be greater than those of research centres and higher education put together. Discrete demand to permanent posts in higher education is likely to be negligible.

When put together with an appraisal of supply trends, this suggests that overall there is likely to be little difficulty in meeting HQM needs at 'junior' level (such as the postdoctoral scientist in his mid- to late twenties). However, significant shortfalls could develop among senior personnel or in a few specialist areas. Areas such as microbial physiology, fermentation and some applied technological specialities could especially suffer. Generally, there is little evidence for any extensive shortage of enabling manpower, but this must be seen in the context of the uncertainties which surround future prospects. The pattern and scale of future UK investment in biotechnology; increased investment, for example, by US R&D establishments; and any fundamental shift in the scale or nature of the UK 'brain drain' are especially important uncertainties[2].

It is also most important to bear in mind that the manpower system within UK biotechnology is focused on a relatively small number of highly qualified scientists/technologists, and even fewer senior personnel in the area of team management. In most aspects of biotechnology, international competitiveness and rapidly developing systems and techniques may mean that a shortfall of a handful of key personnel in a specific specialisation or application could lead to development constraints. Much may depend on the success of manpower initiatives now being taken by SERC and related organisations. If there is currently little evidence of a major shortfall of supply now or in the short term, this should not blind us to the delicacy of this manpower equilibrium. This is perhaps reflected in the difficulties that are currently experienced in recruiting the small numbers of senior staff needed, especially in highly specialised scientific areas such as microbial physiology and more generally in bioprocess engineering.

REFERENCES

1. Parsons, D. J. and Pearson, R. (1984). *Enabling Manpower for Biotechnology in the UK*. Biotechnology Directorate, SERC
2. Pearson, R. and Parsons, D. J. (1984). *The Biotechnology Brain Drain*. Biotechnology Directorate, SERC
3. Royal Society (1981). *Biotechnology and Education*. Report of the Stewart Committee
4. Spinks Report (1980). *Biotechnology: Report of Joint Working Party*. ACARD, ABRC, Royal Society (Spinks report). HMSO, London

37
BIOTECHNOLOGY POLICY INTO THE 1990s

Harry Rothman and Andrew Ashton

INTRODUCTION

In recent years it has become increasingly important to determine methods by which science can be strategically planned. Financial restrictions on governments world-wide have meant that it is essential to observe and detect trends in technology and to discover what is necessary to encourage their proper growth. Technology policy has become the subject of serious research and Technology Policy Units such as that at the University of Aston in Birmingham are now providing valuable insights into the impact of new technologies on society.

Biotechnology is one such emerging technology, and its influence on our society, together with that of 'new' biology may be extensive and long-lasting. The European Commission believes it is necessary to study such impact on a scale of 30 years or more. It is difficult to predict the growth which will have taken place in the 'Bio-Society' by the year 2010. However, by looking at present trends and extrapolating realistic scenarios, it may be possible to assemble sufficient strategic data to assist decision makers and the general public to appreciate the issue and its impact on society.

Recently, techniques have been developed for measuring outputs and advances in research and development, their utility and limitations. Two main techniques that have been examined are Peer Review and Publication Analysis, although neither is ideal for all tastes and each has its advantages and disadvantages for any particular assessment.

One major method for studying present trends in science and technology is Science Mapping — the mapping of research facts based on the use of co-citation clusters, developed by the Institute for Scientific Information (ISI), Philadelphia, who produce the *Science Citation Index (SCI)*.

The Technology Policy Unit (TPU) at Aston has utilised the co-citation technique in order to map biotechnology and has developed literature-based models that have been suggested for strategic planning. This work has formed part of the EEC DGXII FAST programme, sub-programme Biosociety C.I.4 Analysis of Scientific Disciplines Germane to Biotechnology.

AIM OF THE TPU PROJECT

The aim of the project was to provide a greater understanding of growth and internal and external relations of the scientific disciplines and specialities which are regarded as important for, or related to, developments in biotechnology. Thus, a basis of understanding would be available to draw conclusions for the effective development of science and technology policy in biotechnology, with particular reference to the position of the European Community. It was clear that it was impossible to analyse and map the whole of biotechnology with the resources available, so the project made a study of the techniques of science mapping by co-citation, by adopting a case study approach.

CLASSIFICATION OF THE SCIENCES — TRADITIONAL APPROACH

The first problem facing the science planner is to classify the sciences. Traditional schemes of classification are often based on outmoded philosophical or sociological conceptions, which fail to represent the ever-changing diversity and richness of modern science. The boundary zones between the various sciences and fields of research are constantly on the move. An interesting pioneering attempt to 'map' the science disciplines and show the relationships between various technologies was made in 1948 by H. J. T. Ellingham[2]. His chart aimed to illustrate the relations between the branches of natural science and technology by means of a partitioned central panel and an outer border with titles written at various angles to show their proximity to each other as well as the subjects of the central panel (Figure 1). Ellingham himself suggested that the chart would be most effective if wrapped around a sphere[3]. This brave attempt at generating a map indicates the problems faced by present-day science 'cartographers'.

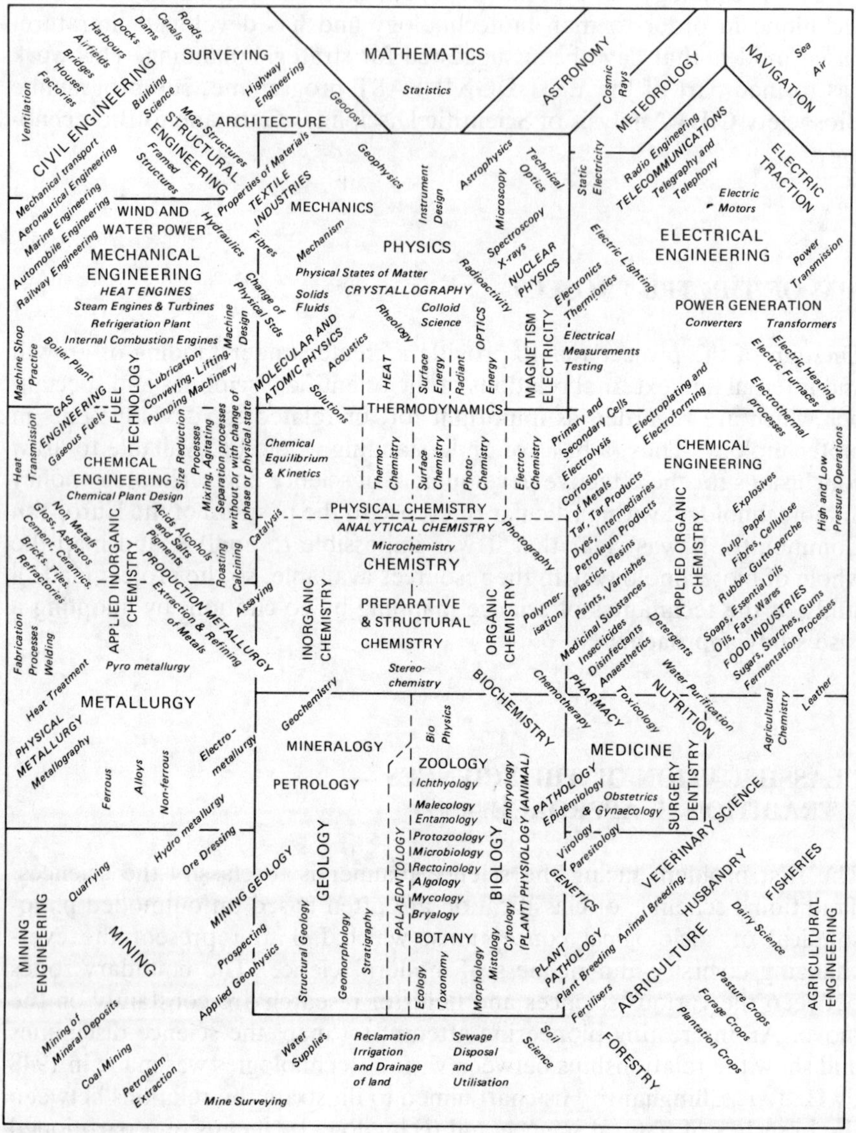
Roads
Canals
Dams
Docks
Harbours
Airfields
Bridges
Houses
Factories
Ventilation
CIVIL ENGINEERING
Building Science
SURVEYING
Highway Engineering
MATHEMATICS
Statistics
Geodosy
ASTRONOMY
Cosmic Rays
METEOROLOGY
NAVIGATION
Sea
Air
ARCHITECTURE
STRUCTURAL ENGINEERING
Moss Structures
Framed Structures
Properties of Materials
TEXTILE INDUSTRIES
Geophysics
Instrument Design
Astrophysics
Technical Optics
Static Electricity
Radio Engineering
TELECOMMUNICATIONS
Telegraphy and Telephony
ELECTRIC TRACTION
Electric Motors
Mechanical transport
Aeronautical Engineering
Marine Engineering
Automobile Engineering
Railway Engineering
WIND AND WATER POWER
Hydraulics
Fibres
MECHANICS
Mechanism
PHYSICS
Microscopy
Spectroscopy
X-rays
NUCLEAR PHYSICS
Radioactivity
Electric Lighting
ELECTRICAL ENGINEERING
Power Transmission
MECHANICAL ENGINEERING
HEAT ENGINES
Steam Engines & Turbines
Refrigeration Plant
Internal Combustion Engines
Physical States of Matter
CRYSTALLOGRAPHY
Solids
Fluids
Colloid Science
Change of Physical Stds
Machine Design
Rheology
OPTICS
Electronics
Thermionics
POWER GENERATION
Converters
Transformers
Machine Shop Practice
Boiler Plant
Lubrication
Conveying, Lifting, Pumping Machinery
MOLECULAR AND ATOMIC PHYSICS
Acoustics
HEAT
Surface Energy
Radiant
Energy
MAGNETISM
ELECTRICITY
Electrical Measurements Testing
Electric Heating
Electric Furnaces
Electrothermal processes
GAS ENGINEERING
Gaseous Fuels
FUEL TECHNOLOGY
Solutions
THERMODYNAMICS
Primary and Secondary Cells
Electrolysis
Electroplating and Electroforming
Heat Transmission
CHEMICAL ENGINEERING
Chemical Plant Design
Size Reduction Processes
Mixing, Agitating
Separation processes without or with change of phase or physical state
Chemical Equilibrium & Kinetics
Thermo-Chemistry
Surface Chemistry
Photo-Chemistry
Electro-Chemistry
Corrosion of Metals
CHEMICAL ENGINEERING
High and Low Pressure Operations
Non Metals
Gloss, Asbestos
Cement, Ceramics,
Bricks, Tiles,
Refractories
Acids Alcohols and Salts
Pigments
Catalysis
PHYSICAL CHEMISTRY
ANALYTICAL CHEMISTRY
Tar, Tar Products
Dyes, Intermediaries
Petroleum Products
Plastics, Resins
Paints, Varnishes
Explosives
Pulp. Paper
Fibres, Cellulose
Rubber Gutto-perche
APPLIED INORGANIC CHEMISTRY
PRODUCTION METALLURGY
Roasting
Calcining
Assaying
Microchemistry
CHEMISTRY
Photography
APPLIED ORGANIC CHEMISTRY
Soaps, Essential oils
Oils, Fats, Wares
FOOD INDUSTRIES
Sugars, Starches, Gums
Fermentation Processes
Extraction & Refining of Metals
INORGANIC CHEMISTRY
PREPARATIVE & STRUCTURAL CHEMISTRY
ORGANIC CHEMISTRY
Polymerisation
Medicinal Substances
Insecticides
Disinfectants
Anaesthetics
Detergents
Fabrication Processes
Welding
Pyro metallurgy
Stereo-chemistry
BIOCHEMISTRY
Chemotherapy
PHARMACY
Toxicology
Water Purification
NUTRITION
Agricultural Chemistry
Leather
Heat Treatment
METALLURGY
Geochemistry
Bio Physics
PHYSICAL METALLURGY
Metallography
Ferrous
Alloys
Non-ferrous
Electro-metallurgy
MINERALOGY
PETROLOGY
ZOOLOGY
Ichthyology
Malecology
Entamology
Protozoology
Microbiology
Rectoinology
Algology
Mycology
Bryalogy
BOTANY
PALAEONTOLOGY
BIOLOGY
Embryology
(PLANT) PHYSIOLOGY (ANIMAL)
Cytology
PATHOLOGY
Epidemiology
MEDICINE
Obstetrics & Gynaecology
SURGERY
DENTISTRY
VETERINARY SCIENCE
Virology
Paresitology
GENETICS
Hydro metallurgy
Ore Dressing
Quarrying
MINING GEOLOGY
GEOLOGY
HUSBANDRY
FISHERIES
Animal Breeding.
Dairy Science
MINING ENGINEERING
MINING
Prospecting
Applied Geophysics
Structural Geology
Geomorphology
Stratigraphy
Ecology
Taxonomy
Morphology
Mistology
PLANT PATHOLOGY
Plant Breeding
Fertilisers
AGRICULTURE
AGRICULTURAL ENGINEERING
Pasture Crops
Forage Crops
Plantation Crops
Mining of Mineral Deposits
Coal Mining
Petroleum Extraction
Water Supplies
Reclamation Irrigation and Drainage of land
Sewage Disposal and Utilisation
Soil Science
FORESTRY
Mine Surveying

Figure 1

MODERN METHODS — PUBLICATION ANALYSIS

It is clear that makers of science maps today must base their work on some tangible data from the science and technology fields. Each year over a million scientific publications of one kind or another are published as part of science's communal process of sharing, peer review and assignment of intellectual property rights. Ostensibly, these publications are written to convey information about scientific theory, techniques and phenomena. However, they also, for a variety of social traditions and cognitive reasons, contain other information — for example, institutional, autobiographical, biographical (citations), keywords, etc. The scientific literature therefore contains an extremely detailed record of its self-activity on intellectual, social, institutional, national, international and many other levels. Of course, it is an imperfect and noisy record, even incomplete, since cogent information may not enter the literature and some of the information that enters may be ambiguous, inaccurate or spurious.

Thus, the literature of science can provide necessary data for both static (snap shot) and dynamic (real over-time) models of science. This technique does not involve traditional library science and information retrieval, but the development of models of science capable of showing how different areas of science relate to each other (science maps) and throwing light on how specialities and research fronts develop. Computer science has made it possible to collect and interpret the literature databases thus provided.

JOURNAL–JOURNAL CITATION MODELS[5]

Journal–journal citation is a technique of relating journals according to their citation patterns. The technique assumes that the more frequently one journal cites another, and vice versa, the more likely they are to be concerned with similar topics. By measuring such citation patterns, it is possible to create a measure of 'relatedness', and cluster journals according to their relatedness and thus make up a map of an area of science. Data required for such modelling can now be obtained for a large number of journals from the ISI *Journal Citation Reports (JCR)*, which has such lists for a specific set of journals — how often a particular journal has been cited in a certain year, what journals have cited it, and how frequently. The inter-citation patterns for a set of journals are entered in a matrix and examined by a computer program devised for cluster analysis, such as Clustan[9].

Case Study — Journal–Journal Citation

Biotechnology —Clustering — Superclusters

In order to compose a case study for biotechnology, 60 biotechnology journals were analysed and the relationships between the journals were first shown as a dendogram (Figure 2). From this it is clear that the journals cluster into distinct groups.

Each cluster which had emerged from the dendogram was then represented on another diagram as a circle (Figure 3). The diameter of the circle indicates the spread of the journals within that cluster. The smaller the circle, the stronger the degree of linkage within the cluster (this is also affected by the number of journals in the cluster). Further, these clusters of journals have been classified and given a subject name. From this map we can see how five basic science clusters and one cluster of interdisciplinary journals feed into a tangled web of applied science and technology clusters.

It is also clear from Figure 3 that 'superclusters' form where many clusters are linked together. Figure 4 analyses one of these clusters in greater detail and shows how this supercluster can have at least five major components. A further enlargement (Figure 5) and breakdown of this supercluster clearly shows the considerable intimate relationship of chemistry to biotechnology. Basic chemistry journals are found at the heart of the supercluster. Three distinct clusters of chemistry journals emerge: clinical, analytical and basic. The highly cited journal *Biotechnology and Bio-engineering* is closely associated with the chemistry clusters in this map.

Advantages and Disadvantages of Journal–Journal Citation

Although journal–journal citation does not provide a definite map of biotechnology, it does serve to illustrate how even such a crude yardstick can allow one to derive a map offering a general overview of the interrelationships between various scientific and technical fields, providing an aid to understanding a particular cognitive network.

The disadvantages of the system are that journals only approximate to actual research fields; they may not always be consistent in their coverage. Moreover, certain journals are multidisciplinary and highly cited and can distort the cluster maps (this can be controlled by seeing how sensitive the clustering is to their presence or absence). A further disadvantage is that the *Journal Citation Reports* produced by ISI contains a limited number of journals (about 3500).

Finally, there are technical limits to the matrix size that can be clustered — that is to say, the data which computers can cope with at present.

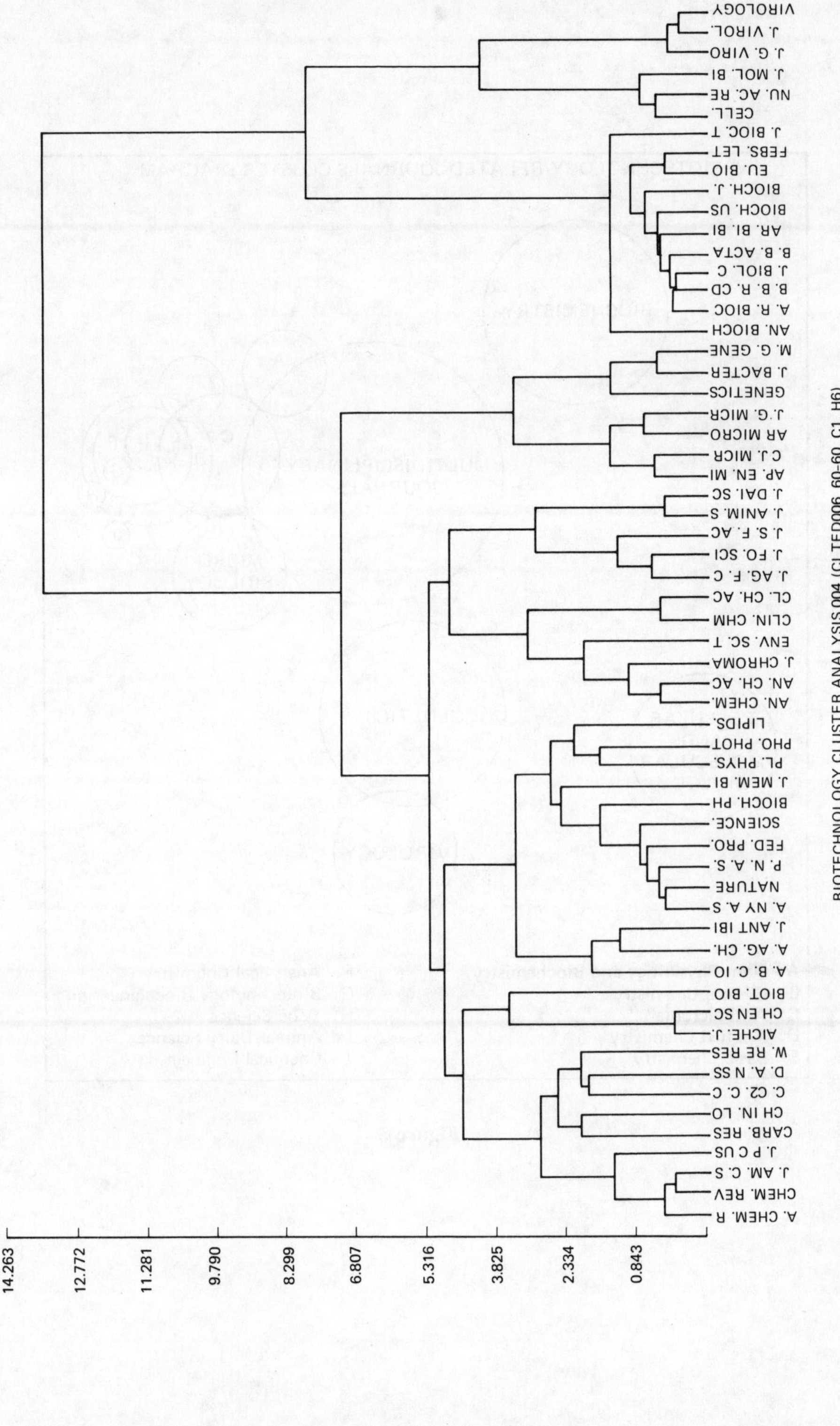
BIOTECHNOLOGY CLUSTER ANALYSIS 004 (CI TED006, 60-60, C1, H6)
VIROLOGY
J. VIROL.
J. G. VIRO
J. MOL. BI
NU. AC. RE
CELL.
J. BIOC. T
FEBS. LET
EU. BIO
BIOCH. J.
BIOCH. US
AR. BI. BI
B. B. ACTA
J. BIOL. C
B. B. R. CD
A. R. BIOC
AN. BIOCH
M. G. GENE
J. BACTER
GENETICS
J. G. MICR
AR MICRO
C. J. MICR
AP. EN. MI
J. DAI. SC
J. ANIM. S
J. S. F. AC
J. FO. SCI
J. AG. F. C
CL. CH. AC
CLIN. CHM
ENV. SC. T
J. CHROMA
AN. CH. AC
AN. CHEM.
LIPIDS.
PHO. PHOT
PL. PHYS.
J. MEM. BI
BIOCH. PH
SCIENCE.
FED. PRO.
P. N. A. S.
NATURE
A. NY A. S
J. ANT IBI
A. AG. CH.
A. B. C. IO
BIOT. BIO
CH EN SC
AICHE. J.
W. RE RES
D. A. N SS
C. C2. C. C
CH IN. LO
CARB. RES
J. P C US
J. AM. C. S
CHEM. REV
A. CHEM. R
14.263
12.772
11.281
9.790
8.299
6.807
5.316
3.825
2.334
0.843

Figure 2

BIOTECHNOLOGY RELATED JOURNALS CLUSTER DIAGRAM

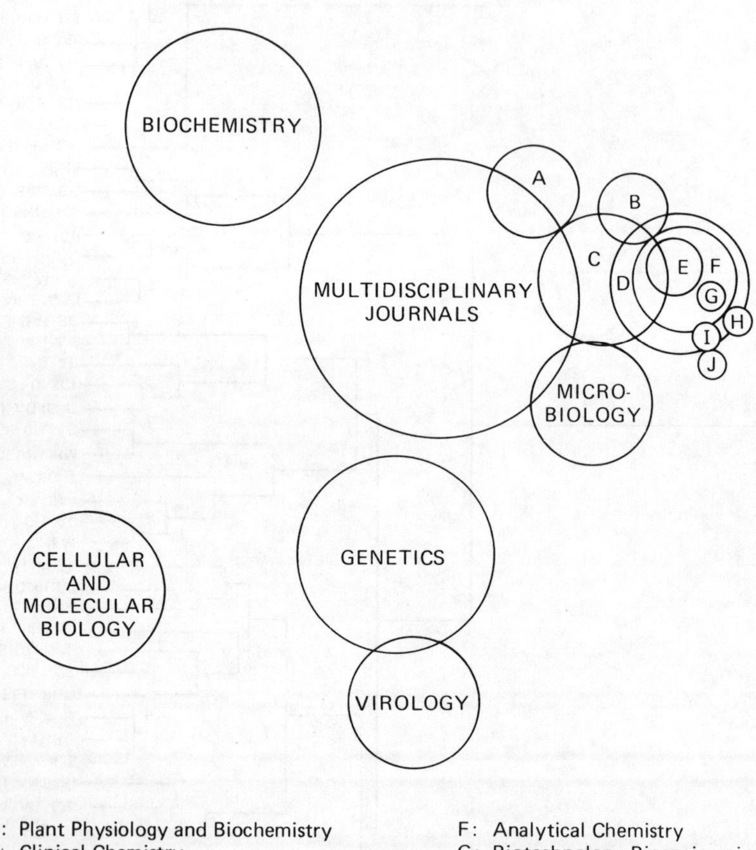

A: Plant Physiology and Biochemistry
B: Clinical Chemistry
C: Antimicrobial
D: Applied Chemistry
E: Basic Chemistry
F: Analytical Chemistry
G: Biotechnology Bioengineering
H: Food Science
I: Animal/Dairy Science
J: Chemical Engineering

Figure 3

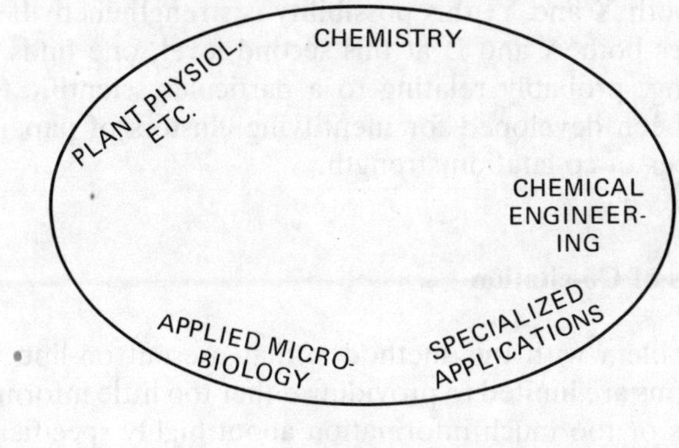

Figure 4

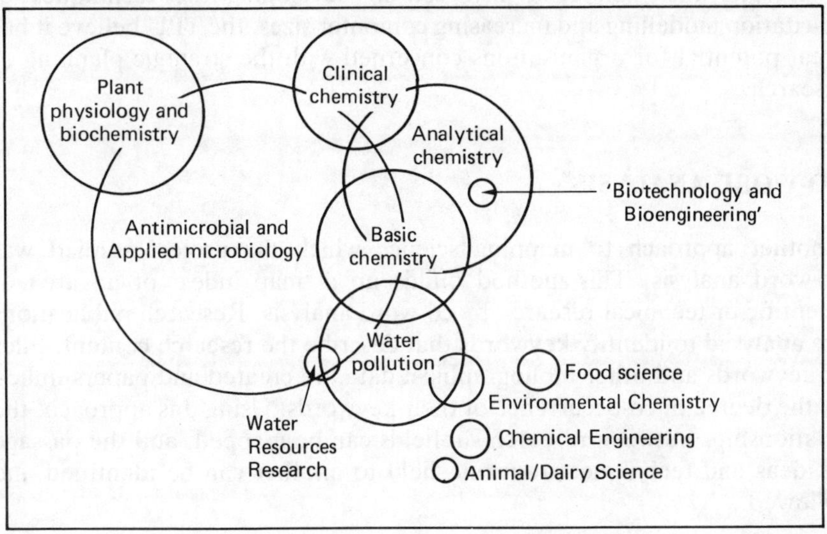

Figure 5

CO-CITATION CLUSTERING[7]

Another technique of science mapping which the TPU at Aston studied was co-citation clustering. Co-citation clusters are groups of highly cited papers that are frequently cited together. Co-citation uses citation to measure common intellectual interests between any two papers. Thus, if paper A cites two papers, X and Y, then both X and Y are co-cited and perhaps have something in common intellectually and/or socially. If paper

B also cites both X and Y, this possibility is strengthened. If yet another paper, C, cites both A and B at this second level, one finds a cluster of papers forming, probably relating to a particular scientific topic. Techniques have been developed for identifying clusters of papers linked by specific degrees of co-citation strength.

Disadvantages of Co-citation

The chief problem with this method is that present on-line information retrieval systems are limited to providing either too little information about many subjects or too much information about highly specific topics. The results produce either a 'coarse-grained' overall picture or a specialised miniature of little value for general science mapping.

Despite the high costs involved in developing the techniques of co-citation modelling and increasing computer sizes, the TPU believe it has great potential for organisations concerned with the strategic planning of research.

CO-WORD ANALYSIS[8]

Another approach to mapping science which the project studied was co-word analysis. This method builds up a 'map index' of an area of scientific or technical research by co-word analysis. Research publications are analysed to identify keywords that describe the research content. Files of keywords and other bibliographical data are created and papers linked by the degree of co-occurrence of their keywords. Using this approach, the relationships between research subfields can be mapped, and the passage of ideas and techniques from one field to another can be identified and followed.

Advantages and Disadvantages of Co-word Analysis

The advantage of this approach over that of co-citation is its flexibility: it is not tied to any specific type or set of documents. It can even analyse official science policy documents or patents — the US Patent Office already plans to put 24 million documents on an electronic database by 1990. It is also less expensive.

However, the system requires an information retrieval base, raising questions of database quality and coverage, although these can be controlled. A second weakness is keywork choice, where a new element of subjectivity is introduced. Only a few validation studies are available for the co-word modelling approach so far.

PATENTS AS INDICATORS OF BIOTECHNOLOGY[4]

The final area of Publication Analysis which was studied was that of patents. Patents are no doubt the most obvious output of Research and Development which can be measured and used to develop indicators of innovative activity. One very valuable aspect of studying patents is the fact that they precede commercialisation of the invention by several years and therefore give early warning of changes in industrial technology which are not available from other sources.

However, in using the information extracted from patents, it is important to remember that some patents are never meant to be used (for example, 'blocking patents') and others may have 'subjective' non-commercial motives (prestige, employee rewards, etc.). Patents taken out by individuals are also utilised to a lesser extent than those assigned to companies, so perhaps they should carry less weight in exercises to predict activity within a particular area.

The study of patents as indicators is considered important, especially along with other 'indicators' to create a balanced picture.

CREATION OF LITERATURE-BASED MODEL[1]

If the *Science Citation Index* is to be a source of useful models of scientific activity, then the body of scientific literature to be analysed must be a statistically representative sample of the field of science being examined and the modelling technique has to be able to pick out signals from noise and control variables which might lead to ambiguous patterns. The co-citation clustering analysis software, developed at the Centre for Research Planning (CRP) in Philadelphia, now meets these requirements.

The basic CRP approach to co-citation modelling of scientific literature involves a series of data analyses software, developed at the Centre for Research Planning (CRP) in Philadelphia, now meets these requirements.

The basic CRP approach to co-citation modelling of scientific literature involves a series of data analyses and computational steps:

(1) One begins the operation by obtaining access to one of the ISI citation databases.

(2) A 'target' literature is identified and extracted.

(3) The citation frequency of cited papers is calculated.

(4) An analytical database is created containing cited papers, a specified citation threshold and current 'source' papers.

(5) Co-cited references are identified and their frequency is calculated.

(6) The co-cited pairs equal to or greater than some specified threshold index of co-citation are clustered.

(7) By creating such co-citation clusters the target literature has been segmented into research front specialities (which one can be named on the basis of a content analysis, the titles of paper citing into the clusters).
(8) Sample research fronts for modelling.
(9) Generate models.
(10) Analyse models.

Figure 6 summarises the technique and the data available for model making. Figure 7 shows the kinds of questions which might be raised in the analysis of the models produced by co-citation analysis.

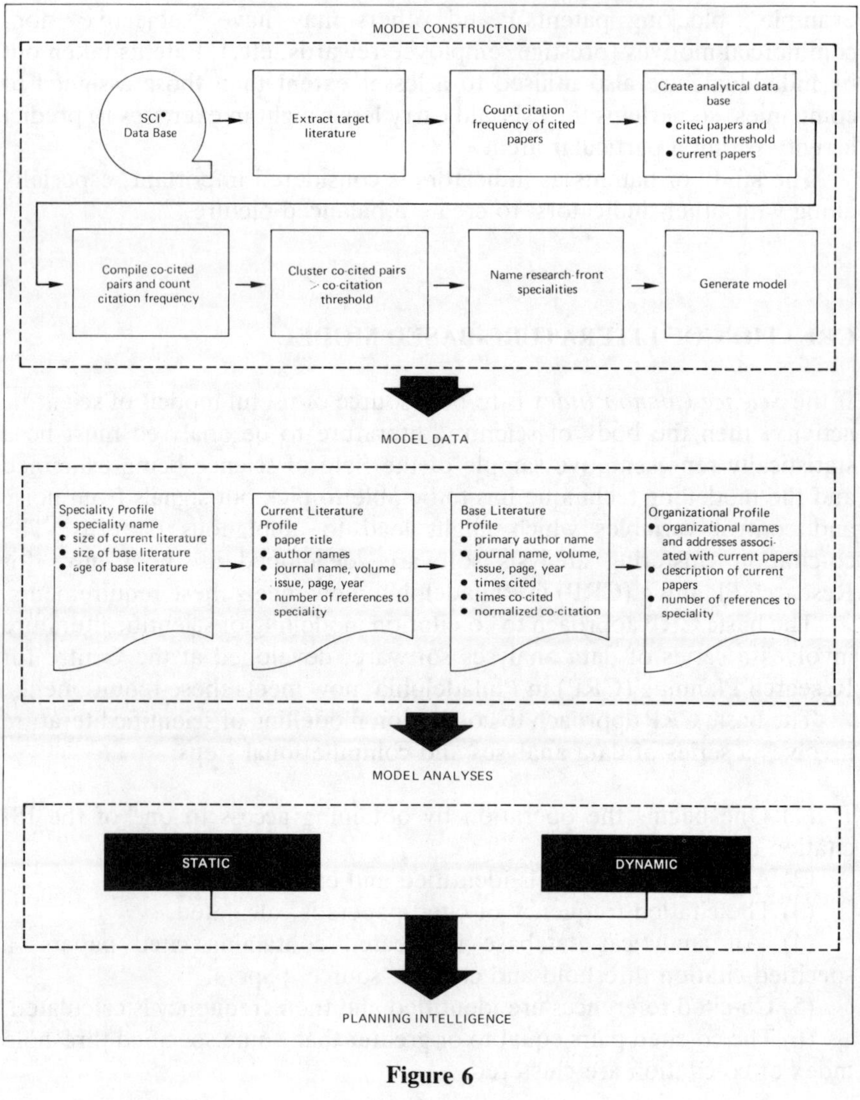

Figure 6

PLANNING INTELLIGENCE

From Static Analysis

- What are the primary regions of activity? What are the primary local points of activity in these regions? What is each concerned with?
- What organizations, people and countries are associated with these local points? Are any dominant? Are any dominant in more than one local point?
- What are the relative levels of activity in the local points? Is there any significant difference from one to the other? If so, which are high and which are low?
- What specialities are involved in each of these local points? Are any being driven by a new set of ideas, which promises rapid advance?
- Is there any interaction between these local points? If so, what are the linking specialities? Are the links what one would expect?
- Are there any relatively isolated specialities in these regions? If so, what focal points are they linked to and where do they connect? Are any of them unexpected? Do any of them represent new directions (as indicated by the age of the ideas driving them)?
- Is there any work being done outside the regions that have been defined? If so, what is the nature of that work? Is any of this work unexpected? Does any of it show higher levels of activity than expected? Does any of it represent new directions (as indicated by unexpected links of new ideas)?
- What people, organizations, and countries are associated with activity features of special interest?

From Dynamic Analysis

- What new speciality areas of activity have developed along the research front, which are the most significant, and in what way?
  - What is the nature of their activity?
    - intellectual locus?
    - interaction with other areas?
  - How big are they?
  - How fast can they be expected to advance?
  - What organizations are most active in them? Do industrial, academic or government organizations dominate?
  - What countries are most active in them?
  - Who are the leading people in them?
- What speciality areas of activity have fallen behind the research front?
  - What is the nature of their activity?
    - intellectual focus?
    - interaction with other areas?
  - How big are they?
  - What are the leading organizations in them? Do industrial academic or government organizations dominate?
- Which of the continuing areas of speciality activity have changed significantly?
  - Which are growing, decaying?
  - Which have changed their intellectual locus? What is the nature of the change?
  - Which have had an infusion of new ideas which can be expected to speed up the rate of advance?
  - Which have had a turnover of dominant organizations? Has the turnover moved the organizational orientation from academic or government to industrial or vice versa?
  - Have any had a change in the geographic orientation of activity?
  - Have any had a turnover of leading people?
- What regional areas of activity have changed significantly?
  - Which are growing, decaying?
  - Which have had a change in direction? What is the nature of the change?
  - Which have had a change in the organizational orientation of activity? What is the nature of the change?
  - Which have had a change in the geographic orientation of activity? What is the nature of the change?

Figure 7

It is clear from these that, for each speciality research front, the models could be used to create sets of 'activity indicators'. However, the nature and significance of activity indicators need study: for example, do changes in the base literature of a specialism signal a change in intellectual focus? Very little work has yet been done on such important technical novelties, which go well beyond the rather crude citation ranking/impact tables of earlier studies or the limited 'maps' of science shown in Figure 3, and it suggests ways of building dynamic models of the progress of scientific specialism[5].

In a study for FAST we did not have the resources to create such models. However, their enormous potential has recently been confirmed by a Science Policy Study commissioned by the Advisory Board for the Research Councils (ABRC) in which five research fields were modelled. One of these was 'protein determination', for which three models were made covering the years 1978, 1980 and 1982. It turned out that they contain a great deal of molecular biology and biochemistry research. One of us (H. R.) is currently examining this utility as an intelligence source for corporate research activity. If the ABRC protein determination models were updated to include the recent research literature, it is our belief that they could make a notable input to strategic thinking about protein engineering research[6].

CONCLUSIONS AND RECOMMENDATIONS OF TPU

General Conclusions

The application and continued growth of science depends on:

(1) Maintaining the health of science by ensuring a proper flow of inputs to the scientific process and protecting it from influences and forces which disrupt its proper process of functioning. The Unit found no sign that long-term retardation of research growth had been caused by, for example, rDNA regulation. The effect of commercial interest on the scientific tradition of openness and free information exchange is a more frequent topic of discussion, especially as the flow of State funding slows down.

(2) The correct evaluation of technical and cost performance. It is essential to avoid *overestimating* the role of 'science push'. The fact that scientists show we can produce X by a new route is not sufficient grounds for utilising that route. There have to be reasons beyond mere novelty — for example, performance and profit, market need and pull.

(3) The requirements of national security and control. Many of the policy recommendations emerging from the FAST Biosociety programme relate to such requirements. Where and to what extent can the use of biotechnology allow Europe to be protected from potential threats by external suppliers or depletion of resources?

The TPU believe that optimal policies for sciences germane to biotechnology will be the result of compromises between these three sets of factors.

The achievement of rational consensus on the nature of such compromises would, it is believed, be aided by the establishment and wide availability of appropriate sets of science, technology and innovation indicators — including modelling strategically significant research fields.

Recommendations

1. *Evaluation Methods for Measuring and Evaluating the Status of Research and Development Fields*

An extended range of publication indicators was applied. The fields selected for study seemed amenable to the use of these techniques and satisfactory results were obtained which threw light on their size, growth rates, interrelationships and national distribution.

There is a need for the development of strategic monitoring of important Research and Development fields. The availability of appropriate science and technology indicators for the community would facilitate this task. Further, the utility of literature-based models of research fields for strategic planning of community Research and Development programmes should be examined.

2. The Identification of Scientific Disciplines at Present Linked to Biotechnology

The findings show that a cluster of basic sciences, including biochemistry, cellular and molecular biology, genetics and virology, now feed into biotechnology, although their interaction among each other is far greater than with biotechnology. Microbiology is closer and more intimately related to biotechnology. Biotechnology itself is a 'supercluster' of many fields, some scientific (plant physiology and biochemistry, chemistry, microbial studies) and others technological (chemical engineering, biochemical engineering, pollution research, food research, etc.).

Some of the 'hot' research specialisms in molecular biology are growing three or four times more rapidly than science as a whole. However, there is evidence of an overall slowing down in the growth rates of biochemistry and microbiology, although they are still growing more rapidly than science as a whole.

'New' biology is highly visible and rightly given great attention in biotechnology policy; however, it should not be at the expense of more traditional research fields such as microbiology, analytical chemistry and chemical engineering, which are more intimately associated with the biotechnology supercluster.

3. Microbiology

The model of biotechnology clusters shows microbiology to be a vital input into biotechnology proper. The European Community member states produce about 30 per cent of the world's literature in four key microbial fields, with the UK producing 40 per cent of the EEC total.

Japan did not produce the high levels of publication output that we had anticipated, and we suspect this is a bias and weakness in the American and Eurocentric information databases that were drawn upon.

Because of microbiology's fundamental relationship to biotechnology, if the European Community seeks to stimulate the latter, they ought to examine the state of their microbiology in research establishments and in higher education, and if necessary ensure increased support.

Improvements in information coverage of Japanese microbiological Research and Development should be sought.

4. Monoclonal Antibody Research

Europe, especially the UK and Switzerland, are strong quantitatively and qualitatively in this field of research. Despite this, Europe was slower than the United States in attempts to commercialise monoclonal antibodies.

The question of US ability to commercialise research fields faster than Europe should be examined to pin-point institutional, cultural and economic causes.

Belated, but significant, steps have been made in several EEC member states to forge novel links between industry and academic research centres. Their experiences should be carefully monitored to evaluate this impact, both on technological innovation and on fundamental research.

5. Biotechnology Patents

Our results appear to show a stronger EEC position than had been reported in earlier studies. Europe was relatively stronger than Japan in the key area of enzyme immobilisation. Most European activity seems to be occurring in well-established fermentation, pharmaceutical and fine chemical companies, responding to rising fuel and raw material costs by improved process engineering. Other significant areas are enzyme reactors, analysis and diagnosis, and biomass. Patterns of 'out-patenting' seem to show that companies within the EEC are in at least as much rivalry with each other as with companies in non-member states.

Supports should be given to develop and encourage patent monitoring based upon analysis of computerised patent information data banks. Such activity could be done on an international basis focusing on fields of strategic significance to the EEC.

REFERENCES

1. Coward, R. H., 'Literature models of science for research planning', and 'Modelling science for research planning intelligence'. Unpublished documents, Centre for Research Planning, Philadelphia, undated
2. Ellingham, H. J. T., Divisions of natural science and technology, in The Royal Society Scientific Information Conference Report, 21 June–2 July 1948, Royal Society, London, pp.477–484
3. Ibid., pp.480–484
4. *Genetic Engineering Biotechnology Patents 1980-1981*, McGraw-Hill, New York, 1982

5. Narin, F., 'Evaluative bibliometrics: the use of publication and citation analysis in evaluation of scientific activity', N. S. F., Washington, D.C., 1976
6. Rothman, H., *ABRC Science Policy Study: Further Studies on the Evaluation and Measurement of Scientific Research.* Report for the ESRC. August, 1985; Healey, P., Rothman, H. and Hoch, P. K., An experiment in science mapping for research planning, *Research Policy*, 1986
7. Small, M. *et al.*, Clustering the *Science Citation Index: Scientometrics,* **7**, 391–409, and **8**, 321–340, 1985
8. Turner, W. A., Courtial, J. P., Bauin, S. and Lampart, C.,'Cartindex des sciences et des techniques', Collogue de la NIDIST, les 24–26 février 1983, Assemblée Nationale, Paris; and Callon, M., Courtial, J. P., Turner, W. A. and Bauin, S., From translations to problematic networks: An introduction to co-word analysis, *Social Science Information,* **22**(2), 191–235, 1983
9. Wishart, D., *Clustan: User Manual*, Programme Library Unit, Edinburgh University, Edinburgh, Scotland, April 1982

BIBLIOGRAPHY

1. *Biotechnology: A Review and Annotated Bibliography.* Rothman, H. et al., London, 1981
2. Op. cit., Monoclonal Antibodies: Another Success for Biotechnology. Parkinson, D.
3. Op. cit., The EEC Policy on Isoglucose. Crott, R.
4. Op. cit., Biotechnology World Patents Analysis. Marstrand, P.
5. Op. cit., Enzyme Technology Patents. Towalski, Z.
6. Op. cit., Microbiology: A Scientometric Analysis. Carson, J.
7. Op. cit., Historical Portrait of a Key Biotechnology Journal: Biotechnology BioEngineering. Rip, A. and van der Es, W.
8. *Op. cit., Biotechnology Company List and Key Research Institutes.* Rothman, H.
9. Op. cit., Measuring European Scientific Capability in Biotechnology. Rothman, H.

38
FUTURES FOR BIOTECHNOLOGY

Jane Llewelyn

Since Biotechnology emerged as a business in the late 1970s the emphasis has shifted from the early predictions[1] that the money makers would be a new generation of raw biochemicals and enzymes to diagnostics and pharmaceuticals, based on the key technologies of monoclonal antibodies and genetic engineering. The over-riding factor here is return on investment.

The short-term emphasis will, therefore, be on low-volume high-value products. Even the large Japanese enzyme and food manufacturers are investing in health care products[2]. In addition to market development and safety clearance costs, the high investment required to produce products from recombinant DNA engineering will initially yield returns in the health care area. Some reports are interestingly predicting that the returns will come first in animal health rather than human health care[3]. Clearly, developments such as growth hormone have applications in both fields.

Commercial goals are leading to a segregation of expertise between DNA engineering and the more traditional biotechnologies, despite the many similarities involved in the development and product marketing. So, apart from companies specialising in the above key technologies, others are looking at applications in food production. For example, Imperial Biotechnology recognise that a whole business can be developed around enzyme applications. Grand Metropolitan's Biocatalysts appear to be following suit[4]. These opportunities are based on high-volume low-value products, relatively speaking. Additionally, companies are beginning to specialise in specific areas of expertise where they see a niche. Agricultural Genetics Co. was launched last year to exploit the Agricultural and Food Research Council's research in genetic engineering in plants[5].

The commercial success of many of the products derived from biotechnology depends on the economics of fermentation. Production of monoclonal antibodies (mabs) on a large scale by continuous fermentation permits a host of applications for pure antibody: from screening blood in immuno-diagnostics, to extracting immune complexes in the blood responsible for disease. Hurdles with cloning have also helped mabs production gain prominence. A lot of the early cloning work has been done in *E. coli*. Unfortunately, this bacterium is not noted for secreting its products into the medium. Therefore, even when a gene sequence coding for a particular enzyme or hormone has been successfully inserted and expressed, it is a further hurdle increasing the yield and then harvesting the final product. Neither is *E.coli* the most attractive choice toxin-wise, so the early models may not be the best 'factories'. There will be a lag period while the vast scientific cloning knowledge now accumulated is applied in commercially feasible operations.

Monoclonal antibody production in continuous air-life fermenters has won favour over batch and stirred tank in this case. Mabs are a particularly tricky species to grow, because they have very rigid temperature and nutrient requirements and are easily damaged in too turbulent a mixture. Bulk production has tremendous advantages over traditional mice breeding and extracting the antibodies from the spleen. Future uses will be not only in preventing human disease, but also in animal disease, animal husbandry and food hygiene. Mabs can be designed like probes to identify harmful bacteria and provide a real time measure, without relying on the results from plating-out, which can take several days at least, resulting in costly losses if the whole production run has to be destroyed[6]. The food industry is naturally very interested in these developments.

Fermentech are developing Protein A, or universal second antibody, which binds to most mammalian IgG molecules. Fermentech believes the market for Protein A to be worth some £10 million a year for purifying immunoglobins from serum[7]. Monoclonals will be used as probes for all types of disease. Centocor has developed two new kits to detect cancers: ovarian cancers, and tumours of the stomach and intestine[8]. Hybritech has developed a pregnancy testing kit which has gained 15–20 per cent of the market in 1983/84 based on mabs[9]. Other companies are developing a range of fertility testing drugs based on mabs which have the advantage of much greater precision, accuracy and ease, and reproducibility of use over existing tests. Boots–Celltech Diagnostics are specialising in diagnostic kits based on mabs, including a number of animal diagnostic products and fertility test kits. Their first product is an immunoradiometric assay kit for the measurement of serum interferon. Celltech's joint venture with Boots is particularly interesting because it combines sound commercial backing with sound scientific expertise and a ready access to MRC technology, from which mabs research at Celltech derives its being. One can expect many interesting products to emerge from this company. Bulk mabs production is the responsibility of Cell Culture Products at Celltech. They claim to have the largest continuous culture for mabs, and include Centocor among their customers.

Gene probes have equal potential in health care. Recent research at the Massachusetts General Hospital and Harvard Medical School heralds a significant step towards developing a gene probe for Huntington's chorea, as they have identified a gene marker for this disease[11]. At the Weizmann–Institute of Science, researchers have isolated the gene responsible for cytoplasmic superoxide dismutase, which is located in human chromosome No. 21. This could be used to develop a probe as a fetal diagnostic test for Down's Syndrome[12]. At Porton Down, CAMR have succeeded in cloning Protein A into *E. coli* but yields of currently 5 per cent of total protein are too low for this route to be commercially attractive yet. Likewise, a probe for hepatitis B virus would have an enormous potential because an estimated 200 million people suffer from the consequences.

A first recombinant product vaccine-wise, Merck Sharp & Dohme's hepatitis B yeast-derived vaccine has just been approved for marketing in W. Germany (Gen-H-B-Vax) and the US (Recombivax H-B)[9]. The advantages of using rDNA derived products are purity and theoretically lower cost. In fact, the production costs for this vaccine are currently greater than for the plasma-derived one. Hopefully, the competition from companies such as Biogen/Ciba-Geigy, Amgen/Johnson, and Johnson and Connaught will reduce the prices, because the medical profession believe that a hepatitis vaccine should be made available to all those at risk. The current three-shots course costing around £60 is prohibitive for most 'at risk'. Smith Kline Beckman also launched its rDNA hepatitis vaccine Energix B in Singapore and Malaysia in June 1986 and Pasteur's rDNA vaccine is expected to be launched in France. Soon a sporozoite vaccine for malaria will be available. At Guy's Hospital in London the 12 amino acid peptide coat of the protozoan has been identified and sequenced[13]. A sporozoite vaccine is about to enter trials in Australia[13a]. Now researchers are working towards a multi-protective vaccine offering total protection throughout the life-cycle of the parasite. Once the antigen is recognised, the gene to produce it can be sequenced and the information spliced into the appropriate host (in this case also a yeast) for commercial production. Other candidates, such as measles, common cold virus and a universal 'flu vaccine should become realities by 1994.

New laser techniques for inserting foreign DNA into host organisms are being developed at the Japanese Institute of Physical Research at Wako Shi, near Tokyo[14]. This is a departure from the use of restriction enzymes. Gene sequencing facilities are becoming widely available with the advent of more sophisticated hardware, but, like computers, each machine is rapidly outdated by the next model. Once the gene sequence has been worked out, it is a matter of stringing the nucleic acid bases together. Many firms and some universities now offer this facility as a customer service. Eventually it should be possible to sequence whole proteins and create new tailor-made molecules.

Should protein engineering become feasible, the patenting possibilities have far-reaching implications. It would be very useful, for example, to be able to isolate and manufacture the functional part of an enzyme. Equally desirable would be the ability to alter the temperature sensitivity or reprogramme the pH tolerance of a catalyst to suit process conditions. Many groups are currently researching into this area and believe success to be achievable inside 20 years.

In recognition of this important area the Government Chemist is setting up a research programme in collaboration with SERC and six commercial companies (Celltech, ICI, Glaxo, Sturge, Tate & Lyle and Unilever) to develop and co-fund a major research programme in protein engineering[15]. It will concentrate on the fundamental properties of enzymes and examine how the detailed structure and folding patterns of these complex molecules affect their catalytic properties and how to improve them by genetic engineering.[16] One of the important features of this work will be computer graphic modelling at the molecular level based on the 3D protein structures elucidated by X-ray crystallography. As in other areas of engineering, computer-aided design will play a vital part in assisting the amalgamation of existing knowledge and prototype development. The first Protein Engineering Conference took place at the University of Oxford in April 1987.

In contrast, some companies are developing simple-to-use biosensors for hospital and chronic patient use. Cambridge Life Sciences have just developed a sugar sensor for diabetes which allows self-diagnosis from a drop of blood and costs around £100[17]. As a result of their own world-wide market research survey, CLS plan to develop other kits along the same lines, marrying Microelectronics and Biotechnology to analyse harmful substances in body fluids and the concentration of therapeutic drugs. Much larger biotech firms with a higher injection of capital have not achieved so much in 18 months.

Genentech has produced and launched recombinant somatrem as Protropin for hypopituitary dwarfism (estimated annual sales $50 million) and is licensed to KabiVitrum for the rest of the world, while Eli-Lilly used recombinant DNA technology to produce somatropin (Humantrope) identical to the natural hormone (estimated annual sales $60 million to $100 million). Genentech and Eli-Lilly have already gained FDA approval for their human insulin development at a cost of around $100 million. Celltech have signed a deal with Serono Labs in the USA to develop HGH[18]. Meanwhile there is still no sign of an earlier rDNA product — rennin. Celltech had reputedly entered into a collaboration with Biotechnica International Inc. to produce rennin, presumably still in *E. coli*[19]. Collaborative Research and Dow Chemical in the USA are producing rennin from yeast[19]. At Biogen clinical trials are under way with rDNA interleukin-2 (IL-2) for AIDS which is due on the market by 1987. Roche has just launched interferon α-2a (Roferon-A) for the treatment of AIDS-related Kaposi's Sarcoma. In the veterinary diagnostic products area, Cytotech, La Jolla, has produced a cytomegalovirus assay which is currently up for FDA approval[19]. Genentech, Biogen and Celltech are considered by many to be the three foremost Biotech companies. It will be interesting to see whether they hold their rank in ten years time. Clearly, in the next decade there will be many diagnostic kits and vaccines developed from mabs and rDNA.

Although early hopes have waned, interferon may still have a role to play in cancer treatment. At the University of Texas interferon has been shown to assist in regulating cell growth. The compound's anti-tumour activity can be transferred from treated to untreated cells in cell culture. Interferon causes a release of secondary messenger which passes between adjacent cells[20]. One of the major future applications of biotechnology in Health Care will be in the field of toxin-linked mabs targeted to specific cancer cells, indicated by current R&D investment in this area.

In the next 20 years, biotechnology will resolve many of the scientific–technical problems which currently hinder its success, but will depend on the 'logistics' (a euphemism for sociopolitical issues) as to whether what is feasible becomes a reality. Much of the current genetic engineering research is subject to legislation. The economics are such that First World commercial concerns are unlikely to invest in Third World health problems. India and some of her neighbours are keenly aware of this fact and are seeking to establish their own expertise in biotechnology as well as the more published nuclear skills. It is interesting to speculate on how these skills will develop in different climates. If one draws the comparison with single-cell protein —the developments of the 1970s seem to have gone underground, beset by various obstacles to commercialisation, except in the USSR. Here well over 2 million tonnes of SCP is being produced at production units all over the Soviet Union[21]. ICI may well find that the best market for Pruteen is sale of expertise to the USSR. Similarly, the biotechnology companies are finding that a lot of R&D work is best sponsored by inviting partners (often end-users) and the more enlightened or adventurous investors to form Research and Development Joint Venture Partnerships. It is equally clear that many developments will not have successful outcomes in such a high risk area.

Apart from the new biotechnology companies, there are many established chemical companies who have been investing in biotechnology for a number of years and whose backroom research will eventually lead to new developments and new markets. Companies like ICI, Hoechst, Bristol

Myers, British Petroleum are all involved in biotechnology, and even Novo, who are notably conservative in their product range, have cloned human gastrin[22] and are developing other rDNA enzymes.

One of the most lucrative areas to be in at the moment is the biotechnology support business and there is still room for expansion. Biotechnology conferences, newsletters and other such topical coverage; dictionaries and catalogues; university and school courses; textbooks and recruitment agencies are all benefiting from the emergence of biotechnology as the latest 'buzz' word and a feeling that everybody should know something about it.

Even the Government has at last made an effort to link the pool of expertise in the UK by concentrating on a vital development — protein engineering. But one wonders how much this is a cosmetic exercise and whether the result will be more than a talking shop. The next vital area is waste treatment. Nevertheless, the multidisciplinary nature of biotechnology does require a clear initiative to bring together all the relevant disciplines and avoid the wasteful duplication of effort, time and money; and in view of the fact that our overseas competitors recognised this need much earlier.

SCP will regain prominence. Phillips Petroleum Co. are growing a yeast on alcohol. 'Provesteen' is 60 per cent protein by weight. MIT are evaluating its human food potential[23]. Rank Hovis McDougall are still developing their fungal food protein ingredient. Concern over the connection with heart disease and diet will increase acceptance of novel proteins with low fat content.

The granting of the second Cohen–Boyer patent to Stanford University and the University of California covering the gene splicing technique has led to hundreds of letters going to biotechnology firms around the world asking for an annual £10 000 licence fee[24]. The future of rDNA technology in its present form may well depend on the majority view, whether or not to uphold the patent claim. Certainly there will be many who will contest this ruling. Patent lawyers have an assured future in the biotechnology business.

Thus, already, attitudes have sobered. There has been much promised, little achieved, and fewer markets are now highlighted in the search for commercial outlets for the new conglomeration of techniques called biotechnology. As well as all the traditional problems of marketing a new product, there are biological factors which cannot be hurried. There is the training of personnel and the development of new sophisticated hardware, and so on. The industry as such is still in its infancy. Likewise, until the requirements of the end-user have been successfully entered into the equation and the potential for the new techniques fully translated, there will continue to be a mismatch. Biotechnology is only a means and should not be viewed as an end in itself.

REFERENCES

1. *European Chemical News* 27.4.81, p.12
2. *Bio/Technology*, April 1984, p.312
3. Scrip No. 645, 23rd November, 1981, p.8
4. *Biotechnology Bulletin*, April 1984, Report No. 47
5. D. Fishlock, Country cousins of biotechnology, *Financial Times*, 12th June 1984, p.14
6. Anon., Recombinant DNA: Economic Study Predicts Limited Future, *Gen. Eng. News*, March/April 1983, p.19
7. *Bio/Technology*, August 1984, p.667
8. *New Scientist* 10.11.83, p.419
9. Scrip No. 1124, 30 July 1986, p.20
10. *New Scientist*, 28 July 1983
11. S. Connor, Gene probes find Huntington's disease, *New Scientist*, Nov. 1983, p.399
12. *Hospital Doctor,* 8 March 1984
13. *Radio 4*, Medicine Now, 19.9.84, 13a Inpharma 2nd Aug 1986:3
14. B. Johnstone, Laser Speeds Up Genetic Engineering, *New Scientist*, 10th Nov. 1983, p.643
15. *The Guardian,* 6 September 1984, p.14
16. *Anon.*, New club beefs up protein research, *New Scientist*, 1st Dec. 1983, p.643
17. *G. Colverd,* A light blue sugar sensor, *Sunday Telegraph*, 16th Oct. 1983, p.12
18. *Chemistry & Industry,* 2.4.84, p.233
19. *European Chemical News* Chemscope, May 1984
20. *Science* 221, p.953
21. Rimmington, *Bulletin of the Biotechnology Centre,* Wales, Article 16:3:8
22. *Financial Times,* 8 March 1983
23. *Times,* 2 August, 1983
24. *New Scientist,* 6 September, 1982, p.7

RECOMMENDED READING

Bu'lock, J. and Kristiansen, B. (1987). *Basic Biotechnology,* Academic Press, New York

Cherfas, J. (1982). *Man-Made Life: A Genetic Engineering Primer.* Blackwell, Oxford

Davies, D. (1986). *Industrial Biotechnology in Europe — Issues for Public Policy.* CEPS–EEC Conference, CUBE DGXII, Commission of the European Communities, Brussels

Dunnil, P. (1981). Biotechnology and industry. *Chemistry and Industry,* **7**, 204

Dunnil, P. and Rudd, M. (1984). *Biotechnology and British Industry.* Report to the Biotechnology Directorate of the Science and Engineering Research Council, Polaris House, Swindon

Elkington, J. (1985). *The Gene Factory.* Century, London

Fishlock, D. (1982). *The Business of Biotechnology.* The Financial Times, London Times Business Information, London

Higgins, I. J., Best, D. J. and Jones, J. (1985). *Biotechnology: Principles and Application,* Blackwell, London

Jacobsson, S., Jamison, A. and Rothman, H. (1986). *The Biotechnological Challenge.* Cambridge University Press, Cambridge

Moo-Young, M. (1985). *Comprehensive Biotechnology,* Vols 1–4, Pergamon Press, Oxford

OECD (1982). *Biotechnology: International Trends and Perspectives.* HMSO, London

OTA (1981). *Impact of Applied Genetics: Micro-organisms, Plants and Animals.* OTA – HR – 132, US Congress Office of Technology Assessment, Washington DC

OTA (1984). *Commercial Biotechnology: An Internal Analysis.* 1 OTA – BA – 218, US Congress Office of Technology Assessment, Washington DC

Prentis, S. (1984). *Biotechnology: A New Industrial Revolution.* Orbis, London

Spinks, A. (1980). *Biotechnology: Report of a Joint Working Party.* HMSO, London

Smith, J. E. (1984). *Biotechnology Principles.* Van Nostrand Reinhold, London

Wheale, P. and McNally, R. (1988). *Genetic Engineering: Catastrophe or Utopia?* Harvester/Wheatsheaf, England, and St Martin's Press, New York

Wiseman, A. (Ed.)(1988). *Principles of Biotechnology,* 2nd edn, Surrey University Press, and Chapman and Hall, New York

Yoxen, E. (1983). *The Gene Business: Who Should Control Biotechnology?* Pan Books, London

INFORMATION AND REFERENCES

1. SERC Biotechnology Directorate, Directory of Research in Biotechnology, Polaris House, North Star Avenue, Swindon
2. The Biotechnology Unit of the Department of Trade and Industry (1984). *Directory of British Biotechnology*. Catermill Publishers, St. Andrews, Fife, Scotland
3. A. Crafts-Lighty (1986). *Information Sources in Biotechnology*, 2nd edn, Macmillan Publishers (Journals Division), London and Basingstoke, and Stockton Press, New York
4. J. Coombs and Y.R. Alston (1989). In *The International Biotechnology Directory: Products, Companies, Research and Organisations* (Eds A. Crafts-Lighty, E. Burak Reed and E. Trawhella), Macmillan Publishers (Journals Division), London and Basingstoke, and Stockton Press, New York
5. AABB and Bio Commerce Data Ltd (1988). *The UK Biotechnology Handbook 1988*, Southwell Press, Camberley
6. M. D. Dibner (1988). *Biotechnology Guide, USA*, Macmillan Publishers (Journals Division), London and Basingstoke, and Stockton Press, New York

Index